CISM International Centre for Mechanical Sciences

Courses and Lectures

Volume 575

The series presents lecture notes, monographs, edited works and proceedings in the field of Mechanics, Engineering, Computer Science and Applied Mathematics. Purpose of the series is to make known in the international scientific and technical community results obtained in some of the activities organized by CISM, the International Centre for Mechanical Sciences.

More information about this series at http://www.springer.com/series/76

Luca d'Agostino · Maria Vittoria Salvetti
Editors

Cavitation Instabilities and Rotordynamic Effects in Turbopumps and Hydroturbines

Turbopump and Inducer Cavitation, Experiments and Design

Springer

Editors
Luca d'Agostino
Civil and Industrial Engineering Department
Pisa University
Pisa
Italy

Maria Vittoria Salvetti
Civil and Industrial Engineering Department
Pisa University
Pisa
Italy

ISSN 0254-1971 ISSN 2309-3706 (electronic)
CISM International Centre for Mechanical Sciences
Courses and Lectures
ISBN 978-3-319-49717-4 ISBN 978-3-319-49719-8 (eBook)
DOI 10.1007/978-3-319-49719-8

Library of Congress Control Number: 2016958494

Printed on acid-free paper

This Springer imprint is published by Springer Nature
The registered company is Springer International Publishing AG
The registered company address is: Gewerbestrasse 11, 6330 Cham, Switzerland

Preface

The attainment of higher power densities in modern hydraulic turbomachinery is invariably obtained by running the impeller at the maximum allowable speed and lower shaft torque. Accordingly, operation under cavitating conditions can occur under special circumstances in hydroturbines and nuclear power plant cooling systems and is often tolerated in liquid propellant rocket feed systems, exposing the machine to the onset of dangerous self-sustained, cavitation-induced fluid dynamic and rotordynamic instabilities. Since these phenomena actually represent the major source of life and reliability degradation of the machine, fundamental information on their nature and behavior is of crucial importance for the effective design of today's high-performance hydraulic turbomachinery. However, the extreme complexity and imperfect understanding of the phenomena involved pose formidable obstacles to the modeling, prediction, and control of cavitation-induced instabilities. For this reason, nowadays theoretical analyses and simulations alone are still of limited value for the solution of specific technical problems and progress in this field must rely on the support of dedicated experimentation.

The objective of this monograph consists in providing the reader with a comprehensive approach to the physics, fluid dynamics, modeling, experimentation, and numerical simulation of cavitation phenomena and their implications on the design and operation of high-performance hydraulic turbomachinery. The material presented herein is, therefore, intended to illustrate some of the most recent advancements concerning both the occurrence of cavitation-induced instabilities and rotordynamic effects in high-performance turbopumps and hydroturbines, as well as the development of more refined and efficient numerical tools for the simulation of the complex cavitating flows involved in these machines.

The book contains most of the material presented in the lectures given by some of the world leading experts at the advanced school Cavitation Instabilities and Rotordynamic Effects in Turbopumps and Hydroturbines, held at the Centre International des Sciences Mecaniques (CISM), Udine, Italy, in July 2014.

The course opens with an introduction by d'Agostino and coworkers to the fundamental aspects of cavitation phenomena and induced flow instabilities, with special emphasis to the operation of hydraulic turbomachinery for space propulsion applications. Next, Ceccio and Mkiharju review the traditional and novel experimental methods for the investigation of hydrodynamic cavitation, and illustrate some of their recent developments in the use of ionizing radiation as a means to visualize cavitating flows. Tsujimoto presents the experimental results, theoretical analyses and numerical simulations he and his collaborators have been using for their pioneering work on cavitation-induced instabilities and rotordynamic effects in inducers, turbopumps and hydroturbines. The fundamental aspects of the preliminary hydraulic design of inducers and centrifugal turbopumps is discussed by d'Agostino et al., which illustrate the development and experimental validation of their reduced-order models capable of jointly providing the geometrical definition and performance prediction of these machines. The last three contributions concern the numerical simulation of cavitating flows. The article by Goncalves presents a comparison of different turbulence and cavitation models, with and without thermal effects, in unsteady RANS simulations, while the contribution by Salvetti focuses on numerical simulation of cavitating flows in complex geometry. Finally, Saurel et al. present a flow model for the numerical simulation of interfacial flows with phase transition and its applications to cavitating nozzle flows.

The monograph is addressed to doctoral/postdoctoral students, researchers, scientists, scholars, and professionals interested in perfecting their knowledge and understanding of cavitating flow phenomena and research occurring in a wide range of applications in aerospace, mechanical, hydraulic, naval and chemical engineering, applied mechanics, applied mathematics, industrial chemistry, and applied physics.

The editors are especially grateful to Prof. Steve Ceccio, Prof. Eric Goncalves, Prof. Richard Saurel, and Prof. Yoshinobu Tsujimoto for their contributions to the present volume. The editors are also very thankful to Prof. B. Schrefler, Secretary General, Prof. A. Soldati, Deputy Secretary General, and to Prof. P. Serafini, Executive Editor, as well as to the staff of the CISM, for their support and assistance in the organization of the Course and in the publication of the present volume.

Pisa, Italy

Luca d'Agostino
Maria Vittoria Salvetti

Contents

An Introduction to Cavitation in Inducers and Turbopumps

Luca d'Agostino, Lucio Torre, Angelo Cervone, Giovanni Pace, Dario Valentini and Angelo Pasini

Abstract After a brief review of the fundamental aspects of cavitation relevant to the operation of high-performance inducers and turbopumps, the article summarizes their application to the analysis of pumping systems, illustrates the scaling of cavitation phenomena from model tests to full-scale operation, describes the occurrence of flow-induced instabilities in turbomachinery, and introduces the concepts of static and dynamic instability of pumping systems and their generalization to cavitating turbopump systems.

1 General Aspects of Cavitation

The term cavitation was first introduced by R. E. Froude to indicate the change of phase occurring within a liquid due to local pressure reductions generated by dynamic actions. In this respect cavitation is closely related to boiling, where phase changes are due instead to the increase of temperature as a consequence of heat transfer to the liquid. In both cases the phase change ideally occurs in the presence of some nucleating agents when the vapor pressure of the liquid locally exceeds its static pressure.

From the physical point of view cavitation represents a complex phenomenon because it involves the simultaneous dynamic and thermodynamic interaction of two phases, it is greatly affected by the uncertain concentration and susceptibility of nucleating agents, it manifests itself in a variety of topological forms, and can be strongly coupled with the fully wetted flow of the liquid phase. Cavitation phenomena in hydraulic machines are, therefore, only partially understood and

L. d'Agostino (✉) · G. Pace · D. Valentini · A. Pasini
Università di Pisa, Pisa, Italy
e-mail: luca.dagostino@ing.unipi.it

L. Torre
Alta S.p.A., Ospedaletto (Pisa), Italy

A. Cervone
T.U. Delft, Delft, The Netherlands

L. d'Agostino and M.V. Salvetti (eds.), *Cavitation Instabilities and Rotordynamic Effects in Turbopumps and Hydroturbines*, CISM International Centre for Mechanical Sciences 575, DOI 10.1007/978-3-319-49719-8_1

predicted by theoretical means and numerical simulations alone. Hence, extensive use of experimentation is still necessary for addressing and solving cavitation problems in engineering applications.

As a consequence of its relevance a vast literature exists on both the more fundamental aspects of cavitation and its technical implications. One can refer, for example, to the reviews of Eisenberg (1961), Acosta (1974), Acosta and Parkin (1975); or to the textbooks of Knapp et al. (1970), Pearsall (1972), Brennen (1995), Franc and Michel (2004), Brennen (2005). More specific aspects of cavitation phenomena have been addressed in the Annual Reviews of Fluid Mechanics, in various specialistic meetings, like the International Symposia on Cavitation (CAV), the conference of the International Association for Hydro-Environment Engineering and Research (IAHR), in the yearly sessions of the ASME Cavitation and Multiphase Flow Forum, in the Proceedings of the International Towing Tank Conference and in other symposia and publications on hydraulic machinery, naval hydrodynamics, marine propulsion, underwater acoustics, etc.

2 Cavitation Nuclei

Early studies have shown (Plesset 1969; Knapp et al. 1970) that the maximum tensile strength that pure liquids can theoretically sustain according to thermodynamic considerations is much larger than observed in practice. The highest tensions attained in experiments with extremely pure water at room temperature are on the order of a few hundred atmospheres. On the other hand, values comparable to the vapor pressure at the liquid temperature are usually observed in cavitation experiments on technical liquids. It has therefore been postulated that the tensile strength of liquids is considerably reduced by the presence of weak spots for the onset of liquid rupture, generically called "nuclei"

It has long been recognized that small bubbles (Fox and Herzfeld 1954; Plesset 1969) and gas pockets contained inside crevices and cracks of a liquid–solid interface (Harvey et al. 1947) act as cavitation nuclei (Holl 1970). The thermodynamic equilibrium of bubbles and/or pockets of gas is influenced by the content of dissolved gas in the liquid, which therefore represents an important parameter in the study of cavitation inception. Local fluctuations of the liquid density of thermostatistical nature or due to the interaction with high energy radiation are also thought to act as cavitation nuclei, although their role is quite negligible in most practical applications where the effects of the abundant supply of weaker nuclei from other sources usually prevail.

Cavitation nuclei can be suspended in the fluid stream or located at the flow boundaries. The relative importance of the cavitation susceptibility of surface and free stream nuclei often determines the form of cavitation developing in a specific flow configuration. There are indications that the effects of free stream nuclei usually prevail in laboratory tests on reduced-scale models and that this situation may sometimes be reversed in full-scale prototype operation.

The concentration of nuclei which become unstable at a given level of tension—synthetically indicated as "liquid quality"—represents an important aspect for predicting the inception and later development of cavitation, as well as for deducing scaling laws capable of extending the results obtained from model tests to full-scale operation.

Since the size of nuclei can be measured more easily than their susceptibility by means of optical, acoustical, and electrical methods (O'Hern et al. 1988), in cavitation research the liquid quality is usually expressed in terms of the nuclei concentration density distribution:

$$n(R_o) = -\frac{dN(R_o)}{dR_o}$$

where $N(R_o)$ is the concentration of nuclei whose equilibrium radius is not smaller than R_o. The nuclei concentration density distributions of many liquids of technical interest (including natural and laboratory waters) spread in practice over a very wide range and in most cases approximately follow an inverse (hyperbolic) power law distribution (Bader 1970) over a relatively large interval of sizes.

The main drawback of using the nuclei concentration density distribution for measuring the liquid quality is that the critical tension that makes each nucleus unstable, as required for cavitation studies, must be deduced indirectly from the measurement of its size. This can only be done with the necessary accuracy in the case of microbubbles. In the nuclei model proposed by Harvey et al. (1947), for example, surface tension effects provide a mechanism capable of stabilizing microscopic pockets of gas trapped in small crevices of hydrophobic solid surfaces with respect to diffusion of their contents in the liquid even at pressures much higher than the saturation value, consistently with the available empirical evidence. However, since in technical liquids the geometry of these crevices and the properties of their surfaces are utterly uncertain, the above method for the characterization of the liquid quality cannot provide accurate predictive indications on the susceptibility of this kind of nuclei. In practice, quantitative information about the actual critical tension of solid particulates, when detected and recognized, must be deduced experimentally.

In principle these limitations are eliminated by making use of standard cavitators, where the intensity of cavitation is suitably monitored under controlled flow conditions, and cavitation susceptibility meters, which in principle should measure the liquid pressure capable of destabilizing individual cavitation nuclei (d'Agostino and Acosta 1991a, b). Unfortunately, also this family of direct methods is affected by significant limitations, mostly introduced by the imperfect control of the nominal flow and by the occurrence of dynamic interactions between nearby nuclei, which prevent the obtainment of accurate liquid quality measurements (d'Agostino and Acosta 1991a, b, c).

3 Forms and Occurrence of Cavitation

Usually the inception of cavitation manifests itself with the sporadic formation of isolated voids in the low pressure regions of the flow when the pressure drops below some critical value, and is associated with the emission of sharp noise bursts due to the sudden collapse of the cavities as they are convected in a region of higher pressure (Holl 1969). Acoustic detection of these bursts is routinely used for the identification of cavitation inception conditions.

Cavitation occurs in different forms as a consequence of various aspects and conditions of the liquid flow. In the absence of separation, it generally takes place either at or close to the boundaries where the minimum pressure occurs, depending on which among surface and freestream nuclei are more susceptible to cavitate. On the other hand, when the flow is separated, cavitation takes place far from the boundaries in the turbulent wake (cloud cavitation) or in the core of shedding vortices (vortex cavitation).

When free stream nuclei are more susceptible than surface nuclei, cavitation develops in the formation of bubbles traveling with the liquid. This form of cavitation is characterized by cavities originating in the low pressure regions of the flow and collapsing further downstream as they are convected in higher pressure regions. When, on the other hand, surface nuclei are more susceptible to cavitate than free stream nuclei, cavitation develops either in a thin recirculation region generated by the separation of the liquid flow (attached cavitation), or as a wake of cavities sequentially released from localized nuclei on the boundary surface (spot cavitation). When developing in regions where the flow is laminar, attached cavities have a quasi-steady glossy appearance in their upstream portion (sheet cavitation) and become increasingly unsteady and fuzzy further downstream as a consequence of the Rayleigh–Taylor destabilization of the liquid/vapor interface. On the other hand, cloud cavitation leads to the formation of large intermittent clusters of bubbles in the fully separated wake of the flow. The latter forms of cavitation are collectively indicated as developed cavitation and, in particular, as partial cavitation when the cavities collapse before reaching the trailing edge of the cavitating body, and supercavitation when a further drop of the pressure leads to the formation of a cavity extending downstream beyond the trailing edge of the body. The above classification should, however, only be considered as indicative since in many practical cases several forms of cavitation can occur simultaneously and variously interact.

4 Classical Theory of Cavitation Scaling

Cavitation scaling consists in correlating the results of cavitation tests on geometrically similar models to full-scale operation. In the classical (Eulerian) theory of cavitation (Holl and Wislicenus 1961) it is assumed in first approximation that:

- cavitation inception occurs as soon as the static pressure p in the flow locally drops below the vapor pressure p_V, corresponding to the liquid temperature;
- pressure changes in the flow are solely due to dynamic (inertial) actions.

Then the fundamental parameter for cavitation scaling is the Euler number (also indicated as cavitation number):

$$\sigma = \frac{p_o - p_V}{\rho_L U_o^2/2}$$

where ρ_L is the liquid density and the subscript o indicates the relevant reference values of the flow pressure p and velocity U. By comparison of the Euler number with the pressure coefficient:

$$C_p = \frac{p - p_o}{\rho_L U_o^2/2}$$

which in the stated assumptions is equal for geometrically similar steady flows, cavitation is therefore predicted when:

$$p_{\min} \leq p_V \quad \Rightarrow \quad \sigma \leq -C_{p\min}$$

The suction performance of turbopumps is usually specified by assigning the admissible value of the Net Positive Suction Pressure:

$$\text{NPSP} = \frac{p_{t1} - p_V}{\rho_L}$$

(or, equivalently, the Net Positive Suction Head NPSH = NPSP/g, where g is the acceleration of gravity) in order to define the required total pressure p_{t1} of the inlet flow to the machine at nominal flow rate for operation under acceptable cavitation conditions (typically at –3% head degradation).

Although more informative for engineering purposes, both NPSP and NPSH are dimensional numbers, unsuited for similarity considerations. Hence, from dimensional analysis, either the design values of σ,

$$\tau = \frac{p_{to} - p_V}{\rho_L U_o^2/2} \quad \text{(cavitation index)}$$

or:

$$\Omega_{SS} = \frac{\Omega\sqrt{\dot{V}}}{\left[(p_{t1} - p_V)/\rho_L\right]^{3/4}} \quad \text{(suction specific speed)}$$

are more efficiently used for scaling the cavitation performance of geometrically similar machines rotating at speed Ω and displacing the volumetric flux $\dot{V}$. Two geometrically similar flows are therefore considered to be equivalent for the purpose of cavitation (i.e., cavitation is predicted to occur in similar form and extent) when they have the same values of σ, τ or Ω_{SS}, as applicable.

The Euler theory of cavitation scaling provides in first approximation an adequate correlation of cavitation phenomena between full-scale and model flows. However, in specific cases violation of the underlying assumptions as a consequence of:

(1) changes of the fully wetted flow field due to viscosity, gravity, and compressibility, including boundary layer development, separation, transition, turbulence, reattachment, fully separated shear and vortical flows, turbulent pressure fluctuations;
(2) changes of the geometric similarity, including surface roughness;
(3) changes of the cavity growth, due to inertia, heat transfer, surface tension, liquid viscosity and compressibility, transport phenomena between dissolved and noncondensable gas, susceptibility and concentration of cavitation nuclei.

can lead to significant discrepancies, collectively indicated as "cavitation scaling effects" (Arndt 1981). In particular, the available evidence indicates that the latter set 3) of scaling effects usually prevails in developed cavitation.

Scaling effects from various sources often overlap and influence each other, making it very difficult to recognize their ultimate causes and to predict their intensity. The literature on cavitation scaling effects is therefore quite vast and intricate. Not all of it is, however, of immediate relevance to the analysis of the steady and unsteady performance of hydraulic turbomachinery, which is mostly concerned with fully developed cavitation. Thus, here we limit our attention to this form of cavitation, which is the decisive factor for the occurrence of major difficulties in the operation of high power density turbopumps.

5 Cavitation and Bubble Dynamics

Whatever the initial nature and size of cavitation nuclei may be, one can in first approximation visualize their growth into larger cavities (and their later collapse) with reference to the idealized case of a spherical bubble in an unbounded incompressible liquid subject to a change of its far field pressure. In practice there can be significant departures from this simplified model, especially in the initial stages of growth and in the final stage of collapse, or as a consequence of various causes such as the proximity of the boundaries, interactions with other bubbles, flow shearing effects, instabilities of the cavity interface, etc. However, consideration of spherical cavities greatly simplifies the analysis, while still retaining the most important aspects of the physical phenomena governing the interactions of the bubbles with the surrounding liquid. Under these assumptions the bubble radius

R is determined by the Rayleigh–Plesset equation (Knapp et al. 1970; Plesset and Prosperetti 1977):

$$R\ddot{R}+\frac{3}{2}\dot{R}^2+4\nu_L\frac{\dot{R}}{R}+\frac{2S}{\rho_L R}=\frac{p_B-p_L}{\rho_L}$$

Here dots indicate Lagrangian derivatives with respect to time t, S is the surface tension, $\nu_L=\mu_L/\rho_L$ is the kinematic viscosity of the liquid, p_L is the far field liquid pressure and p_B is the bubble internal pressure, which consists of the partial pressures of vapor p_V and noncondensable gas p_G. When mass diffusion effects within the bubbles are neglected, p_V is constant and p_G is expressed by the perfect gas equation:

$$p_G=p_{Go}\left(\frac{R_o}{R}\right)^3\frac{T_B}{T_{Bo}}$$

where p_{Go} is the gas partial pressure at the initial radius R_o, T_{Bo} is the corresponding temperature and T_B is in general determined by the simultaneous solution of the energy equation for the thermal contact of the bubble with the surrounding liquid. In this case, if the initial state (indicated by the subscript o) is a mechanical equilibrium state, then the corresponding pressure in the liquid is:

$$p_{Lo}=p_{Go}+p_{Vo}-\frac{2S}{R_o}$$

For a bubble containing a fixed amount of gas behaving isothermally $(T_B=T_L)$, the quasi-static stability condition (Blake 1949; Knapp et al. 1970):

$$\frac{d}{dR}\left(\frac{2S}{R}\right)\geq\frac{dp_G}{dR}$$

shows that no stable mechanical equilibrium exists when the tension applied to the bubble exceeds the critical value expressed by:

$$(p_V-p_L)_{cr}=\frac{4S}{3R_o}\left[3\left(1+\frac{p_{Lo}-p_V}{2S/R_o}\right)\right]^{-1/2}$$

In this case, neglecting thermal and diffusive effects, the bubble growth rate is determined by the inertia of the liquid and, when p_V-p_L is constant, for large times approaches the value:

$$\dot{R}=\sqrt{\frac{2(p_V-p_L)}{3\rho_L}}$$

Conversely, when thermal effects dominate (Plesset and Zwick 1954):

$$\dot{R} = \sqrt{\frac{3}{\pi}} \frac{k_L(T_{Lo} - T_R)}{q_{LV}\rho_V\sqrt{\alpha_L t}}$$

where k_L is the thermal conductivity of the liquid, T_R is the temperature at the bubble interface (assumed equal to the bubble internal temperature T_B), T_{Lo} is the unperturbed temperature of the liquid, q_{LV} is the latent heat of vaporization, ρ_V is the vapor density, α_L is the thermal diffusivity of the liquid and t is the time from the beginning of the bubble growth. On the other hand, in the presence of surface tension free bubbles in an infinite liquid are unstable with respect to diffusion of their noncondensable gas into the liquid (Epstein and Plesset 1950) because a reduction of the bubble radius increases their pressure, driving the gas into solution and resulting in a further reduction in size. The opposite happens when the bubble radius increases. The situation is, however, different when a finite amount of liquid is considered and the concentration of the dissolved gas in the liquid is appreciably affected by the diffusion from the bubble (Mori et al. 1977). The rate of bubble diffusion in liquids is especially relevant in cavitation loop experiments when bubble resorption is longer than the recirculation time of the liquid and the bubbles created by the earlier collapses of cavities can accumulate and become a significant source of nuclei. Various stabilization mechanisms have, therefore, been proposed in order to explain the low tensile strength of waters where bubbles are not expected nor observed to remain in suspension. Among these mechanisms, the model proposed by Harvey et al. (1947) where a gas pocket is stabilized against diffusion by surface tension in a crevice of an hydrophobic solid surface, seems to be the most consistent with experimental observation (Apfel 1970).

6 Thermal Cavitation Effects

Thermodynamic or thermal effects on cavitation originate from the difference between the vapor pressure in the cavities and the vapor pressure corresponding to the bulk temperature of the liquid. Scaling of thermal effects is very important in the attempt to extrapolate the results from cavitation tests in cold water to cavitation in liquids with relatively high vapor pressures, such as light hydrocarbons and cryogenic fluids, where testing is impractical, dangerous, or expensive. In general, thermal effects are not restricted to the case of travelling bubble cavitation alone, but also occur in attached cavitation and supercavitation.

The detailed physical mechanism originating thermal effects is, however, different in the two cases. In travelling bubble cavitation thermal effects occur as a consequence of the evaporation rate required to sustain the individual bubble growth (or collapse) produced by the mutual interaction of the cavities with the rest of the flow field. In developed attached cavitation and in supercavitating flows thermal effects originate from the evaporation rate necessary to balance the flux of vapor entrained in the wake of the cavity. In the absence of entrainment and diffusion of noncondensable, gas the pressure in a steady attached cavity would be equal to the vapor pressure of the liquid at its bulk temperature and no thermal effects would occur. Thus, in principle, a comprehensive scaling theory of thermal effects should also account for the type of cavitation occurring in a specific application. This, however, proved at first exceedingly complex, and therefore simplified scaling theories have initially been developed on the basis of averaged homogeneous flow models, which do not account for any specific topological form of cavitation.

The first method to scale thermal effects in cavitating turbopumps has been proposed by Stahl and Stepanoff (1956), and later by Stepanoff (1961, 1964). They suggested to correlate the cavitation characteristics of different liquids using the parameter B, defined as the vapor to liquid volume ratio consequent to the flashing of various liquids subject to the same "dynamic depression" produced by the operation of the machine:

$$B = \frac{\rho_L}{\rho_V} \frac{c_{pL} \Delta T}{q_{LV}}$$

Here ΔT is the temperature drop due to the phase change and c_{pL} is the specific heat of the liquid. Implicit in this method is the assumption that phase changes take place adiabatically (Acosta and Hollander 1959) and that the same "dynamic depression" is produced in geometrically similar machines independently on the nature of the cavitating liquid, thus neglecting both the dynamic aspects of cavitation and its mutual interactions with the liquid flow. From the relative value of the parameter B in various liquids it would then be possible to infer, for example, whether a certain NPSH reduction is possible for a pump operating at design conditions in a specific flow application. The temperature and vapor pressure drops are related by the Clausius–Clapeyron equation:

$$\Delta p_V = \left(\frac{dp_V}{dT} \right)_o \Delta T = \frac{q_{LV}}{T(v_V - v_L)} \Delta T$$

where $v_V = 1/\rho_V$ is the specific volume of the vapor, $v_L = 1/\rho_L$ is the specific volume of the liquid, and, for small changes Δp_V and ΔT, dp_V/dT is computed along the saturation line at the reference unperturbed conditions, indicated by the subscript o. Then, neglecting the specific volume of the liquid with respect to the specific volume of the vapor, the corresponding positive correction to the cavitation index is expressed by

$$\Delta \sigma_V = B \left(\frac{\rho_V}{\rho_L} \right)^2 \frac{q_{LV}^2}{T_{Lo} c_{pL} U_o^2}$$

The parameter B for given values of U_o and $\Delta\sigma_V$ varies greatly in different liquids and correctly indicates the observed trends of resistance to cavitation. The results of Salemann (1959) indicate that the values of B for many commonly used liquids at cut-off conditions are well correlated by an approximately hyperbolic variation with vapor pressure. Thus, for example, at the two opposite sides of the spectrum, very high values of B characterize water at room temperature with respect to liquid hydrogen at 20 K, thus suggesting that cavitation tests in cold water are quite conservative when compared to most other liquids, as confirmed by experience.

The method of Stahl and Stepanoff has later been generalized by Gelder et al. (1966), Moore and Ruggeri (1968), Ruggeri and Moore (1969), Hord et al. (1972), Hord (1973a, b), Hord (1974), Moore (1974), to include the influence of the flow characteristics on thermal effects. In their work a plausible functional dependence of the B-factor on the flow parameters is initially postulated. Then, by comparison with experimental results, semi-empirical correlations are obtained for various liquids (particularly cryogenic fluids) in relatively simple flow configurations such as venturi tubes, ogives, hydrofoils as well as in the more complex case of turbopumps (Ruggeri and Moore 1969; Hord 1974; Moore 1974).

Most of the above analyses only predict the minimum NPSH necessary for pump operation without cavitation or the NPSH corresponding to some standard cavitation conditions, usually the 3% efficiency degradation point. On the other hand, the method proposed by Ruggeri and Moore (1969), Moore (1974) predicts the suction performance throughout the pump's cavitation range from the knowledge of two reference suction characteristics under different thermal cavitation conditions. This method claims a high accuracy, but is very sensitive to the choice of the reference characteristics.

Some useful semi-empirical correlations of thermodynamic effects on the minimum NPSH of cavitating turbopumps have been proposed by Zika (1984). Also these correlations, however, are essentially based on the B-factor method and do not provide any additional understanding of the mechanisms responsible for the occurrence of thermodynamic effects.

In travelling bubble cavitation some of the most important cavitation scaling effects can be identified and interpreted using the Rayleigh–Plesset equation. Any positive value of the right hand side will determine the growth of the bubble. Of particular interest for turbomachinery application is the case of developed travelling bubble cavitation, which involves large changes of the bubble volume. In this situation the effects of surface tension and gas content quickly become negligible as the radius increases. Hence, the growth of the bubble can be controlled either by the inertial effects due to the acceleration of the surrounding liquid or by the decrease of the pressure differential that drives the motion. One possible way the latter situation can occur is through an increase of the far field pressure of the bubble as a consequence of the interactions of cavitation with the rest of flow field. Clearly, this effect cannot be accounted for by the equation of an isolated bubble, where the external pressure in the liquid must be assigned as a known input of the dynamic problem. Another possible limitation of the bubble growth is a decrease of the bubble internal pressure. Since the effect of gas content quickly becomes negligible, this implies a drop of the vapor pressure due to the inability of the heat transfer from the liquid to the bubble surface to sustain the evaporation rate required to keep pace with the increase of the bubble volume (thermal effects). This type of cavitation is usually indicated as vaporous cavitation and can be viewed as an intermediate regime between inertially controlled cavitation and boiling.

As previously mentioned, the asymptotic growth of the bubble radius is linear in time when controlled by inertial effects and is proportional to the square root of time when controlled by thermal effects with purely conductive heat transfer. This situation is modified, but not fundamentally changed, by the simultaneous presence of convection due to the relative motion of the bubble in the liquid. One important consequence is that, the other conditions being the same, a bubbly mixture will display a larger tensile strength when the growth of the cavities is controlled by thermal rather than inertial effects. This phenomenon provides a qualitative interpretation of the lower cavitation susceptibility and more gradual performance breakdown commonly observed in pumps operating in travelling bubble cavitation conditions with liquids closer to the boiling point (Chivers 1969).

In view of the above considerations, inertial effects tend to be important in the initial stages of bubble growth, especially if the vapor pressure is small, but at some later time thermal effects will inevitably prevail. In some situations, for example when the vapor pressure is large and therefore the required evaporation rate is high, inertial effects are never significant and the bubble growth will be controlled by thermal effects alone. Whether in a practical case the bubble growth will be controlled by thermal effects rather than by inertial effects clearly depends on the relative importance of the two phenomena during the largest portion of bubble growth.

A quantitative description of this phenomenon clearly requires the determination of the behavior of the bubble pressure p_B through the simultaneous solution of the Rayleigh–Plesset equation and the energy equations for the two phases. The resulting mathematical problem is quite complex due to its nonlinear nature and to the strong coupling between mechanical and thermodynamic effects. This analysis for a stream of bubbles travelling with velocity U_o in a liquid stream at temperature T_{Lo} has been outlined by Brennen (1973). It is assumed that the pressure and temperature of the vapor inside the bubble are uniform and that the temperature boundary layer at the liquid–vapor interface is small with respect to the bubble radius (Plesset and Zwick 1954; Zwick and Plesset 1955). The drop of the liquid temperature at the bubble surface $\Delta T_R = T_R - T_{Lo}$ is related to the corresponding drop of the vapor pressure inside the bubble $\Delta p_V = p_V - p_{Vo}$ using the Clausius–Clapeyron equation. Then, by only retaining the prominent terms in the energy equations and neglecting inertial effects in the momentum equation for the bubble growth, it follows that vaporous cavitation will occur when

$$\left(\frac{U_o^3\sigma_V}{L}\right)^{1/2} << \frac{2q_{LV}^2\rho_V^2}{\alpha_L^{1/2}c_{pL}T_{Lo}\rho_L^2} = \Sigma(T_L)$$

In this equation L is the typical length of the reduced pressure region crossed by the bubbles and $\Sigma(T_L)$ is a very strong dimensional function of the flow temperature T_L (Brennen 1973) which represents the combined thermodynamic properties of the liquid and the vapor with respect to the insurgence of vaporous cavitation. Therefore, the above equation implicitly defines a transitional temperature of the flow depending on the value of the parameter $\sqrt{U_o^3\sigma_V/L}$. Vaporous cavitation will occur when the flow temperature is higher than the above transitional temperature, oth-erwise inertially dominated cavitation will take place. Assuming, for example, $U_o = 100\,\mathrm{m/s}$, $L = 0.5$ m, and $\sigma_V = 0.1$, the transitional temperatures of water and liquid oxygen, respectively, are 60 °C and 70 K ca., while, on the other hand, cavitation in liquid hydrogen would virtually always be of the vaporous type.

Experimental evidence also suggests the existence of a correlation between the relative magnitude of thermal effects with respect to inertial effects and the topological form of cavitation. In particular, travelling bubble cavitation has been consistently observed in connection with the predominance of thermal effects, whereas attached cavitation is increasingly more likely when inertial effects prevail (Salemann 1959; Sarosdy and Acosta 1961). Quantitative evaluation of thermal effects in travelling bubble cavitation is difficult because the departure of the vapor pressure from its unperturbed value depends on various uncertain factors, including the size and the growth time of the bubble. As a first orientation, the maximum change of the vapor pressure according to Plesset (1957) is expressed by

$$\Delta p_V = \frac{R_{\max}}{3\sqrt{\alpha_L t_{\max}}}\frac{q_{LV}}{c_{pL}}\frac{\rho_V}{\rho_L}\frac{dp_V}{dT}$$

where $R_{\max}$ is the maximum radius reached by the individual bubble during its growth time $t_{\max}$. Due to the different physical characteristics of the problem, the above analysis of thermal effects in travelling bubble cavitation is not directly applicable to developed attached cavitation. In this case entrainment theories have been proposed by various authors (Brennen 1969; Billet 1970; Holl et al. 1975, Billet et al. 1978) on the assumption that the cavity behavior and the hydraulic performance degradation in geometrically similar cavitating turbomachines are only functions of the cavitation index based on the cavity pressure p_C:

$$\sigma_C = \frac{p_o - p_C}{\frac{1}{2}\rho_L U_o^2}$$

In steady-state conditions, equating the heat flux $\dot{q} = q_{LV}\dot{m}_V$ necessary to evaporate the mass flow rate $\dot{m}_V$ of vapor entrained at the trailing edge of the cavity to the heat flux $\dot{q} = h_L A_C(T_{Lo} - T_C)$ transferred from the liquid to the cavity results in the following expression (Billet et al. 1981) for the temperature drop in the cavity:

$$\Delta T = \frac{C_M}{C_A}\frac{\text{Pe}}{\text{Nu}}\frac{\rho_L}{\rho_V}\frac{q_{LV}}{c_{pL}}$$

where $\text{Pe} = U_o D/\alpha_L$ is the Péclet number based on the characteristic transversal dimension D of the cavity and $\text{Nu} = h_L D/k_L$ is the Nusselt number based on the heat transfer coefficient h_L and thermal conductivity k_L of the liquid. The cavity surface area coefficient $C_A = A_c/D^2$ and, in particular, the vapor entrainment coefficient $C_M = \dot{m}_V/\rho_V D^2 U_o$ depend in a complex way on the specific flow configuration. For simple geometries C_A can sometimes be estimated from free streamline theories of cavity flows as a function of the cavitation index. No reliable theoretical methods exist, on the other hand, for evaluating the coefficient C_M, which has to be determined experimentally.

The above expression of the temperature drop in attached cavitation can be considered as a generalization of the B-factor method where

$$\Delta T = B\frac{\rho_V}{\rho_L}\frac{q_{LV}}{c_{pL}}$$

More complex empirical correlations, which include the effects of viscosity, turbulence, gravity, surface tension and geometry of the cavity, are obtained upon comparison with experimental results by assuming a plausible functional dependence of ΔT on the Reynolds number $\text{Re} = U_o D/\nu_L$, Prandtl number $\text{Pr} = \nu_L/\alpha_L$, Froude number $\text{Fr} = U_o/\sqrt{gD}$, Weber number $\text{We} = U_o\sqrt{\rho_L D/S}$ and on the cavity aspect ratio L/D. The relatively large number of free parameters makes these correlations rather flexible for representing available experimental results and, not surprisingly, satisfactory agreement has been obtained in several simple flow configurations. For the same reason, however, the above correlations are of little use in the more general case when trying to extrapolate thermal effects observed in a specific application to geometrically different flow configurations. Thus, for example, the experimental information currently available on thermal effects in simple flow geometries cannot be safely generalized to the more complex case of cavitating turbomachines.

7 Steady Cavitation in Turbomachines

As mentioned earlier, in the majority of technical applications cavitation takes place long before other constraints (structural, dynamic, etc.) become critical and therefore represents the most severe limitation to the increase of the specific power of hydraulically operating turbomachines. Among the various undesirable effects of cavitation, such as erosion, vibrations, noise, operation irregularities, instabilities, etc., the progressive head and efficiency degradation and final breakdown are the most crucial ones from the strict fluid dynamic point of view.

Cavitation occurs when the local pressure in the liquid stream drops below the inception value as a consequence of a combined decrease of the inlet liquid pressure and a simultaneous increase of the rotational speed and volume flux. When investigating cavitation in turbomachines it is, however, more convenient to consider the situation where the inlet pressure is quasi-statically reduced, while the rotational speed and volume flow rate are kept constant. The resulting variation of the head rise is the cavitation performance curve (or cavitation characteristics) of the turbomachine. The development of cavitation in turbomachines can ideally be divided in four steps: inception, usually first appearing as a laminar separation cavity at the leading edges of the blade tips, or as tip vortex cavitation; blade cavitation, a more extensive form of cavitation in some cases occurring on alternate blades only; developed cavitation, characterized by larger, periodically pulsating cavities; cavitation breakdown, where the cavities greatly interfere with the flow and produce a rapid drop of the turbomachine's performance. This rather idealized situation is complicated in practice by the simultaneous presence of separation, tip clearance vortex cavitation, secondary and leakage flows, real fluid effects, and other interfering phenomena. The steady cavitation problem in turbomachines essentially consists in the prediction of head and efficiency degradation and final breakdown as a function of the inlet pressure at constant flow rate and rotational speed.

The theoretical analysis of steady cavitation in turbomachines has mostly relied on the application of free streamline theory on simplified models of the cavitating flow. A classic example is the work of Gongwer (1941) who used the theory of Betz and Petersohn (1931) for flat plate cavitating cascades (essentially derived from the corresponding theories for isolated hydrofoils) to predict the cavitation performance of centrifugal impellers. The same approach has been applied to axial machines and, more specifically, to helical inducers by Stripling and Acosta (1962). Neglecting radial velocity components, the flow in the inducer is assumed to occur on cylindrical surfaces, which are then developed on a plane. Cavitation in the resulting cascade of flat plate hydrofoils is then represented by a free streamline model, where flow separates at the leading edges and reattaches downstream on fictitious surfaces, parallel to the hydrofoils, whose only function is to guide and stabilize the flow in the wake. At the end of these surfaces the flow undergoes a sudden expansion, with the simultaneous occurrence of diffusion losses and pressure recovery. The cavity is assumed to be much shorter than the blade passages (partial cavitation), so that the uniform downstream flow is perfectly guided and not affected by the occurrence of cavitation upstream. The height of the cavity relative to the blade spacing determines the flow blockage and therefore the diffusion losses in the wake, which crucially contribute to the performance degradation and breakdown of the turbomachine. The solution clearly depends on the cavitation number and on other geometric and kinematic parameters: inlet flow velocity, angle of attack, blade angle and spacing. In their paper Stripling and Acosta extensively document the variation of diffusion losses and cavity height, length, and aspect ratio as functions of the cavitation number (or other equivalent cavity pressure parameter) for various values of the attack and blade stagger angles.

The above theory has been extended and modified by Stripling (1962) in order to predict head rise performance and breakdown conditions of cavitating inducers as a function of the inlet pressure. The results are first expressed in terms of the NPSH and all radially varying parameters are reduced to blade tip station. Head breakdown is assumed to occur when the cascade reaches supercavitation conditions. In this situation the cavity length becomes infinite and the cavitation number attains its lowest possible value, which only depends on the angle of attack and on the blade stagger angle. The maximum value of the cavitation number at breakdown, which determines the suction performance of the inducer, is achieved when the angle of attack is equal to one half the blade stagger angle measured from the circumferential direction. Comparison with experimental results shows, however, that the theory significantly underestimates the inducer NPSH at breakdown. Stripling speculates that this discrepancy is due to real fluid displacement effects, which effectively reduce the available through flow area in the blade passages. Consequently he introduces an empirical correction coefficient of the through flow area accounting for the difference between the ideal and actual blockage of the cavity. Small changes of this correction coefficient have a large influence on the suction performance of the inducer. Stripling shows that the minimum NPSH at breakdown and the blockage correction coefficient of a helical inducer can be expressed as a function of a single parameter, which only involves the inducer's flow coefficient, the angle of attack and the blade angle. The minimum NPSH at breakdown and the whole cavitation performance curve predicted by the modified theory are in substantial agreement with the experimental results from various helical inducers of different geometric characteristics. In his paper Stripling also discusses the effects of different leading edge shapes on the inducer cavitation characteristics. Theoretical considerations (Stripling and Acosta 1962) suggest that superior suction performance should be obtained when the blade leading edge is fully contained in the cavity (i.e., when the suction side does not interfere with the separating streamline) and the actual flow more closely follows the ideal free streamline model. The experimental results presented by Stripling (1962) and relative to inducers with various leading edge shapes confirm this conclusion, indicating that maximum suction performance and better agreement with the theory are achieved when the suction side, rather than the pressure side, is filed in order to obtain a sharp corner for fixed separation to occur from the pressure side without any flow rotation around the leading edge. The modified theory by Stripling (1962) proved to be fairly successful in predicting the cavitation characteristics of helical inducers in water and represents a very useful tool in inducer design analysis. However, caution must be exercised in generalizing its results to liquids with important thermodynamic effects, where cavitation is more homogeneous and the free streamline model may be misleading. The same consideration holds for the extension of Stripling's theory to inducers of significantly different geometries, such as aerodynamically designed inducers with variable cross-sectional area and spiraling leading edges, where the assumption of cylindrical flow is less applicable and tip vortex cavitation may play a larger role with respect to cavitation on the blade suction surfaces.

An attempt to incorporate the influence of thermodynamic properties of the liquid in the free streamline model of inducer flow proposed by Stripling and Acosta has been carried out by Jakobsen (1964). He starts from the observation that in liquids with significant thermodynamic effects cavitation takes place in a rather homogeneous way, with the presence of a bubbly stream of cavities dispersed in the liquid phase. Therefore, Jakobsen postulates that the cavity collapse in the wake can then be represented as the steady propagation of a shock wave in the compressible bubbly mixture, whose speed of sound depends on the individual properties of the two phases and on their volume ratio. Head breakdown is assumed to occur when sonic conditions are reached in the inducer passages, corresponding to a certain critical volume ratio of the two phases. By identifying this volume ratio with the *B*-factor for the correlation of thermodynamic effects in cavitating flows, the maximum dynamic depression (and therefore the minimum NPSH) possibly occurring in the bubbly mixture in adiabatic conditions can be calculated. The actual dynamic depression will be lower due to the heat transfer between the two phases. The form of the functional dependence of the NPSH on the *B*-factor at breakdown conditions is then estimated from qualitative consideration of the thermal and dynamic problem of the cavitation bubbles. The theory seems to correlate successfully experimental data for inducers operating at design conditions in water and liquid oxygen. At higher flow coefficients the agreement is, however, less satisfactory, although still within the intrinsic indetermination of available test data.

8 Flow-Induced Instabilities in Turbomachines

In high-performance hydraulically operating turbomachines, the development of higher power densities through the increase of the rotational speed has aggravated, especially in space applications with stringent space and weight limitations, the local instability problems generated by the interaction of the flow with the rotating and static parts. Instability phenomena in turbomachines can affect the rotordynamic operation of single components or the overall behavior of the flow system. A typical example of rotordynamic instability is represented by the forced or self-excited orbital (whirl) motion of turbomachine impellers under the influence of destabilizing forces. These forces can be of mechanical or fluid dynamic origin. Among the destabilizing forces of mechanical origin are those due to internal damping and hysteresis of the rotor and shaft (Kimball 1924), anisotropic stiffness of the rotating parts and bearings (Grabowski 1982), dynamic imbalance (Gunter et al. 1982), interference of stationary and rotating parts (Goggin 1982; Matsushita et al. 1982), system nonlinearities (Wachel and Szenasi 1980), and possibly other sources. Destabilizing forces of fluid dynamic origin are, among others, those produced by seals (Childs 1983) and lubricated bearings (Lund 1974), liquid trapped between the rotating and stationary parts (Ehrich 1967), flow asymmetries (Colding-Jorgensen 1980), unsteady phenomena and, in many cases, cavitation. The global instabilities of a flow system depend instead on the interactions among the various components and therefore vary widely in the specific applications. Apart from the case of turbomachines operating in the positively sloped part of their characteristic curve, the large majority of the global instabilities observed in turbomachines are related to the interaction of the inertia and the compressibility of the flow system or of some of its components, in the presence of an excitation source capable to initiate the instability.

Excluding the case of acoustic phenomena (Abbott et al. 1963; Jeager 1963), in liquid pumping systems unsteady cavitation, especially near rotating parts, often represents the main source of fluid volume changes and can be responsible for the onset of operational instabilities similar to rotating stall and surge phenomena commonly observed in compressors. The first type of unsteady cavitation is usually indicated as rotating cavitation and consists in a blade to blade variation of the extent of cavitation, which rotates with respect to the inducer (or the impeller) of the machine. This type of instability involves significant perturbations of the inlet pressure and generates circumferential flow distortions similar to those observed during rotating stall phenomena (Taylor et al. 1969; Kamijo et al. 1977; Brennen 1994; Hashimoto et al. 1997; Zoladz 2000; Pace et al. 2015), but is not associated to time-dependent fluctuations of the output pressure and mass flow rate of the machine. The second type of unsteady cavitation phenomena consists, instead, of perturbations of volume fluxes and pressures at the input and output of the machine as a consequence of axial fluctuations of the flow caused by simultaneous changes in the volume of the cavities. This type of instability, called auto-oscillation or surge, is generally more severe, since it often involves rather large and violent fluctuations of the flow parameters with respect to their steady-state values (Sack and Nottage 1965; Young et al. 1972; Hobson and Marshall 1979). In turbopumps belonging to more complex systems, unsteady cavitation can also be strongly coupled with other related components capable to provide positive feedback and therefore sustain joint oscillations. A typical example is the POGO instability in liquid propellant rockets (Fashbaugh and Streeter 1965; Rubin 1966; Zielke 1969). Here the longitudinal vibrations of the rocket structure generate inertial pressure fluctuations in the vertical lines of the fuel feed system, which can interfere with the occurrence of cavitation in the turbopumps. The resulting unsteady perturbations of the fuel and/or oxidizer fluxes to the rocket engine can in turn induce thrust fluctuations capable to excite the structure and make the oscillations self-sustained.

The occurrence of cavitation-induced auto-oscillations depends greatly on the specific characteristics of the pumping system in addition to the characteristics of the pumping elements alone. If the conditions are favorable, auto-oscillations initiate when a critical value of the cavitation number is reached and can seriously compromise the overall performance of a pumping system even before the occurrence of large scale performance breakdown due to chocking and flow blockage effects (Braisted 1979). The cavitating element of the hydraulic system is responsible for initiating the auto-oscillations, while the other components only influence their development. The onset of instabilities is greatly facilitated by the presence of the compliant volume due to the presence of the cavities, in conjunction with the inherently unsteady nature of cavitation, which provides a natural excitation and triggering mechanism (Kamijo et al. 1977). The unsteady character of cavitation is particularly enhanced when either the length of the cavity approaches the chord of the inducer blades (Barr 1967) or when the trailing edge of the cavity just enters the passage formed by the adjacent blade (Etter 1970). As anticipated earlier, the problem of cavitation-induced instabilities has, in general, both fluid mechanical and dynamic aspects, since it manifests itself as a time-dependent perturbation of the flow, which generates in turn unsteady forces on rotating and/or stationary elements of the pumping system.

9 Flow Stability of Pumping Systems

In consideration of their widespread occurrence and great practical importance instability problems in turbomachinery have been the subject of intense studies in the past (Greitzer 1981). The final objectives of these studies is twofold, consisting in the identification of the conditions which will lead to the onset of instabilities (stability criteria) and in the prediction of the system behavior once the instability has been initiated, in order to control its effects. The first step in the analysis of a system consists in specifying its nature and in defining a suitable model to represent it. For pumping systems this analysis is normally carried out using a "lumped parameter model," where the various components are identified, ideally isolated from each other and their behavior is represented by a suitable set of discrete parameters. Distributed parameter models are also used in some cases, when necessary. Then a definition of stability must be introduced. According to the commonly accepted definition, an operational point of a dynamic system is considered stable when any small disturbance will decay in time and the system will tend to return to its original state. With reference to pumping systems, two different kinds of instability can be identified. The former, normally indicated as static instability, implies the divergence of the system response to perturbations from the standpoint of purely quasi-steady operational properties and therefore only requires consideration of the characteristic curves of both the pump and the system resistance. The latter kind of instability, usually called dynamic instability, also involves the action of unsteady parameters such as inertances and capacitances and manifests itself with the occurrence of diverging oscillations of the system away from the original operational point. The conditions for dynamic stability are usually more stringent than those for static stability.

The relationships between static and dynamic stability can be illustrated in a more formal way following the approach outlined by Greitzer (1981) if it is assumed, as often is the case in practice, that the system under examination can be described by a second order linear system:

$$\ddot{X} + 2\alpha\dot{X} + \beta X = 0$$

with α and β constant. By assuming solutions of the form e^{st} for the controlled variable X as a function of time t, one is reduced to the determination of the roots of the associated characteristic equation:

$$s^2 + 2\alpha s + \beta = 0$$

for the complex parameter s. Then, the transient response of the system is expressed by:

$$X(t) = c_1 \exp\left[\left(-\alpha + \sqrt{\alpha^2 - \beta}\right)t\right] + c_2 \exp\left[\left(-\alpha - \sqrt{\alpha^2 - \beta}\right)t\right]$$

with the two constants c_1 and c_2 depending on the initial conditions. Thus, clearly, the system will be statically and/or dynamically stable when either of the following two conditions is satisfied:

(1) $0 < \beta < \alpha^2$ with $\alpha > 0$, leading to converging exponential growth;
(2) $\beta > \alpha^2$ with $\alpha > 0$, leading to exponentially converging oscillations.

Viceversa, the system will be statically and/or dynamically unstable in any of the complementary situations, i.e., when

(3) $\beta > \alpha^2$ with $\alpha < 0$, leading to exponentially diverging oscillations;
(4) $0 < \beta < \alpha^2$ with $\alpha < 0$, leading to diverging exponential growth;
(5) $\beta < 0$ with any α, leading to diverging exponential growth.

Thus, in general, a system will be stable when both α and β are positive and will be unstable when the reverse is the case. On the other hand, from the standpoint of quasi-static considerations the system reduces to a first order system with $\alpha = 1$ and the condition for static stability simply requires $\beta > 0$, i.e., that the system response to any perturbation is of restoring type. Note that static stability ($\beta > 0$) is necessary for dynamic stability (conditions 1 and 2), but that a statically stable system can still be dynamically unstable (conditions 3 and 4).

The above approach can easily be applied to a simple liquid pumping consisting of a pump, a pressurized storage vessel, a throttle and the necessary pipelines for connection to the inlet and outlet constant pressure reservoirs. Assuming that the liquid is inviscid and incompressible and upon linearization for small perturbations from the steady-state equilibrium conditions, the following characteristic equation is obtained (Greitzer 1981):

$$\frac{d\Delta p_T}{d\dot{m}} s^2 \left[\frac{\gamma p_1}{\rho_L V_G} - \frac{A_o}{L}\frac{d\Delta p_P}{d\dot{m}}\frac{d\Delta p_T}{d\dot{m}}\right] s + \left[\frac{d\Delta p_T}{d\dot{m}} - \frac{d\Delta p_P}{d\dot{m}}\right]\frac{\gamma p_1 A_o}{\rho_L V_G L} = 0$$

Here p_1 is the inlet liquid pressure, $d\Delta p_P/d\dot{m}$ and $d\Delta p_T/d\dot{m}$ are, respectively, the slopes of the steady-state characteristic curves of the pump and of the throttle, V_G is the volume of the pressurizing gas in the vessel, which behaves isoentropically with specific heat ratio γ, and L is the effective inertial length of the liquid piping system (including the pump and the inlet and outlet lines), expressed by

$$L = A_o \int_{x_1}^{x_2} \frac{dx}{A(x)}$$

where A_o is a convenient reference area, $A(x)$ is the duct cross-sectional area at station x, and the integral extends from the inlet to the outlet sections x_1 and x_2. Since the slope $d\Delta p_T/d\dot{m}$ of throttle characteristics is always positive, from the above equation the well-known condition for static stability of a pumping system is obtained:

$$\frac{d\Delta p_T}{d\dot{m}} < \frac{d\Delta p_P}{d\dot{m}}$$

As indicated earlier, dynamic stability requires that the coefficient of s in the characteristic equation be positive. In most applications the slopes of the throttle characteristics and the volume of gas in the pressurized vessel are comparatively large, therefore this condition is usually verified near the peak of the pump characteristic curve.

A similar approach can also be applied to the analysis of the stability of a cavitating pumping system with respect to the occurrence of fluctuations of the input and output mass flow rates and pressures as a consequence of unsteady changes of the cavitation volume. Let the pumping system consist of a liquid reservoir at constant pressure (station 0) connected by a pipeline to the inlet section (station 1) of a cavitating pump, whose outlet section (station 2) is in turn connected through a throttle to a second reservoir at constant pressure (station 3). With respect to the noncavitating case previously examined, volumetric changes in the system are now due to cavitation inside the pump rather than to the external pressurized storage tank. In a simplified approach the analysis is only concerned with small in-phase departures (here indicated with the symbol δ) from the steady-state operational conditions (Greitzer 1981). Then, the pressure perturbation at the pump inlet is

$$-\delta p_1 = \frac{L_1}{A_1} \frac{d\delta\dot{m}_1}{dt}$$

where A_1 is a suitable reference cross-sectional area and L_1 is the corresponding inertial length of the pump inlet duct. Neglecting the effects of changes of the rotational speed, it is then assumed that the unsteady cavitating performance of the pump is only a function of its inlet pressure p_1 and mass flow rate $\dot{m}_1$. Therefore:

$$\delta p_2 - \delta p_1 = \frac{\partial \Delta p_P}{\partial p_1}\delta p_1 + \frac{\partial \Delta p_P}{\partial \dot{m}}\delta \dot{m}_1 - \frac{L_P}{A_1}\frac{d\delta \dot{m}_1}{dt}$$

$$\delta V_C = \frac{\partial V_C}{\partial p_1}\delta p_1 + \frac{\partial V_C}{\partial \dot{m}}\delta \dot{m}_1$$

and, assuming $\rho_G \ll \rho_L$:

$$\delta \dot{m}_2 - \delta \dot{m}_1 = \rho_L \frac{d\delta V_C}{dt}$$

where L_P is the inertial length of the pump (from the inlet to the outlet section) and V_C is the cavitation volume. Finally, the linearized pressure drop in the throttle is expressed by

$$\delta p_2 = \frac{L_2}{A_2}\frac{\partial \Delta p_T}{\partial \dot{m}}\delta \dot{m}_2$$

In the linearized quasi-steady approximation the derivatives appearing in the above equations are known quantities computed from the steady-state operational conditions. The characteristic equation for the system of previous perturbation equations is

$$\frac{L_1}{A_1}\frac{\partial V_C}{\partial p_1}s^2 - \left[\frac{1 + \frac{\partial \Delta p_P}{\partial p_1} + \frac{L_P}{L_1}}{\frac{\rho_L A_1}{L_1}\frac{\partial \Delta p_T}{\partial \dot{m}}} + \frac{\partial V_C}{\partial \dot{m}}\right]s - \frac{1}{\rho_L}\left[1 - \frac{\frac{\partial \Delta p_P}{\partial \dot{m}}}{\frac{\partial \Delta p_T}{\partial \dot{m}}}\right] = 0$$

In addition to the steady-state slopes of the pump and throttle characteristics, the above equation also involves the derivatives of the cavitation volume with respect to the inlet pressure and mass flow. These parameters, respectively called compliance ($-\partial V_C/\partial p_1$) and mass flow gain factor ($-\partial V_C/\partial \dot{m}$), characterize the behavior of cavitation in the pump with respect to changes of the independent inlet variables p_1 and $\dot{m}_1$. Since the pressure versus flow characteristics of the pump at the conditions of interest and the pump compliance are both negative, the system will become unstable when the mass flow gain factor exceeds a critical value:

$$-\frac{\partial V_C}{\partial \dot{m}} > \frac{1 + \frac{\partial \Delta p_P}{\partial p_1} + \frac{L_P}{L_1}}{\frac{\rho_L A_1}{L_1}\frac{\partial \Delta p_T}{\partial \dot{m}}}$$

The above analysis can be made nondimensional by introducing the usual expressions for the flow and head coefficients and for the cavitation number of the pump, based on the impeller tip speed Ωr_T and pump inlet area A_1. In this form the above characteristic equation for the nondimensionalized frequency $sL_P/\Omega r_T$ becomes independent on the pump speed. Therefore, contrary to the results for the noncavitating case, the point of critical stability in dimensionless coordinates should be independent on the pump speed. The same analysis also indicates that the natural frequency of the cavitation-induced flow oscillations should be proportional to the pump speed, a result which seems to be in agreement with the experimental observations of Braisted (1979), Brennen and Braisted (1980).

10 Flow Stability of Cavitating Turbopump Systems

The above simplified treatment of the unsteady performance of a cavitating turbopump clearly indicates the importance of the cavitation compliance and, in particular, of the mass flow gain factor for the onset of auto-oscillations in cavitating pumping systems. Early theoretical attempts to describe the dynamic behavior of cavitating turbopumps used the slopes of the steady-state characteristics of head rise versus flow rate and cavitation number in conjunction with an empirical compliance coefficient in order to account for the volume changes due to cavitation (Rubin 1966; Wagner 1971; Vaage et al. 1972; Farrel and Fenwick 1973; Brennen and Acosta 1973; Brennen 1973). A more quantitative and systematic approach to the problem has been proposed by Brennen and Acosta (1976) and their co-workers (Ng et al. 1976; Ng and Brennen 1978; Brennen 1978a, b; Brennen et al. 1982). In the linear limit for small perturbations the unsteady performance of each component of a pumping system is represented by a transfer matrix which relates the unsteady fluctuations of the inlet variables to their corresponding outlet values. The inlet pressure and mass flow rate are used as the independent variables for each component. In order to account for the possibility of phase lags, periodically fluctuating quantities (denoted by primes) are represented in complex exponential form:

$$p'(t) = \mathrm{Re}\{\hat{p}e^{-i\omega t}\}; \quad \dot{m}'(t) = \mathrm{Re}\{\hat{\dot{m}}e^{-i\omega t}\}$$

by means of their complex amplitudes, here indicated by hats. Therefore, for the generic component of the pumping system:

$$\begin{pmatrix} \hat{p}_2 - \hat{p}_1 \\ \hat{\dot{m}}_2 - \hat{\dot{m}}_1 \end{pmatrix} = \begin{bmatrix} Z_{11} & Z_{12} \\ Z_{21} & Z_{22} \end{bmatrix} \begin{pmatrix} \hat{p}_1 \\ \hat{\dot{m}}_1 \end{pmatrix}$$

where the indexes 1 and 2 refer to the inlet and outlet conditions. In another slightly different representation the pressure and mass flow fluctuation amplitudes at discharge are expressed as functions of the corresponding inlet values:

$$\begin{pmatrix} \hat{p}_2 \\ \hat{\dot{m}}_2 \end{pmatrix} = \begin{bmatrix} Y_{11} & Y_{12} \\ Y_{21} & Y_{22} \end{bmatrix} \begin{pmatrix} \hat{p}_1 \\ \hat{\dot{m}}_1 \end{pmatrix}$$

Clearly the two transfer matrices are related by

$$Y = Z + 1$$

where 1 is the unitary matrix of order two. In general, the coefficients of the transfer matrices $\boldsymbol{Z}$ and $\boldsymbol{Y}$ are complex and are expected to depend on the frequency, the mean operating point and the extent of cavitation. In the limit for low frequencies the above equations reduce to the steady-state operational characteristics.

Of particular interest is the case of a cavitating turbopump. In a more generalized treatment also the rotational speed could be included in the modeling of the unsteady flow through a turbopump, leading to 2×3 matrix representation. However, usually the fluctuations of the pump rotational speed in response to fluctuating loads on the impeller blades are insignificant, being effectively limited by the large inertia of the rotating parts. The influence of the pump rotational speed can only become important where there exists the potential for significant interactions between the performance of the pumping system and the operational speed of the pump itself. Conceivably this situation may occur when the pump is driven by a turbine powered by fluid bled from the pump output, a configuration that is typical of low pressure turbopumps (prepumps) of liquid propellant rocket engine feed systems, like in the Space Shuttle Main Engine. In general, Brennen and Acosta (1976) have shown that angular speed changes have a negligible effect on the dynamics of the pump as long as their relative value is small compared with the relative change of the cavitation number. In cavitating turbopumps Z_{11} is the change of the pressure rise versus inlet pressure, which is expected to be zero in the absence of cavitation (or of a second gaseous component of the flow), since in this case the pressure rise produced by the pump does not depend on the value of the inlet pressure. Clearly, for very low frequencies Z_{11} reduces to the slope of the cavitation performance curve versus the cavitation number of the pump. The element Z_{12} is the change of the pressure rise versus inlet mass flow rate and therefore represents the impedance of the pump, comprising a resistive and an inertial component At low frequencies the pump resistance simply approaches the slope of the steady-state characteristic curve of the pump. The elements Z_{21} and Z_{22} account for the effects of the inlet pressure and mass flow rate on the cavitation volume of the pump and therefore on the instantaneous difference between the inlet and outlet mass fluxes. Hence, Z_{21} and Z_{22} ,respectively, represent the pump compliance and mass flow gain factor. Neglecting the effects of liquid compressibility and the deformability of the pump casing, Z_{21} and Z_{22} are only due to the occurrence of cavitation or to the presence of a second gaseous component in the flow. Therefore, the pump compliance and mass flow gain factor are expected to vanish either in fully wetted flow conditions or in the limit for very low frequencies, when the cavitation volume (if any) remains constant in time.

Similar considerations can be made for other typical elements of an hydraulic system. For example, the transfer matrix of an incompressible flow in a rigid duct of reference cross- sectional area A and inertial length L is given by:

$$Y = \begin{bmatrix} 1 & -R + i\omega\frac{L}{A} \\ 0 & 1 \end{bmatrix}$$

where $I = R - i\omega L/A$ is the impedance of the duct, consisting of a resistive contribution R due to the hydrodynamic dynamic losses, and of an inertial component $-i\omega L/A$, which depends on the oscillation frequency ω. On the other hand, the transfer matrix of a simple compliant volume Va containing a gas at pressure p is given by:

$$Y = \begin{bmatrix} 1 & 0 \\ i\omega\frac{\rho_L V_G}{\gamma p_1} & 1 \end{bmatrix}$$

Once the transfer matrix of each element of a pumping system is known it is then possible to determine by simple multiplication the transfer matrix of the whole flow line.

The stability analysis of the pumping system can then be carried out using energy considerations, as proposed by Brennen and Braisted (1980). Since any form of instability requires the supply of a certain amount of energy in order to sustain the oscillations, Brennen and Braisted suggested to consider as an instability parameter the net output change of the mechanical energy flux developed by the system in an oscillation cycle:

$$\Delta E = \int_t^{t+T} \frac{\dot{m}_2' p_2' - \dot{m}_1' p_1'}{\rho_L g} dt$$

which they called dynamic activity parameter. When applied to an individual component, this parameter would indicate whether that component would have a potentially stabilizing or destabilizing effect for the onset of auto-oscillations. Each component would be characterized as active or passive with respect to the occurrence of instabilities depending on the positive or negative value of its dynamic activity parameter. When applied to the whole system, the dynamic activity parameter would provide a measurement of the overall tendency to develop instabilities, higher positive values being again characteristic of increasingly unstable systems. Clearly the stability condition is

$$\Delta E \leq 0$$

corresponding to no net positive energy addition to the oscillation phenomenon in a cycle. This criterion is of rather general applicability. In a noncavitating system, for example, it reduces to

$$\mathrm{Re}\{I_1 - Z_{12} + I_2\} \geq 0$$

where I_1 and I_2 are, respectively, the impedances of the inlet and outlet lines and Z_{12} is the only nonzero term of the transfer matrix of the pumping element. Thus, the stability condition simply requires that the resistance of the pumping element must be smaller than the combined resistances of the inlet and outlet lines. Violation of this condition can lead to surge both in compressors (Pearson and Bowmer 1949; Greitzer 1976a, b) and in pumps (Stepanoff 1948). In the cavitating case all of the pump transfer matrix elements are in general different from zero and the condition for stability becomes formally more complex (Braisted 1979):

$$\mathrm{Re}\left\{\frac{I_1(1+Z_{11})+I_2(1+Z_{22})-Z_{12}-I_1I_2Z_{21}}{1+Z_{22}-I_1Z_{21}}\right\}\geq 0$$

The above two equations confirm that stability is a global characteristics of the pumping system, involving the combined properties of each individual component. It is also evident from the first of these equations that only the pumping element(s) can actively induce instability in a system and that increasing the (positive) resistances of the remaining elements have a favorable stabilizing effect. In general, the above equations indicate that the system stability can be improved by properly modifying and/or combining the characteristics of its components. Attempts to this effect have actually been made for example in order to suppress auto-oscillations in the Space Shuttle Main Engine (Murphy 1969; Dorian 1977).

In general, the dynamic activity parameter is expected to depend on the cavitation number, the mean operational point and the frequency of the auto-oscillation. Specifically, the observations of Braisted (1979) indicate that the dynamic activity parameter goes through a maximum for a certain value of the oscillation frequency and increases at low cavitation numbers. These considerations, therefore, suggest a simple method for the determination of both the frequency of auto-oscillation and of the cavitation number at which the auto-oscillation will occur. In particular, given the mean operational conditions of the pumping system, one would expect that auto-oscillations would initially develop with frequency corresponding to the peak in the spectral distribution of the dynamic activity parameter and at the highest value of the cavitation number for which the peak becomes positive. The analysis also indicates that the auto-oscillation frequency is proportional to the pump rotational speed for given values of the flow coefficient and of the cavitation number. Another noteworthy result is that the stability of the system is most sensitive to changes of the mass flow gain factor and, to a lesser degree, of the pump impedance (Braisted 1979). Braisted and Brennen (1980) have used the above stability analysis in conjunction with experimentally determined characteristics of the pumping system and found reasonable in agreement with direct observation of the auto-oscillations. Experimental data also confirm that the auto-oscillation frequency tends to decrease with the cavitation number, as indicated by the stability analysis.

Thus, within the limits of a linearized analysis, the dynamic problem of a cavitating turbopump consists in the determination of the possibly complex elements of its transfer matrix. Transfer matrices of cavitating pumps have been investigated both theoretically (Brennen and Acosta 1973; Brennen and Acosta 1976; Brennen 1978a; Brennen and Braisted 1980) and experimentally (Ng 1976; Ng et al. 1976; Ng and Brennen 1978; Brennen 1978b; Braisted 1979; Brennen et al. 1982;

Franz 1989; Bhattacharyya 1994; Bhattacharyya et al. 1997). Theoretical investigations have been based on quasi-steady analysis of the flow, using two different approaches to model cavitation in the pump inducer. In the first approach (Brennen and Acosta 1973, 1976) radial velocities are neglected in the inducer flow and linear free streamline theory is applied to the resulting two-dimensional cavitating cascade problem in order to evaluate the cavity volume at each radial location from the axis of rotation. The results are then integrated from the hub to the tip radius of the inducer, possibly accounting for the radial variations of several geometrical parameters (blade and attack angles, blade thickness), in order to estimate the total volume of the cavity. Finally, the compliance and the mass flow gain factor are computed from the response of the cavity volume to small quasi-steady changes of the inlet pressure and mass flow rate. The remaining elements of the transfer matrix are deduced from steady-state operational conditions. This approach involves rather crude simplifying assumptions with respect to the observed features of the inducer flow. However, it is relatively simple from the mathematical standpoint and has the distinct advantage of being virtually self-contained, i.e., capable to derive both the compliance and the mass flow gain factor of the inducer from purely theoretical considerations. Considering its limitations, the results of this model are remarkably consistent with the known dynamic characteristics deduced from some tests on rocket engine turbopumps (Brennen and Acosta 1973, 1976).

A second theoretical approach to the description of the dynamic transfer matrices of cavitating turbopumps has been proposed by Brennen (1978a) as a possible interpretation of some of the unexpected features of the experimental transfer matrices reported by Ng and Brennen (1978).

Direct observation had shown that cavitation in high suction specific speed inducers mostly occurs in the form of a bubbly mixture on the suction side of the blades (especially in liquids close to boiling), rather than as a relatively steady fully developed attached cavity. Thus, in Brennen's approach cavitation in the blade passages is modeled as a one-dimensional homogeneous bubbly mixture originating from the blade leading edge. The length and the uniform average void fraction of the cavitation region depend on the cavitation number and the flow coefficient and can be estimated from experimental data. The dynamic transfer matrix of the inducer is then described in terms of two additional free parameters, the compliance and the mass flow gain factor of the cavitation region, whose values must also be estimated from experience or obtained from comparison with experimentally determined transfer matrices. Contrary to the previous approach by Brennen and Acosta (1976), the present model with its four free parameters has a distinct semi-empirical character and therefore is not capable to derive the dynamic transfer matrix of a cavitating turbopump from purely theoretical considerations. However, with proper choice of these parameters, good qualitative and quantitative agreement is obtained with the complex transfer matrices measured in the experimental tests by Ng and Brennen (1978), thus strongly suggesting that the most significant dynamic effects observed in cavitating turbomachinery are indeed consequences of the physical mechanisms and phenomena intimately connected with the compliance and the mass flow gain factor of the cavitation volume.

11 Conclusions

As anticipated in the introduction and illustrated later, cavitation is an extremely articulated and complex phenomenon. These complexities arise as a consequence of a variety of aspects:

- the simultaneous presence of two phases in dynamic and thermodynamic equilibrium;
- the stochastic nature of the inception process, due to the interaction of nuclei with the local static pressure field, either in a Lagrangian frame for free stream nuclei or in an Eulerian frame in the case of surface nuclei;
- the uncertain dynamic behavior of cavitation nuclei, often involving the unknown and uncontrollable interfacial and diffusion properties of microscopic and submicroscopic impurities or inhomogeneities in the liquid and/or at the flow boundaries;
- the extremely rapid speed of cavity growth and collapse especially in the initial and final stages, which control cavitation inception and erosion effects;
- the often unpredictable form of cavitation form and development in response to the specific flow conditions;
- the strong interactions of the cavities with the fully wetted flow and with neighboring cavities;
- the already involved fluid dynamic field of the liquid phase where cavitation occurs, often characterized by complex phenomena due to viscosity, boundary layers, flow separation, turbulence, etc.

The analysis of three-dimensional unsteady cavitating flows in turbomachines is therefore a complex and challenging problem. No comprehensive treatment is yet available for capturing all of the above effects in a general way.

Theoretical analyses are limited to simplified situations (idealized geometry, reduced dimensionality, linearized dynamics, etc.). They can be useful for the study of specific problems, but are difficult to integrate in a global picture because they cannot model many of the existing interactions between different phenomena.

Simulation methods for turbomachines do not suffer, in theory, from such limitations, however, insufficient space and time resolution, computational stability and numerical stiffness constraints have so far prevented their widespread application to three-dimensional unsteady cavitating turbomachinery flows, except for a few exploratory research studies. This limitation may, however, be eventually lifted, since progress in this field is very rapid and the potential for practical application is enormous.

The further inclusion of cavitation adds one more dimension to the complexity of the problem. In this respect the most crucial implications of cavitation for the performance and survival of high power density cryogenic turbopumps are:

- auto-oscillations
- system coupled oscillations (i.e., POGO instability)
- cavitation-induced forces on rotors and stators

As of today only linearized theories of auto-oscillations and system coupled oscillations (POGO instability) are available. They seem to predict with reasonable accuracy the onset of instability (stability limits), but they are unable to describe its following development beyond the linear range (limit cycles) and do not provide any quantitative indication on how to control the system once the oscillation has started.

Cavitation-induced forces represent another area where reliable information is, at the same time, vital and almost completely lacking in the open literature. Fluid forces, in general, are often responsible for very serious problems in high-performance turbopumps, ranging from long-term fatigue failure of the supports and bearings to the onset and development of explosive rotordynamic instabilities. Cavitation forces due to mass and/or blade load imbalance can represent a real danger for themselves and at the same time a powerful source of excitation for the onset of rotordynamic instabilities. Therefore, further research in the following cavitation related areas appears to be particularly significant for progress in this field.

Because of the extreme complexity and articulation of the phenomena involved cavitating and noncavitating flows in turbomachines are only partially understood and imperfectly predicted by theoretical means alone. Hence, today one must recur extensively to experimentation in addressing and solving specific technical problems. For the same reasons it is not likely that significant progress in the understanding of cavitating turbomachines may be achieved only through further theoretical studies, either analytical or numerical, without the orientation, the guidance and, most importantly, the validation obtained from direct observational experience. For the foreseeable future the most viable and profitable strategy for progress in the fluid mechanical design of high-performance cavitating turbomachines must forcibly rely on the careful combination of guided intuition, simplified theoretical analyses, numerical simulations, and experimental results.

References

Abbot, H. F., et al. (1963). Measurements of auto-oscillation in hydroelectric supply tunnel and penstock system. *ASME Journal Basic Engineering*, *85*, 625–630.

Acosta, A. J. (1974). Cavitation and fluid machinery. In *Cavitation* (pp. 386–396), Conference held at Heriot-Watt University, Edinburgh, Scotland, September 1974, London Institution of Mechanical Engineers.

Acosta, A. J., and Hollander, A. (1959). *Remarks on Cavitation in Turbomachines*. California Institute Technical Report No. 79.3, Pasadena, CA, USA.

Acosta, A. J., & Parkin, B. R. (1975). Cavitation inception—a selective review. *Journal of Ship Research, 19,* 193–205.

Apfel, R. E. (1970). The role of impurities in cavitation-threshold determination. *Journal of the Acoustical Society of America*, *48*(5), 1179–1186, Part 2.

Arndt, R. E. A. (1981). Cavitation in fluid machinery and hydraulic structures. *Annual Review of Fluid Mechanics*, *13*, 273–328.

Barr, R. A. (1967). Study of instabilities in high tandem row inducer pumps. Hydronautics Interim Technical Report 703-1.
Bader, H. (1970). The hyperbolic distribution of particle sizes. *Journal of Geophysical Research, 75*(15), 2822–2830.
Betz, A., & Petersohn, E. (1931). *Anwendung du Theorie der Frien Strahlen*. In Ingenieur Archiv, Band II.
Bhattacharyya, A. (1994). *Internal Flows and Force Matrices in Axial Flow Inducers*. Report No. E249.18, California Institute of Technology, Pasadena, USA.
Bhattacharyya, A., Acosta, A. J., Brennen, C. E., & Caughey, T. K. (1997). Rotordynamic forces in cavitating inducers. *ASME Journal of Fluids Engineering, 199*(4), 768–774.
Billet, M. L. (1970). *Thermodynamic efects on developed cavitation in water and freon 113*. M.S. thesis, Department of Aerospace Engineering, Pennsylvania State University, University Park, PA, USA.
Billet, M. L., et al. (1978). Correlations of thermodynamic effects for developed cavitation. In *Polyphase flow in turbomachinery* (pp. 271–289). ASME Special Publications.
Billet, M. L., et al. (1981). Correlations of thermodynamic effects for developed cavitation. *ASME Journal of Fluids Engineering*, *103*, 534–542.
Blake, F. G. Jr. (1949). *The Onset of Cavitation in Liquids*. In Harvard Acoustics Research Laboratory, TM 12.
Braisted, D. M. (1979). *Cavitation Induced Instabilities Associated with Turbomachines*. Report No. E184.2, California Institute of Technology, Pasadena, CA, USA.
Brennen, C. E. (1978a). Bubbly flow model for the dynamic characteristics of cavitating pumps. *Journal of Fluid Mechanics, 89,* 223–240.
Brennen, C. E. (1978b). On unsteady dynamic response of phase changes in hydraulic systems. In *Proceedings of International Seminar of Center for Heat and Mass Transfer in Two-Phase and Chemical Systems*, Dubrovnik, Yugoslavia, Hemisphere Publishing Corporation.
Brennen, C. E. (1969). The Dynamic balances of dissolved air and heat in natural cavity flows. *Journal of Fluid Mechanics, 37,* 115–127.
Brennen, C. E. (1994). *Hydrodynamics of pumps*. Oxford Engineering Science Series 44. Oxford University Press.
Brennen, C. E. (1995). *Cavitation and bubble dynamics*. Oxford University Press.
Brennen, C. E. (2005). *Fundamentals of multiphase flows*. Cambridge University Press.
Brennen, C. E. (1973). The dynamic behavior and compliance of a stream of cavitating bubbles. *Transactions of the ASME Journal Fluids Engineering*, 533–541.
Brennen, C. E., & Acosta, A. J. (1973). Theoretical, quasi-static analysis of cavitation compliance in turbopumps. *Journal of Spacecraft, 10*(3), 175–179.
Brennen, C. E., & Acosta, A. J. (1976). The dynamic transfer function for a cavitating inducer. *ASME Journal of Fluids Engineering, 98,* 182–191.
Brennen, C. E., & Braisted, D. M. (1980). Stability of hydraulic systems with focus on cavitating pumps. In *Proceedings of IAHR Conference*, Tokyo, Japan.
Brennen, C. E., et al. (1982). Scale effects in the dynamic transfer functions for cavitating inducers. *ASME Journal of Fluids Engineering*, *104*, 428–433.
Childs, D. W. (1983). Finite length solutions for rotordynamic coefficients of turbulent annular seals. *ASME Journal of Lubrication Technology*, *105*, 437–444.
Chivers, T. C. (1969). Cavitation in centrifugal pumps. In *Proceedings of the Institution of Mechanical Engineers*, *184*(2), Pt. l.
Colding-Jorgensen, J. (1980). Effect of fluid forces on rotor stability of centrifugal compressors and pumps. In 1*st Workshop on Rotordynamic Instability Problems in High Performance Turbomachinery*, Texas A&M University, TX, USA.
d'Agostino, L., & Acosta, A. J. (1991a). A cavitation susceptibility meter with optical cavitation monitoring—part one: design concepts. *ASME Journal of Fluids Engineering, 113*(2), 261–269.
d'Agostino, L., & Acosta, A. J. (1991b). A cavitation susceptibility meter with optical cavitation monitoring—part two: experimental apparatus and results. *ASME Journal of Fluids Engineering, 113*(2), 270–277.

d'Agostino, L., & Acosta, A. J. (1991c). Separation and surface nuclei effects in a cavitation susceptibility meter. *ASME Journal of Fluids Engineering, 113*(4), 695–699.
Dorian, H. H. (1977). *Space Shuttle POGO Prevention*. SAE, 770969.
Ehrich, F. F. (1967). The influence of trapped fluids in high speed rotor vibration. *ASME Journal of Engineering for Industry*, 806–812.
Eisenberg, P. (1961). Mechanics of cavitation. In V.L. Streeter (Ed.), *Handbook of fluid dynamics* (pp. 12.2–12.24). New York: McGraw-Hill.
Epstein, P. S., & Plesset, M. S. (1950). On the stability of gas bubbles in liquid-gas solutions. *Journal of Chemical Physics, 18,* 1505–1509.
Etter, R. J. (1970). *An Investigation on Tandem Row High Head Inducers*. Hydronautics Technical Report 703-9.
Farrel, E. C., & Fenwick, J. R. (1973). *Pogo Instabilities Suppression Evaluation*. NASA Report CR-134500.
Fashbaugh, R. H., & Streeter, V. L. (1965). Resonance in liquid rocket engine systems. *Transactions of the ASME, Series D, 87*(4).
Fox, F. E., & Herzfeld, K., F. (1954). *Gas Bubbles with Organic Skin as Cavitation Nuclei*. Journal of the Acoustical Society of America, *26*, 984.
Franc, J. -P., & Michel, J. -M. (2004). *Fundamentals of cavitation*. Kluwer Academic Publishers.
Franz, R. J. (1989). *Experimental Investigation of the Effect of Cavitation on the Rotordynamic Forces on a Whirling Centrifugal Pump Impeller*. Ph.D. Thesis, California Institute of Technology, Pasadena, USA.
Gelder, T. F., Ruggeri, R. S., & Moore, R. D. (1966). *Cavitation Similarity Considerations Based on Measured Pressure and Temperature Depressions in Cavitated Regions of Freon 114*. In NASA TN D-3509.
Greitzer, E. M. (1976a). Surge and rotating stall in axial flow compressors—part I: Theoretical compression system model. *ASME Journal of Engineering for Power*, *98*, 190–198.
Greitzer, E. M. (1976b). Surge and rotating stall in axial flow compressors—part II: Experimental results and comparison with theory. *ASME Journal of Engineering for Power*, *98*, 198–211.
Greitzer, E. M. (1981). The stability of pumping systems. *ASME Journal of Fluids Engineering*, *103*, 193–242.
Goggin, D. C. (1982) Field experiences with rub induced instabilities in turbomachinery. In *2nd Workshop on Rotordynamic Instability Problems in High Performance Turbomachinery*, Texas A&M University, TX, USA.
Gongwer, C. A. (1941). A theory of cavitation flow in centrifugal pump impellers. *Transactions of the ASME*, *63*, 29–40.
Grabowski, B. (1982). Shaft vibrations in turbomachinery excited by cracks. In *2nd Workshop on Rotordynamic Instability Problems in High Performance Turbomachinery*, Texas A&M University, TX, USA.
Gunter, E. J., Springer, H., & Humphris, R. (1982). Influence of unbalance on the stability characteristics of flexible rotor bearing systems. In *ASME/Applied Mechanics, Bioengineering and Fluids Engineering Conference* (Vol. 55), Houston, TX, USA, AMD.
Harvey, E. N., McElroy, W. D., & Whiteley, A. H. (1947). On cavity formation in water. *Journal of Applied Physics*, *18*, 162.
Hashimoto, T., Yoshida, M., Watanabe, M., Kamijo, K., & Tsujimoto, Y. (1997). Experimental study of rotating cavitation of rocket propellant pump inducers. *AIAA Journal of Propulsion and Power, 13*(4), 488–494.
Hobson, D. E., & Marshall, A. (1979). Surge in centrifugal pumps. In *Proceedings of 6th Conference on Fluid Machinery* (Vol. 1), Hungarian Academy of Sciences, Budapest.
Holl, J. W. (1969). Limited cavitation. *ASME, Cavitation State of Knowledge*, 26–63.
Holl, J. W. (1970). Nuclei and cavitation. *ASME Journal of Basic Engineering*, *92*, 681–688.
Holl, J. W., & Wislicenus, G. F. (1961). Scale effects on cavitation. *ASME Journal of Basic Engineering*, 385–398.
Holl, J. W. et al. (1975). Thermodynamic effect on developed cavitation. In: *Cavity Flows, ASME Fluids Engineering Conference* (pp. 101–109), Minneapolis, MN, USA.

Hord, J., Anderson, L. M., & Hall, W. J. (1972). *Cavitation in Liquid Cryogens. I. Venturi*. In NASA CR-2054 (83 pp).
Hord. J. (1973a). *Cavitation in Liquid Cryogens. II. Hydrofoil*. In NASA CR-2156 (157 p).
Hord, J. (1973b). *Cavitation in Liquid Cryogens. III. Ogives*. In NASA CR-2242 (235 pp).
Hord, J., (1974). *Cavitation in Liquid Cryogens, Volume IV—Combined Correlations for Venturi, Hydrofoil, Ogives and Pumps*. In NASA CR-2448.
Jakobsen, J. K. (1964). On the mechanism of head breakdown in cavitating inducers. *ASME Journal of Basic Engineering*, *86*, 291–304.
Jeager, C. (1963). The theory of resonance in hydropower systems: Discussion of incidents and accidents occurring in pressure systems. *ASME Journal of Basic Engineering*, *85*, 631–640.
Kamijo, K., Shimura, T., Watanabe, M. (1977). *Experimental Investigation of Cavitating Inducer Instability*.
Kimball, A. L., Jr. (1924). Internal friction theory of shaft whirling. *General Electric Review*, *27*, 244–251.
Knapp, R. T., Daily, J. W., & Hammit, F. G. (1970). *Cavitation*. New York: McGraw-Hill.
Lund, J. W. (1974). Stability and Damped Critical Speeds of a Flexible Rotor in Fluid-Film Bearings. *ASME Journal of Engineering for Industry*, 509–517.
Matsushita, O., et al. (1982). Rotor vibration caused by external excitation and rub. In *2nd Workshop on Rotordynamic Instability Problems in High Performance Turbomachinery*, Texas A&M University, Texas.
Moore, R. D., & Ruggeri, R. S. (1968). *Predictions of Thermodynamic Effects of Developed Cavitation Based on Liquid Hydrogen and Freon 114 Data in Scaled Venturis*. NASA TN D-4899.
Moore, R. D. (1974). Prediction of pump cavitation performance. In B. Lakshminarayana, W. R. Britsch & W. S. Gearhart (Eds.), *Symposium on Fluid Mechanics, Acoustics and Design of Turbomachinery—Part II*, NASA SP-304 (pp. 733–755).
Mori, Y., Hijikata, K., & Nagatani, T. (1977). Fundamental Study of Bubble Dissolution in Liquid. *International Journal of Heat and Mass Transfer*, *20*, 41–50.
Murphy, G. L. (1969). Pogo Suppression Analysis of the S-11 and S-IVB LOX Feed Systems. Summary Report ASD-ASTN-1040, Brown Engineering Co., Huntsville, AL, USA.
Ng, S. L. (1976). *Dynamic Response of Cavitating Turbomachines*. PhD thesis, California Institute of Technology, Pasadena, CA, USA.
Ng, S. L., Brennen, C. E., and Acosta, A.J. (1976). *The Dynamics of Cavitating Inducer Pumps*. In *Conference on Two-Phase Flow and Cavitation*, Grenoble, France.
Ng, S. L., & Brennen, C. E. (1978). Experiments on the dynamic behavior of cavitating pumps. *ASME Journal of Fluids Engineering*, *100*, 166–176.
O'Hern, T. J., d'Agostino, L., & Acosta, A. J. (1988). Comparison of holographic and coulter counter measurements of cavitation nuclei in the ocean. *ASME Journal of Fluids Engineering*, *110*, 200–207.
Pace, G., Valentini, D., Pasini, A., Torre, L., Fu, X., & d'Agostino, L. (2015). Geometry Effects on Flow Instabilities of Different Three-Bladed Inducers. *ASME Journal of Fluids Engineering*, *137*(4)/011102-1, 041304.
Pasini, A., Torre, L., Cervone, A., & d'Agostino, L. (2011a). Characterization of the rotordynamic forces on tapered axial inducers by means of a rotating dynamometer and high-speed movies. In *Proceedings of WIMRC 3rd International Cavitation Forum 2011*, University of Warwick, United Kingdom.
Pasini, A., Torre, L., Cervone, A., & d'Agostino, L. (2011b). Continuous spectrum of the rotordynamic forces on a four-bladed inducer. *ASME Journal of Fluids Engineering*, *133*(12).
Pearsall, L. S. (1972). *Cavitation*. London: Mills & Boon.
Pearson, H., & Bowmer, T. (1949). Surging of axial compressors. *Aeronautical Quarterly*, *I*, 195–210.
Plesset, M. S. (1957). Physical effects in cavitation and boiling. In *Proceedings of 1st Symposium on Naval Hydrodynamics* (pp. 297–323), Washington, D.C., USA.
Plesset, M. S. (1969). The tensile strength of liquids. In *ASME Fluids Engineering and Applied Mechanics Conference*, (pp. 15–25), Evanston, Illinois.

Plesset, M. S., & Prosperetti, A. (1977). Bubble dynamics and cavitation. *Annual Review of Fluid Mechanics, 9*, 145–185.

Plesset, M. S., & Zwick, S. A. (1954). The growth of vapor bubbles in superheated liquids. *Journal of Applied Physics, 5*(4), 493–500.

Rosenmann, W. (1965). Experimental Investigations of Hydrodynamically Induced Shaft Forces With a Three Bladed Inducer. In *Proceedings of ASME Symposium on Cavitation in Fluid Machinery.*

Rubin, S. (1966). Longitudinal instability of liquid rockets due to propulsion feedback (POGO). *Journal of Spacecraft and Rockets, 3*(8), 1188–1195.

Ruggeri, R. S., & Moore, R. D. (1969). *Method for Prediction of Pump Cavitation Performance for Various Liquids, Liquid Temperatures and Rotative Speeds*. In NASA TN D-5292, Washington, D.C., USA.

Sack, L. E., & Nottage, H. B. (1965). System oscillations associated to cavitating inducers. *ASME Journal of Basic Engineering*, *87*, 917–924.

Salemann, V. (1959). Cavitation and NPSH requirements of various liquids. *ASME Journal of Basic Engineering*, *81*, 167–180.

Sarosdy, C. R., & Acosta, A. J. (1961). Note on observations of cavitation in different fluids. *ASME Journal of Basic Engineering*, *3*, pp. 399–400.

Stahl, H. A., & Stepanoff, A. J. (1956). Thermodynamic aspects of cavitation on centrifugal pumps. *ASME Transactions*, *78*, 1691–1693.

Stepanoff, A. J. (1948). *Centrifugal and axial pumps; theory, design and application*. Wiley.

Stepanoff, A. J. (1961). Cavitation in centrifugal pumps with liquids other than water. *ASME Transactions, Series A, Journal of Engineering for Power*, *83*(1), 79–90.

Stepanoff, A. J. (1964). Cavitation properties of liquids. *ASME Journal of Engineering for Power*, *86*, 195–200.

Stripling, L. B. (1962). Cavitation in turbopumps, part 2. *ASME Journal of Basic Engineering*, *84*(3), 329.

Stripling, L. B., & Acosta, A. J. (1962). Cavitation in turbopumps—part 1. *ASME Journal Basic Engineering*, *84*, 326–338.

Taylor, W. E., et al. (1969). *Systematic Two-Dimensional Cascade Tests, Volume I—Double Circular Arc Hydrofoil*. In NASA CR-72498.

Torre, L., Pasini, A., Cervone, A., Pecorari, L., Milani, A., & d'Agostino, L. (2010). Rotordynamic forces on a three bladed inducer. In *Proceedings of Space Propulsion Conference*, San Sebastian, Spain.

Torre, L., Pasini, A., Cervone, A., & d'Agostino, L. (2011a). Continuous spectrum of the rotordynamic forces on a four-bladed inducer. In *ASME/JSME/KSME Joint Fluids Engineering Conference*, Hamamatsu, Japan.

Torre, L., Pasini, A., Cervone, A., & d'Agostino, L. (2011b). Experimental Characterization of the Rotordynamic Forces on Space Rocket Axial Inducers. *ASME Journal of Fluids Engineering*, *133*(10).

Torre, L., Cervone, A., Pasini, A., & d'Agostino, L. (2011c). Experimental characterization of thermal cavitation effects on space rocket axial inducers. *ASME Journal of Fluids Engineering*, *133*(11).

Vaage, R. D., et al. (1972). *Investigation of Characteristics of Feed System Instabilities*. Final Report MCR-72-107, Martin Marietta Corporation, Denver, CO, USA.

Valentini, D., Pace, G., Pasini, A., Torre, L., & d'Agostino, L. (2015). Influences of the operating conditions on the rotordynamic forces acting on a three-bladed inducer under forced whirl motion. *Journal of Fluids Engineering*, *137*(7), 071304.

Wachel, J. C., & Szenasi, F. R. (1980). Field verification of lateral-torsional coupling effects on rotor instability in centrifugal compressors. In *1st Workshop on Rotordynamic Instability Problems in High Performance Turbomachinery*, Texas A&M University, TX, USA.

Wagner, R. G. (1971). *Titan II Engine Transfer Function Test Results*. Report No. TOR-0059 (6471)-9, Aerospace Corporation, El Segundo, CA, USA.

Young, W. E., et al. (1972). *Study of Cavitating Inducer Instabilities*. Pratt and Whitney Report PWA FR-5131.

Zielke, W., et al. (1969). Forced and self-excited oscillations in propellant lines. *ASME Journal of Basic Engineering*, *91*, 671–677.

Zika, V. J. (1984). Correlation of cavitating centrifugal pumps. *ASME Journal of Fluids Engineering*, *106*, 141–146.

Zoladz, T. (2000). Observations on rotating cavitation and cavitation surge from the development of the fastrac engine turbopump. In *36th AIAA/ASME/SAE/ASEE Joint Propulsion Conference*, Huntsville, AL, USA.

Zwick, S. A., & Plesset, M. S. (1955). On the dynamics of small vapor bubbles in liquids. *Journal of Mathematics and Physics, 33,* 308–330.

Experimental Methods for the Study of Hydrodynamic Cavitation

Steven L. Ceccio and Simo A. Mäkiharju

Abstract A review of traditional and novel experimental methods for the investigation of hydrodynamic cavitation is presented. The importance of water quality is discussed, along with its characterization and management. Methods for the direct and indirect experimental determination of cavitation inception are presented. Along with traditional optical visualization, methods of measuring developed cavitation are described, including point and surface electrical probes, optical bubble probes, acoustic measurements, and indirect measurements of noise and vibration. Recent developments in the use of ionizing radiation as a means to visualize cavitating flows are also discussed.

1 Introduction

Hydrodynamic cavitation can occur in a variety of important liquid flows, including those associated with turbines, pumps, and other turbomachinery, ship propulsors, ventricular assist devices, fuel injectors, and other macro and micro fluidic systems. The presence of cavitation can degrade the performance of these devices, and can lead to excessive noise, vibration, and erosion. However, cavitation can be used to enhance the performance of some systems, such as high-speed underwater vehicles (Ceccio 2010).

Given the complexity of many cavitating flows, engineers have often resorted to experimental testing in order to reveal the presence of cavitation and its effect on system performance. And, for physically large systems such as turbines and propulsors, testing of scale models is often the only practical means of developing an optimized

S.L. Ceccio (✉)
Department of Naval Architecture and Marine Engineering, University of Michigan, Ann Arbor MI, USA
e-mail: ceccio@umich.edu

S.A. Mäkiharju
Department of Mechanical Engineering, University of California, Berkeley, CA, USA
e-mail: makiharju@berkeley.edu

L. d'Agostino and M.V. Salvetti (eds.), *Cavitation Instabilities and Rotordynamic Effects in Turbopumps and Hydroturbines*, CISM International Centre for Mechanical Sciences 575, DOI 10.1007/978-3-319-49719-8_2

design before its manufacture at full scale. Experimental testing has also been used to illuminate the basic processes of cavitating flows, usually through the examination of canonical flows that may be used to study specific cavitating flow processes, such as flows with variable area ducts and Venturis, and over headforms and hydrofoils.

The goal of this chapter is to review some experimental methods that have been successfully employed to study hydrodynamic cavitation (e.g., cavitation produced by flowing liquids). However, many of the methods are also useful for examination of cavitation produced by acoustic fields as well as other gas–liquid multiphase flows, including boiling flows. And some are also applicable to gas–solid and liquid–solid flows as well. The focus will be on the experimental methods that are available to researchers as they examine the flow processes that lead to cavitation inception, its development, and its effect on system performance, and its erosive potential rather than the test facility itself. But, a brief summary is provided here.

Experimental examination of hydrodynamic cavitation is often performed in a dedicated test facility. These can be broadly classified into flow loops and towing tanks. In the former, a prime mover delivers liquid (usually trough flow conditioners and a contraction) to the inlet of a test section where the cavitating flow will be examined. The flow is then returned (usually after passing through a diffuser where the pressure rises) to the prime mover to continue recirculation. Examples of modern cavitation tunnels are the Grand Tunnel Hydrodynamique (GTH) in France (Lecoffre et al. 1987), the Large Cavitation Channel (LCC) in the U.S.A. (Etter et al. 2005), and the Hydrodynamics and Cavitation Tunnel (HYKAT) in Germany (Wetzel and Arndt 1994a; Wetzel and Arndt 1994b). Turbomachinery may be tested with closed recirculating flow loops as well, as described by Avellan et al. (1987). Cavitation may also be studied by towing a test article in a stationary liquid, and facilities have been developed to allow for the ambient pressure over the free surface to be varied such that cavitation can form under a variety of flow conditions, such as the Depressurized Wave Basin in The Netherlands (Van der Kooij and De Bruijn 1984). The International Towing Tank Committee (ITTC) offers a catalog of many of these facilities, and Brandner et al. (2007) describe the recent design of a modern cavitation tunnel.

The capabilities and quality of the test facility is, of course, of central importance to the conduct of experimental investigations of cavitating flows. Many of the criteria used to assess a cavitation test facility are identical to those used for any aerodynamic or hydrodynamic test apparatus, including the uniformity and quality of the freestream flow, including the level of freestream turbulence, and the precision and range over which flow speed and pressure may be fixed. Many of the design requirements and approaches of subsonic wind tunnels presented by Rae and Pope (1984) apply equally to conventional water tunnels.

The acoustic characteristics of a cavitation flow facility may also be important to manage, as the acoustic emission of cavitation may be an important aspect of the testing program. The recently developed cavitation flow facility of the Japan Defense Agency is a modern example of a channel developed with these acoustic considerations (Sato et al. 2003). The tunnel was designed to reduce the amount of background noise and reverberation to improve the signal-to-noise ratio of for both noncavitating and cavitating flows. Finally, an important consideration for cavitating flow is management of the freestream water quality, which will be discussed below.

2 Characterization and Management of Water Quality

The inception and development of cavitation can be strongly related to the amount of free and dissolved gas within the cavitating liquid. Liquid that is supersaturated with dissolved gas and has many large free gas nuclei would be considered "weak", and cavitation may form where the liquid pressure falls below vapor pressure. Conversely, liquid that is undersaturated with dissolved gas and has few free gas nuclei can sustain pressures below vapor pressure (e.g., can be in tension), and is considered "strong". Determination and control of the liquids cavitation "water quality" or "susceptibility" is an important consideration for many experimental studies of cavitating flows. Discussions of the importance of water quality with regard to the conduct and interpretation of cavitation experiments are found in Lecoffre and Bonnin (1979), Kuiper (1985), Gindroz and Billet (1998), Arndt (2002), and Atlar (2002).

Cavitation inception occurs when a reduction in the liquid pressure results in a local pressure below the vapor pressure, and the liquid begins to change phase into vapor. Homogeneous nucleation process can occur in the bulk of the liquid as a result of inclusions that naturally form due to the random motion of the liquid molecules (Brennen 1995). Homogeneous nucleation typically requires a significant level of liquid tension, and ultraclean water can sustain tensions of over 30 MPa at room temperature (Mørch 2007). Yet, for many practical situations, cavitation incepts as a result of heterogeneous nucleation whereby nucleation sites within the bulk of the fluid or at solid boundaries grow when exposed to sufficiently strong tension. The characterization of the fluid's susceptibility to nucleation is an important aspect of many cavitation studies, especially those concerned with inception. In turn, the susceptibility of the flow is related to both the free and dissolved gas content as well as the nature of potential surfaces and flow-borne nucleation sites.

Water quality can also affect developed cavitation. For example, the presence of many freestream nuclei can lead to the suppression of sheet cavitation through the formation of traveling bubbles upstream of the cavity separation location (Li and Ceccio 1996; Keller 2001), and diffusion of dissolved gas into a developed tip vortex can significantly alter its core size (Gowing et al. 1995). As a consequence, it is incumbent upon the cavitation research engineer to adequately characterize and, if possible, manage the facility's water quality before conducting experimental studies and scale testing.

2.1 Dissolved Gas Content

Henry's law states that the equilibrium amount of a dissolved gas in a liquid at a given temperature is related to the partial pressure of the gas. When the gas concentration is at equilibrium, this is the saturated condition. Then, a reduction in the liquid pressure would result in supersaturation, and outgassing can occur. Likewise, if the pressure of the liquid is increased, the undersaturated liquid can dissolve more free gas. Hence, as a saturated liquid flows into regions of low pressure and cavitates, it is possible that significant amounts of outgassing may accompany any vaporization. Moreover, the level of gas saturation will influence the stability of free gas nuclei, which will be discussed below. The total dissolved gas content can be determined using a van Slyke apparatus developed for measurement of blood gas (Simoni et al. 2002). The van Slyke apparatus uses a vacuum placed over the liquid sample to produce outgassing, and it is quite accurate. Traditionally, the vacuum was created by the movement of a mercury manometer, which has led to its replacement by other methods that do not require the manual manipulation of mercury reservoirs.

For many practical applications, the cavitating liquid is water, and the dissolved gases are the main components of air, molecular nitrogen, and oxygen. But, other dissolved gases may be of interest, especially noble gases (Rooze et al. 2013). Measurement of dissolved gas content can be achieved with a Total Dissolved Gas Pressure (TDGP) probe. A sample of the liquid is placed in a probe beneath a headspace of gas at a known pressure that is separated from the liquid by a permeable membrane. Over a period the transfer of dissolved gas into or out of the headspace will change the gas pressure, and this change can then be related to the original dissolved gas concentration of the liquid sample. In order to relate the pressure change to the gas concentration, the chemical composition of the dissolved gases must be known, and some systems combine the probe with a separate instrument to sample and characterize the composition of the gas released into the headspace. For air dissolved in water, it is often assumed that the ratio of the dissolved gases track the ratio of nitrogen and oxygen in air at standard conditions (Yu and Ceccio 1997; Lee et al. 2016). In fact, in many test facilities, only the dissolved oxygen is measured, and it is assumed that the level of oxygen saturation parallels the overall dissolved air concentration. Dissolved oxygen (DO) probes employ a measurement technique similar to that of pH meters. Two electrodes are suspended in a liquid electrolyte, which is separated from the test sample by a semipermeable membrane. A low DC voltage is applied between the electrodes within the electrolyte, and when oxygen molecules from the test liquid cross the membrane, the magnitude of current between the electrodes will change.

2.2 Free Gas Content and Cavitation Nuclei

Measurement of the free gas content is more difficult. The intent is to characterize the nucleation sites in the freestream liquid, and a variety of methods have been proposed and evaluated (Lecoffre and Bonnin 1979; Oldenziel 1982; d'Agostino and Acosta 1991; Ceccio et al. 1995; Billet 1985; Pham et al. 1999). Freestream nuclei are any inclusion in the fluid that will cavitate when exposed to a sufficient tension (as opposed to surface nucleation sites which reside on flow boundaries). An ideal nucleus is a clean gas bubble, and the nucleation characteristics of such a bubble can be readily predicted. However, a wider variety of nucleation sites exist in the flow, including gas pockets on the surface of particles and bubbles that have significant surface contamination. Nevertheless, it is useful to review the basic nucleation process of a clean bubble in order to give us a reference to compare with practically observed nuclei.

Consider a nucleus that is a clean gas bubble with a radius R_N that contains vapor and noncondensable gas. The pressure inside the bubble, $P_B = P_V + P_G$ is the sum of the partial pressures of the vapor and noncondensable gas, respectively. This pressure is balanced by the liquid pressure on the bubble surface, P_∞, such that

$$P_\infty = P_V + P_G - \frac{2S}{R_N},$$

where S is the surface tension. When a nucleus experiences a drop in the surrounding liquid pressure, the radius may grow quasi-statically from its initial equilibrium radius to a larger equilibrium radius. However, if a tension is applied below a critical value, $P_V = P_C - P_\infty$, the radius will grow unboundedly. This critical tension is given by

$$\frac{4S}{3R_N} < P_V - P_\infty < \frac{2S}{R_N}$$

(Brennen 1995). Note that since the fluid is in tension, $P_V > P_\infty$, and that the static pressure is in fact negative. While not all nuclei are clean bubbles, the fluids nuclei content is often reported as a distribution of nuclei with a given critical radius, R_C. This is the radius that corresponds to the required critical pressure, P_C, needed to produce explosive growth of the nucleus. Therefore, the nuclei content of a liquid is typically reported as a Nuclei Number Density Distribution (*NNDD*) as a function of R_C, where $NNDD(R_C)$ has units of [Number]/[Length]$^{-4}$. Then, the number of nuclei over a range of critical radius ΔR_C is given by $NNDD(R_C) \cdot R_C$. Alternatively, if the bin size of the distribution is fixed, the Nuclei Number Distribution (*NND*) with units of [Number]/[Length]$^{-3}$ can be presented, as shown in Fig. 1 (Chang et al. 2011). The typical range of nuclei critical radii in test facilities spans 1 micron $< R_C <$ 500 micron, and the critical tensions range from $0 > P_C > -100$ kPa. Nuclei concentrations can range widely, with order 1 per cubic centimeter for the smallest nuclei and order 10 per cubic meter for the largest (Gindroz 1998).

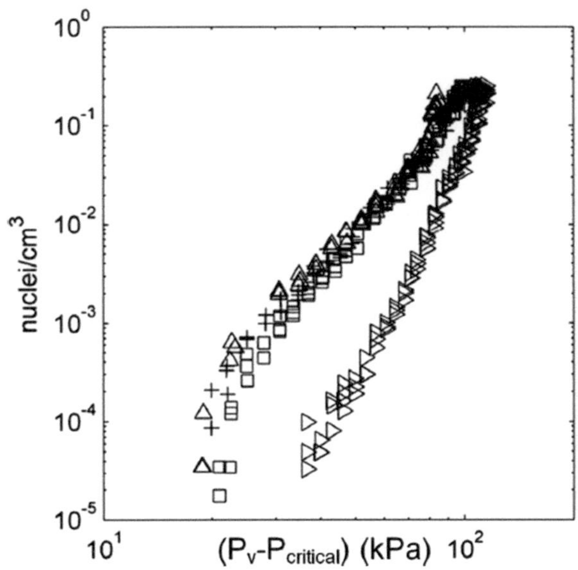

Fig. 1 Example nuclei concentration measured with a centerbody Venturi (Chang and Ceccio 2011). The different spectra are for varying freestream dissolved oxygen levels and concentrations of long-chain polymer

2.3 Direct Measurement of the Cavitation Nuclei Distribution

Measurement of the nuclei distribution can be accomplished with both direct and inferred means. In direct methods, a sample of the flow is exposed to a known tension, resulting in the cavitation of nuclei that incepts at (or above) that critical pressure. These devices are collectively known as Cavitation Susceptibility Meters (CSMs), as first discussed by Schiebe (1972). The most common CSM consists of a simple venture through which the liquid is passed. The pressure in the throat is measured or inferred, the occurrence of single cavitation events are detected optically, electrically, and/or acoustically. CSMs of this type have been developed by Oldenziel (1982), Chahine and Shen (1986), d'Agostino and Acosta (1991), Chambers et al. (2000), for example. During typical operation, a sample of the freestream liquid is drawn from the flow facility and delivered to the CSM inlet. Care must be taken to ensure that the sampling process does not significantly alter the nuclei distribution itself. Then, the flow is passed through the CSM. Changing the flow rate then varies the throat pressure, and the number of cavitation inception events is counted at each level of throat tension. Knowledge of the flow rate and throat tension yields the concentration of nuclei that incept at or above the given tension. This data is then converted into the nuclei number distribution.

One limitation of Venturi-based CSMs is the limited volume of the throat. If the nuclei distribution is too dense, multiple bubbles may simultaneously form in the throat, creating a blockage. A solution is to use a centerbody Venturi (Lavigne 1991), as shown in Fig. 2. Now the high-tension region is an annulus of fluid around the centerbody, increasing the volume of liquid that is in tension. Keller (1987) developed

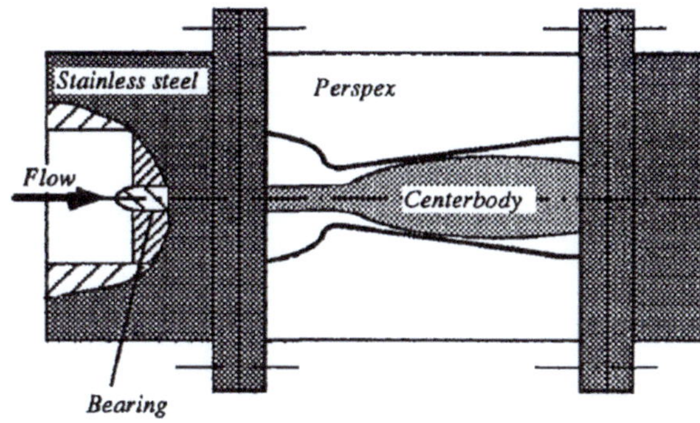

Fig. 2 Schematic diagram of a centerbody venturi cavitation susceptibility meter (Gindroz and Billet 1998)

a CSM that used a swirling flow passed through a vortex to create the region of low pressure.

While the operation of a CSM can be somewhat cumbersome, it is a device that directly measures the number of cavitation nuclei and their critical tension. It is important to note that not all nuclei are clean gas bubbles. In fact, particulates with small gas pockets on their surface can readily act as nuclei, and free gas bubbles can be coated by an organic skin, effectively modifying their interfacial properties. (Mørch 2007). Hence, a measurement of the size of the nucleus in the flow may not necessarily yield an accurate measure of its critical tension.

2.4 Indirect Measurement of the Cavitation Nuclei Distribution

Indirect measurement of the nuclei distribution can be operationally advantageous, especially if online or in situ measurement is required. Unlike CSMs, indirect measurements ascertain some aspect of suspected nuclei, such as its light or acoustic scattering properties. From this, the critical tension of the detected nucleus is inferred. As noted above, this relationship may not be easily determined. Nevertheless, the advantages of indirect methods have motived their development and use.

Bubble populations in liquids can be determined via acoustic scattering. Duraiswami et al. (1998) and Chahine and Kalumuck (2003) report on a method that employs the dispersion of sound passing through a bubbly medium. Both the attenuation and phase velocity are measured, and analytical relationships are used to invert these data into the bubble population. Once the bubble population is determined, the cavitation susceptibility can be inferred.

Light scattering can be used to detect the presence of nuclei in the flow (Keller 1972). Mie scattering by small spherical nuclei can be detected as they pass through a focused region of laser light, for example. A somewhat more sophisticated method employs Phase Doppler Anemometry (PDA), where multiple detectors are used to record the light scattered from a particle passing through the probe volume made by two crossed laser beams (Tanger and Weitendorf 1992). PDA systems can more readily determine the radius and velocity of the presumably spherical nucleus passing through the control volume. Care must be taken to relate the measured event rate to the actual nuclei density, since the effective measurement volume may not be easily determined given that nuclei may not pass directly through the measurement volume, for example. And, it is important to discriminate between bubbles and particles as they pass through the probe volume.

A direct optical measurement of the nuclei distribution can be made using holography (Katz et al. 1984; Katz and Sheng 2010). Holographic imaging can yield the absolute nuclei distribution in a volume of fluid, and with proper resolution, can be used to distinguish between bubbles, particulates, and other contaminants. Hence, holography is often used as the calibration standard for other nuclei measurement systems. Holographic systems have been used to measure nuclei distributions both in the laboratory and in the environment (Katz and Acosta 1981; O'Hern et al. 1985). Kawanami et al. (2002) used laser holography to study the structure of a cloud shed from a hydrofoil and estimated the bubble size distribution. Holography is not typically used as a routine nuclei measurement method, but recent advances in both camera technology and digital processing have made its everyday use more feasible.

2.5 Management of Water Quality

Characterization of the freestream dissolved and free gas content can be an essential component of any experimental test and evaluation effort. Moreover, it may be advantageous to actively manage these quantities through the addition or removal of dissolved gas and freestream nuclei. Besides filtering, the most basic method to control the water quality is through degassing the bulk of the test liquid, and deaeration is a common practice to reduce both the free and dissolved gas during testing. Typically, the dissolved gas concentration will be reduced to levels below 50%.

Control of the water quality can be improved via the active management of the free and dissolved gas content. The major flow facilities that have implemented these systems include the Grand Tunnel Hydrodynamique in France (Lecoffre et al. 1987) and the Australian Maritime College Cavitation Tunnel (Brandner et al. 2007). Figure 3 shows an image of latter facility that highlights the means of gas control. The facility is equipped with bulk degassing systems to control the average dissolved gas concentration. Additionally, small gas bubbles can be controllably injected directly into the flow upstream of the test section, while both small and large gas bubbles are removed downstream of the test section via gravity separation, coalescence, and resorption. In this way, the nuclei distribution can be prescribed and maintained during an experiment.

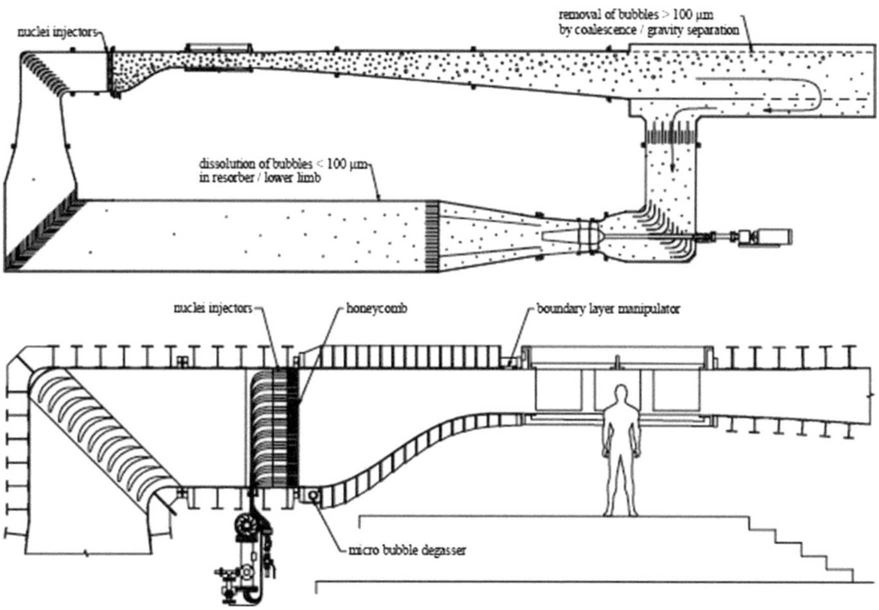

Fig. 3 A schematic diagram of the water quality management systems of the Australian Maritime College Cavitation Tunnel (Brandner et al. 2007). Nuclei can be injected upstream of the test section, and the tunnel is designed to remove gas by separation, coalescence, and resorption

3 Detection and Measurement of Incipient Cavitation

Cavitation inception occurs when cavitation is first observed in the flow, and the determination of inception conditions is important characterization of the flow itself as well as an important consideration for the scaling of the performance of model hydraulic systems. Inception usually occurs when the first freestream or surface nuclei encounter sufficient tension in the flow field to cavitate. Since a distribution of nuclei exists in the flow, and the pressure field producing the tension can often have important contributions from flow unsteadiness, inception is usually a stochastic process. Therefore, the average flow conditions under which inception is determined is, many cases, subjective. Moreover, the extent of cavitation chosen as necessary to call inception can vary widely. In some cases, such as the characterization of naval systems, only a minimal amount of cavitation is required to call inception, and the cavitation may not be easily visible to the naked eye. Conversely, limited cavitation may not be of practical interest to the operators of industrial hydraulic systems, and inception would be called only when the amount of cavitation begins to alter the performance of the device. And finally, proper use of inception observations in the scaling of model hydraulic systems to full scale is also of vital importance. Acosta and Parkin (1975) and Rood (1991) review the basic elements of cavitation inception for a variety of cavitation forms.

3.1 Detection of Inception with Acoustic, Vibration, and Force Measurements

Since the presence of incipient cavitation will many times be accompanied by emission of sound from the cavitating nuclei, the detection of inception is often accomplished through acoustic means. Hydrophones can be placed directly within the flow field (Ran and Katz 1994), within the cavitating test article itself (Ceccio and Brennen 1991; Kuhn de Chizelle et al. 1995) or in acoustically coupled chambers that are separated from the flow by an acoustic window (Choi and Ceccio 2007). Since the acoustic impedance of acrylic (3.1×10^6 Pa s / m^3) is only about twice that of

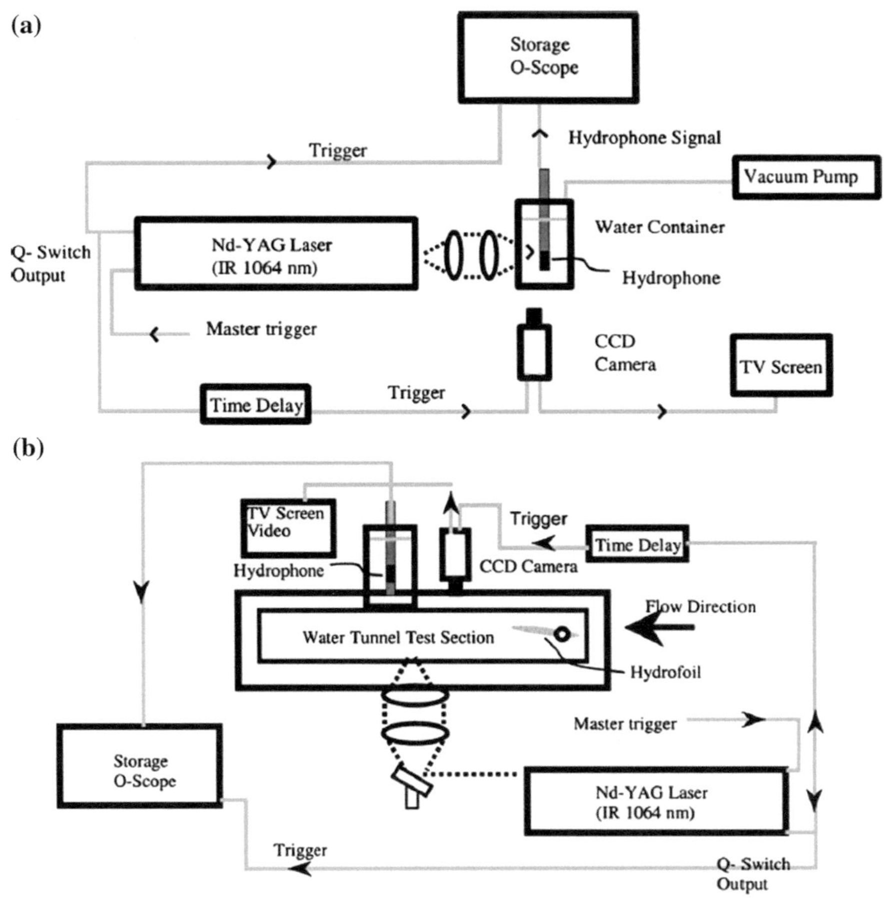

Fig. 4 Schematic diagram of the quiescent laser-induced cavitation bubble experiment (**a**) and a bubble–vortex interaction experiment (**b**). A single laser pulse is used to create a cavitation bubble in the bulk of the fluid from Oweis et al. (2004). The acoustic emission of the bubble is captured with a hydrophone within the flow (**a**) and in an external chamber through an acoustic window (**b**)

water (1.5×10^6 Pa s / m^3), it is commonly used as a rigid acoustic flow boundary. (A metal boundary would be much more reflective, with impedance mismatches in excess of ten times that of water.) Fig. 4 shows two typical setups from Oweis et al. (2004).

The noise produced by incipient cavitation bubbles often takes to form of discrete bursts or pulses, and example sound traces are shown in Fig. 5 from Chang and Ceccio (2011). In this case, the bubbles were formed in the cores of stretched vortices,

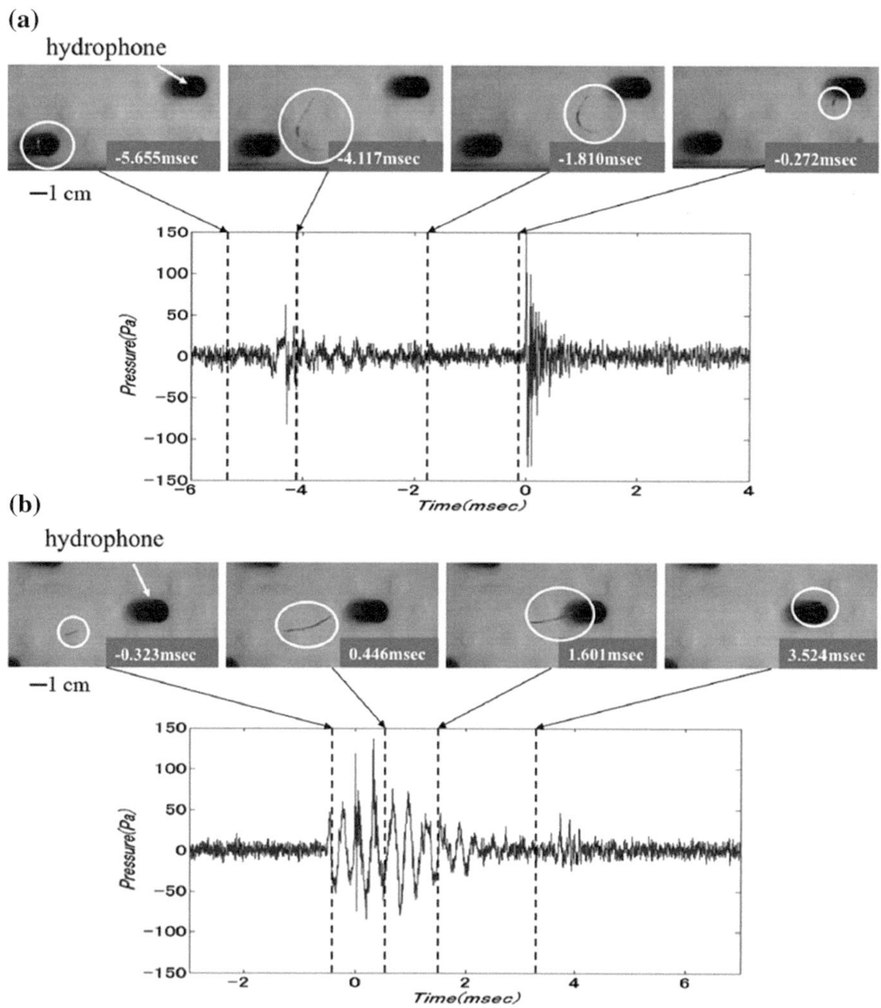

Fig. 5 Images and sound traces of a growing and collapsing vortex cavitation bubble in the secondary vortex producing an **a** acoustic "pop" and **b** a "chirp". The broadband acoustic pulse was abrupt lasting approximately 1 ms (Chang and Ceccio 2011)

and the sounds recorded from individual bubbles could be a pulse ("pop") or a periodic tonal burst ("chirp"). The hydrophone array can be seen in the image, and, in this case, an array of hydrophones was used to localize the sound source (Chang and Dowling 2009).

As the cavitation develops, the amount of sound emitted and the cavitation's effect on the overall flow begins to increase. Hence, measurement of vibration and changes to the system performance (e.g., lift coefficient, flow coefficient, efficiency) can be used to call inception. See, for example, Arndt (1981), McNulty and Pearsall (1982), Shen and Dimotakis (1989), and Koivula (2000). Indirect methods of inception detection are often calibrated against visual observations when optical access to the incepting flow is available. Escaler et al. (2006) report on a comprehensive study that illustrates how cavitation inception can be detected using measurements of structural vibrations, acoustic emissions, and hydrodynamic pressures measured in turbomachinery where optical access may be limited or unavailable.

3.2 Optical Measurement and Light Scattering for Inception Detection

Direct visual observations of incepting nuclei are often used to call inception. Traditionally, the flow is illuminated with stroboscopic lighting, and a human observer is tasked with determining which conditions have produced detectable and sustained cavitation. The availability of high-speed video systems have enabled the digital recording and analysis of the incepting flow, making inception calls with the naked human eye less common. However, at the first moments of inception, the cavitation bubbles may be quite small and difficult to locate; and they may occur infrequently. Such limited event rate cavitation inception is difficult to discern with visual detection alone, and the camera systems can be synchronized with acoustic detection systems (see, for example, Gopalan et al. (1999) and Chang and Ceccio (2011)). If the location of inception is known a priori, then focused light scattering can be used to detect the onset and rate of bubble formation. Keller (1972) developed such a system by directing a focused light source into the inception region of a headform and then into a photodetector. The presence of the bubble in the measurement volume would block the light to produce a signal.

4 Optical Measurement of the Cavitating Flow Field

Many optically based methods that have been developed for fluid measurements can also be effectively employed to study cavitating flows. General reviews of optical methods are provided by Goldstein (1996) and Tropea et al. (2007), and these methods and their applications are wide raging and varied. With this in mind, this section will concentrate on the use of optical methods in cavitating flows.

4.1 High-Speed Imaging

The need to study the dynamics of cavitation has stimulated the development of high-speed imaging. From the early work of Benjamin and Ellis (1966), Kling and Hammitt (1972), and Lauterborn and Bolle (1975), high-speed photography has played a key role in understanding bubbly dynamics and cavitation. A recent review by Thoroddsen et al. (2008) provides a good summary of the history and recent methods. As the resolution and frame rate of high-speed digital imaging systems has improved, the ability for detailed examination of cavitating flows has significantly improved. Frame rates of order 1 Khz with spatial resolution of $10^3 \times 10^3$ pixels are now commonly available, and cameras with much higher imaging speeds are commercially available. Ultrahigh-speed imaging systems have been created as well, with frame rates in excess of 1 million per second. Such systems can resolve the fine details of cavitation bubble dynamics, as shown by Lauterborn and Hentschel (1985), Ohl et al. (1995), and Obreschkow et al. (2006).

The study of high-speed bubble dynamics often requires the controlled creation of single or multiple bubbles. In many cases cited above, a laser pulse is focused to create the bubble in the imaging region of the camera. The bubble creation can then be synchronized with the imaging system. This technique can be used to controllably place nuclei into the freestream flow as well, as shown by Choi and Ceccio (2007). Figure 6 presents a time series from high-speed imaging of a laser-induced bubble from Ohl et al. (1998), including a schematic of the setup and an example of an aspherically collapsing bubble with detected luminescence. And, Fig. 7 shows images of a vortex cavitation bubble formed after a laser-induced nucleus was created within the upstream core of the vortex.

4.2 Laser Doppler Velocimetry and Light Scattering Methods

Laser Doppler Velocimetry (LDV) (also known as Laser Doppler Anemometry) is a well-established technique to measure local flow velocity. In this method, two beams of laser light are crossed within the flow domain to create a probe volume consisting of an interference pattern of light. As flow-borne particles pass through the probe volume, the scattered light from the fringe pattern is detected. The frequency of the light "burst" is related to the fringe spacing (and fringe velocity if one laser beam is frequency shifted) and to the velocity of the particle. If it can be assumed that the particle travels at the local flow speed, inference of the particle velocity yields a nonintrusive measurement of the flow velocity. Laser beams with multiple colors (wavelengths) can be used to measure two or three components of the flow speed in the same probe volume. A comprehensive review of this method can be found in Durst (1982) and Goldstein (1996).

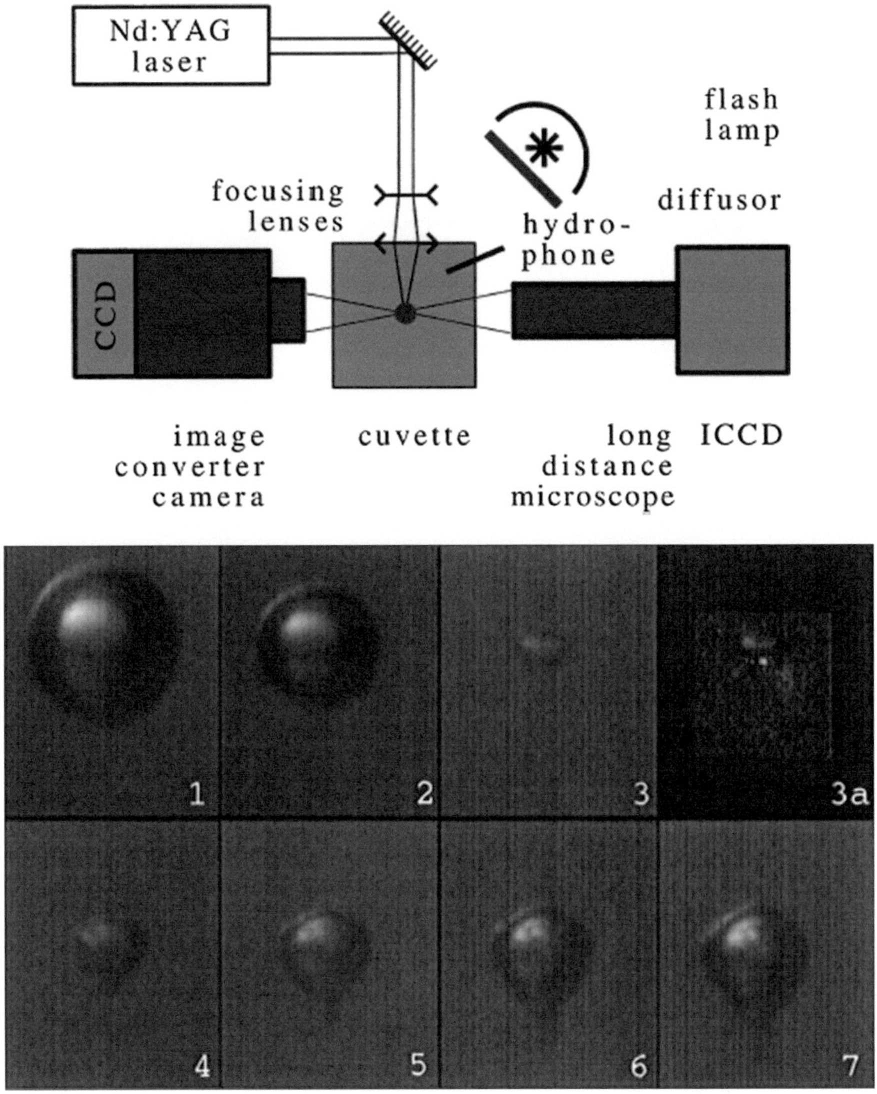

Fig. 6 High-speed imaging of a laser-induced bubble from Ohl et al. (1998). A schematic of the setup used to create the laser-induced bubbles is shown in along with images of an aspherically collapsing bubble with detected luminescence

LDV can be a useful method to examine cavitating flows. Kubota et al. (1989) used conditional sampling of the LDV for velocity measurements in a flow around a shedding partial cavity, as shown in Fig. 8. More recently, Roth et al. (2002) used similar methods to examine the cavitating flow in a fuel injector. In these cases, LDA was used to measure the velocity of the liquid flow around and outside the

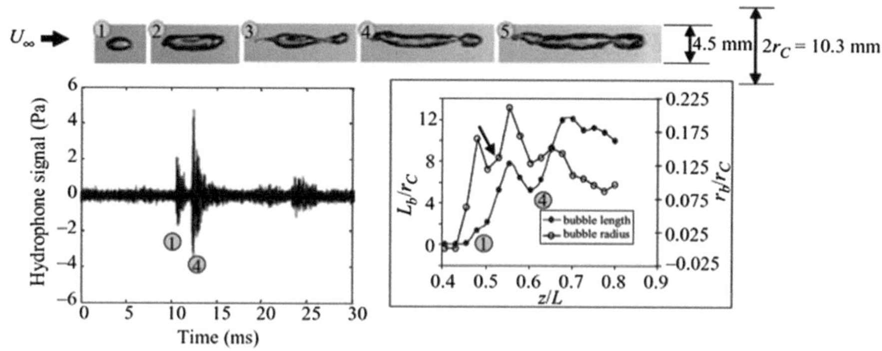

Fig. 7 Images of a vortex cavitation bubble created by a laser-induced nucleus; the images were used to compute the length and average radius of the bubble as a function of position within the Venturi. Also shown is the corresponding acoustic signal detected from a hydrophone Choi and Ceccio (2007)

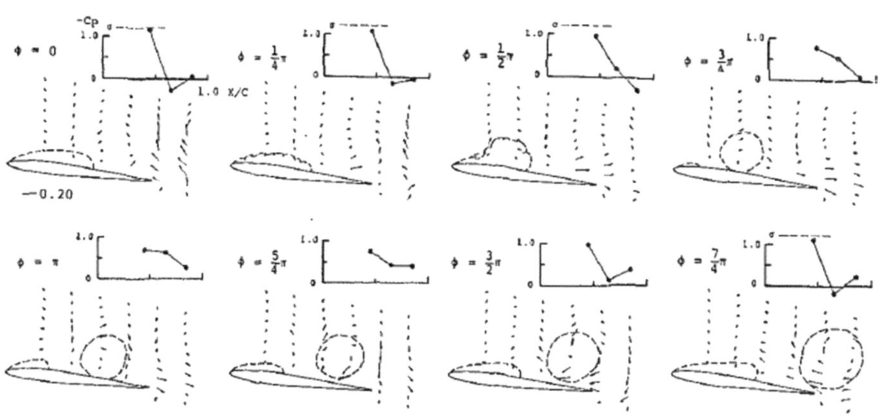

Fig. 8 Local measurements of flow velocity acquired with Laser Doppler Velocimetry were conditionally sampled and correlated to the measured surface pressure during the shedding of cloud cavitation on a hydrofoil (Kubota et al. 1989)

region of developed cavitation. When employing LDV in this way, it is important to determine if the flow tracers are small seed particles or small bubbles generated by the cavitation itself since large bubbles may not necessarily behave as Lagrangian flow tracers, especially in regions of high turbulence and shear.

4.3 Particle Imaging Velocimetry

Particle Imaging Velocimetry (PIV) is a full-field method used to measure two or three components of the flow velocity in a plane or volume. In this method, the flow is densely seeded with flow tracers, and a plane or volume of the flow is illuminated with pulsed laser light. Two or more images of the flow tracers are compared to determine the motion of the tracers over a known time interval, and the motion of many particles are analyzed to determine the spatial distribution of velocities within the illuminated flow region. Since PIV's development in the 1990s there have been significant advances in its development and use, and a general review is provided by Adrian and Westerweel (2011). The use of PIV to study dispersed multiphase flows has also been progressing. In these cases, care is taken to distinguish the tracer particles (which are intended to be nonintrusive) from those that constitute the dispersed phase itself, such as larger particles or bubbles. Hassan et al. (1992), Lindken and Merzkirch (2002), and Balachandar and Eaton (2010) provided examples of PIV applied to such dispersed multiphase flows.

As in the case of LDV, PIV can be used to study the low void fraction flow around developed cavitation. Examples include the work of Vogel and Lauterborn (1988), Leger and Ceccio (1998), Leger et al. (1998), Gopalan and Katz (2000), Laberteaux and Ceccio (2001a), Laberteaux and Ceccio (2001b), Dular et al. (2005), and Foeth et al. (2006). Figure 9 presents results from Foeth et al. (2006), who examined the dynamic flow around shedding partial cavities. The figure shows the steps needed to separate the images of the PIV tracer particles from the cavity and resulting bubbly flow. Synchronization of the image acquisition with periodically shedding cavity flows can yield phase-averaged flow data.

For limited cavitation, PIV can be used to interrogate the flow in and around the cavitation bubbles. Examples include Ran and Katz (1994), Iyer and Ceccio (2002), and Straka et al. (2010) who examined inception and bubble–vortex interactions in shear flows, and Wosnik et al. (2003) who examined the bubbly wake of supercavities. Figure 10 presents an example of the use of PIV in limited cavitating flows from Iyer and Ceccio (2002) who examined the influence of cavitation on the dynamics of a planar shear layer. The setup and an example image of the cavitating shear layer are shown in (a), and the mean flow and vorticity profiles are shown in (b) for increasing levels of cavitation (void fraction) in the shear layer. Recent advances in PIV systems include digital holographic PIV, micro-PIV, tomographic (volume) PIV, and high frame-rate cinemagraphic PIV. All of these methods have the potential to bring new insights into our understanding of cavitating flows.

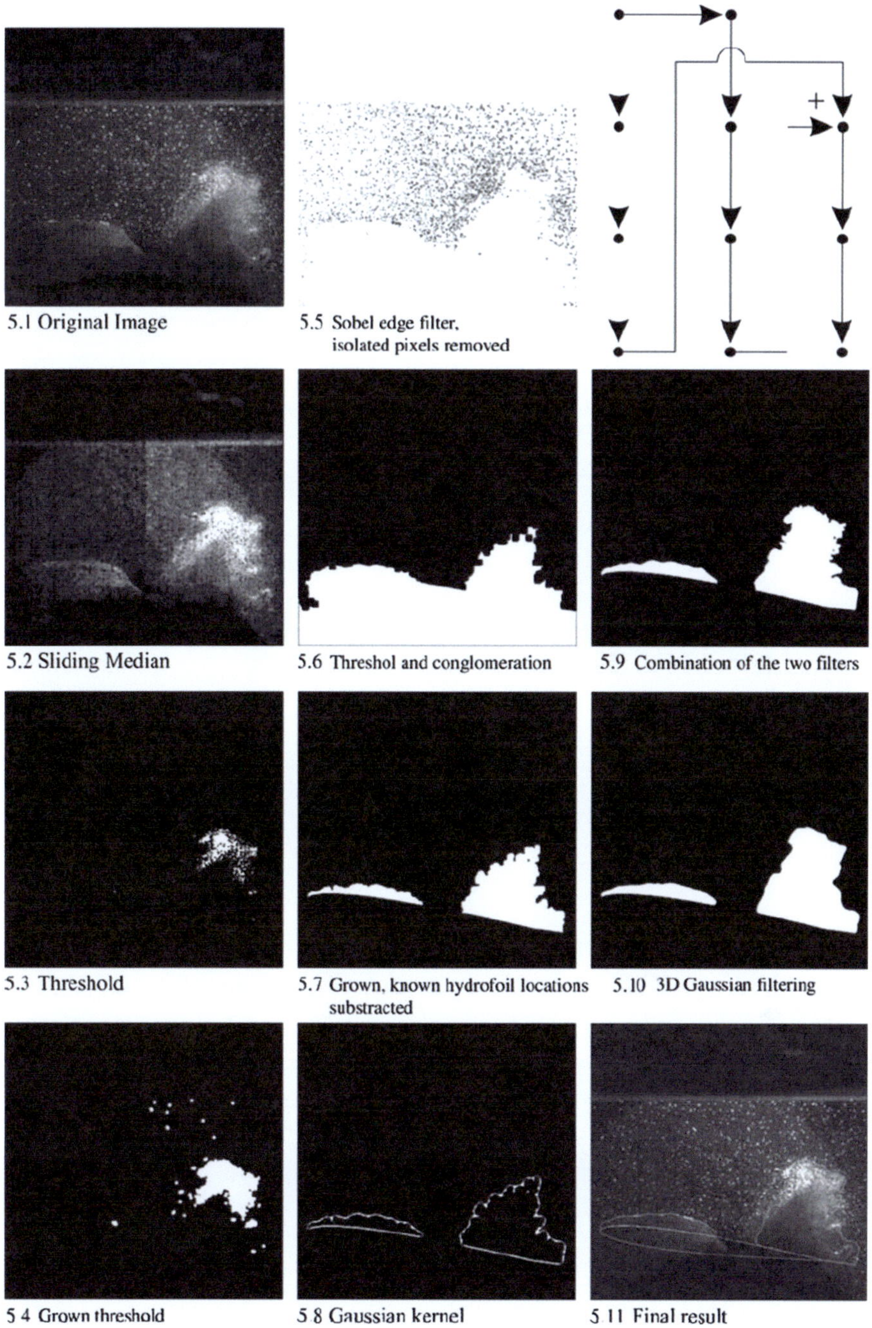

Fig. 9 The image processing steps employed by Foeth et al. (2006) to determine the flow field around a developed cavitation forming on a twisted hydrofoil

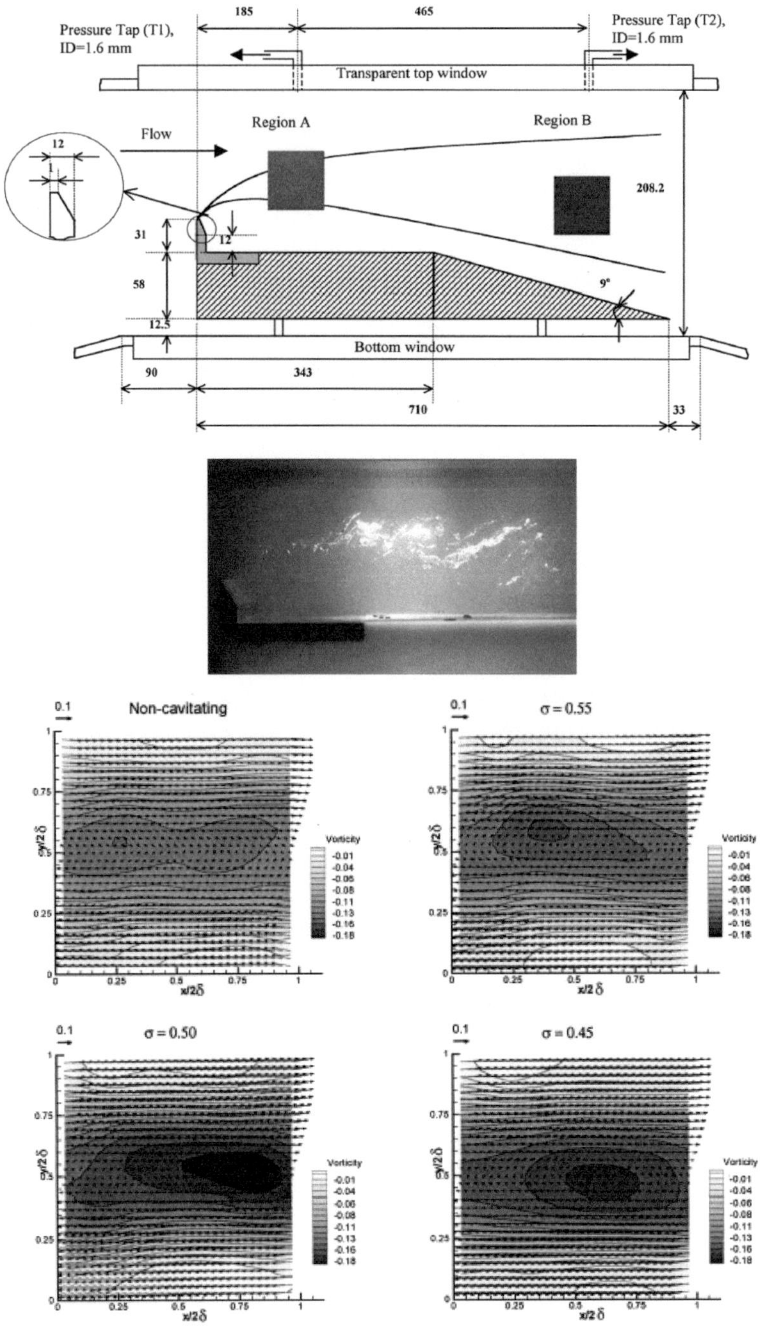

Fig. 10 Examination of the influence of cavitation on a shear layer using PIV (Iyer and Ceccio 2002). The setup and an example image of the cavitating shear layer are shown in (**a**) and (**b**), and the mean flow and vorticity profiles are shown in (c) for increasing levels of cavitation (void fraction) in the shear layer

5 Measurement of Cavity Flows with High Void Fraction

As the void fraction of the cavitating flow begins to exceed a few percent, the bubbly flow becomes opaque, and optical methods such as LDV and PIV begin to fail due to multiple scattering of the incident and reflected light. For very high void fraction flows, alternative techniques must be used, as described below.

5.1 Surface Pressure, Acceleration, and Forces

Cavitating flows with high void fraction often occur near solid boundaries, and it is therefore possible to place instruments close to or flush against the cavitating surface. The most common flush-mounted instruments measure the static and dynamic pressure. These measurements can be combined with local or average acceleration and forces on the test article to help elucidate the underlying flow. For example, Kjeldsen et al. (2000) measured the static and dynamic pressures on the surface of a cavitating hydrofoil, along with the time-varying lift force. Callenaere et al. (2001) employed dynamic and ultrasonic transducers to measure the reentrant flow beneath a partial cavity, and Le et al. (1993) employed arrays of dynamic pressure transducers to examine the unsteady pressures developed by partial cavitation. Escaler et al. (2006) illustrated how a range of nonoptical methods can be used to detect and quantify cavitation in turbomachinery, including external vibrations and noise signatures.

5.2 Electrical Impedance Probes

Since the gas and liquid phases of the cavitating flow have quite different electrical properties, measurements of the local or average mixture impedance can be used to infer the void fraction of the flow, and a review of these general methods is provided by Ceccio and George (1996). Measurement of the mixture impedance can be accomplished with flush-mounted electrodes or intrusive probes, albeit the latter may excessively perturb the cavitating flow. The electrodes can be made large enough to measure the bulk-averages void fraction, or small enough to measure the local void fraction or the passage of individual gas pockets. A high-frequency alternating current can be applied as the probing signal, making the temporal response of the transducers very rapid. However, as the path lines of the applied current are strongly affected by the presence of the gas phase, it is not always possible to fix the measurement volume of the probe. And, the relationship between the measure bulk impedance and the void fraction may not be straightforward. For bubbly flows, a mixture relationship can be developed that can relate the bulk impedance to the void fraction and electrical properties of the liquid and gas components (Hewitt 1978;

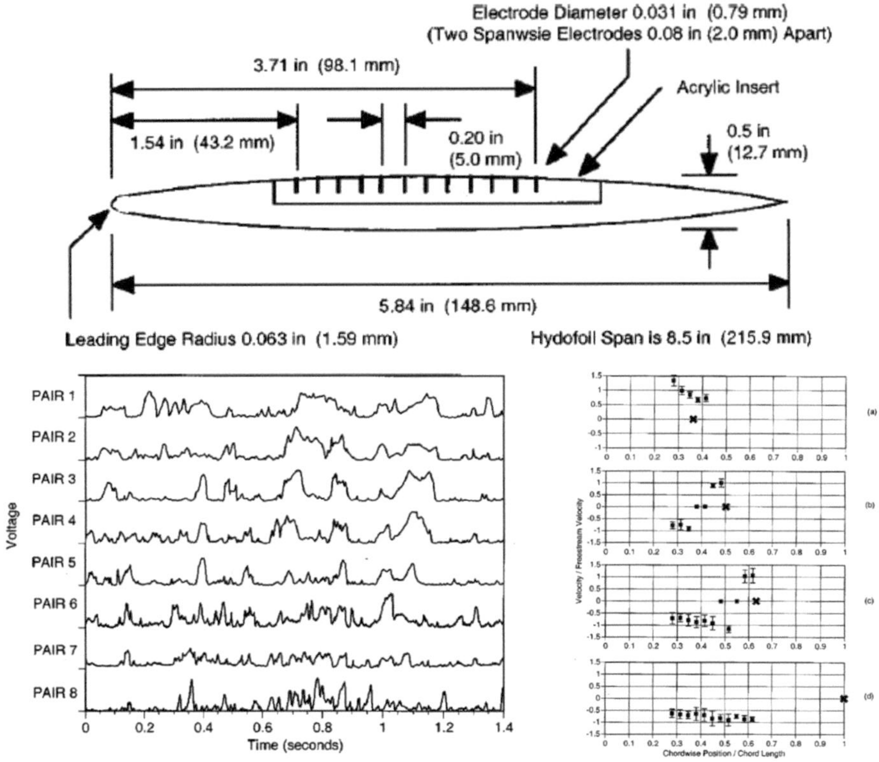

Fig. 11 The near-surface gas-phase velocity beneath a partial cavity measured with flush-mounted electrical impedance probes (George et al 2000a). The electrode locations on the hydrofoil (**a**), sample voltage signal transduced from the probes (**b**), and the gas-phase velocity determined through cross-correlation of the signals (**c**)

George et al. 2000b). But such relationships will generally fail for highly stratified flows.

Flush-mounted impedance probes have been used successfully to quantify cavitating flows. Examples include the work of Ceccio and Brennen (1991) and Kuhn de Chizelle et al. (1995) for the study of traveling bubble cavitation; and Ceccio and Brennen (1992), Pham et al. (1999), and George et al (2000a) for the study of partial cavitation. Cross-correlation of signals from multiple electrodes can be used to determine the gas-phase interface velocity, as reported by George et al (2000a) (Fig. 11). Intrusive conductivity probes with one or more electrode pairs have also been developed for gas–liquid flows, as discussed by Wu and Ishii (1999), Lucas and Mishra (2005), and Elbing et al. (2013).

5.3 *Fiber Optic Probes*

Fiber optic probes can be used to detect the presence of bubbles and gas–liquid interfaces via the difference in the index of refraction between the liquid and gas. The end of the fiber is placed in the flow, and light is directed toward the sharpened tip. If the tip is immersed in the pure liquid, the majority of the light will be transmitted into the fluid. But, if gas is present at the tip, the light will internally reflect within the fiber and return to its source to be detected. Fiber optic probes have been successfully employed to measure bubble size populations, average void fraction, phase speed, and interfacial area density, and a review is provided by Boyer et al. (2002) and Chang et al. (2003). The performance of optical and conductivity probes for measurement of bubbly flow was compared by Le Corre et al. (2003). The use of intrusive optical probes in cavitating flow has been limited due to the probes' delicate construction and their ability to perturb the flow, but such probes were successfully used to study the dynamics of partial cavitation by Stutz and Reboud (2000) and Stutz (2003). Figure 12 presents the probe employed by Stutz and Reboud (2000) to measure the bubbly flow within a partial cavity, the time traces from two probes that can be used to measure the local volume fraction and phase speed, and an example data set showing the average volume fraction within the cavity. As in the case of electrical impedance probes, care must be taken to carefully determine how the signal transduced from the probe relates to the flow quantity of interest (e.g., bubble size and velocity) and what the influence volume of the probe may be. Single point probes have shown the best results when they are oriented in a flow with a strong rectilinear velocity and when the probe tip is small compared to the bubbles to be measured (Cartellier 1992; Mäkiharju et al. 2013).

Images of a high void fraction bubbly cavitating flow were acquired by Coutier-Delgosha et al. (2006) by traversing an endoscope within a partial cavity flow. They were able to demonstrate that the bubbly flow within the cavity often consists of highly distorted gas bubbles, as shown in Fig. 13.

5.4 *Ionizing Radiation*

The use of X-ray and gamma-ray densitometry and tomography for the study of multiphase flows has been well established, and reviews are provided by Kumar et al. (1997), George et al. (2000b), and Heindel (2011). These methods have also been applied for the study of high void fraction cavitating flows.

The underlying principal of these systems relies the material and density dependent attenuation coefficient of high-energy photons as they pass through the multiphase mixture. When a beam of high-energy X-ray photons, for example, encounters the mixture, a fraction of the photons passes through without scattering or absorption, and this fraction depends on the mixture's attenuation coefficient, μ, density, ρ, and thickness, x. For a beam encountering a domain with N distinct materials, the

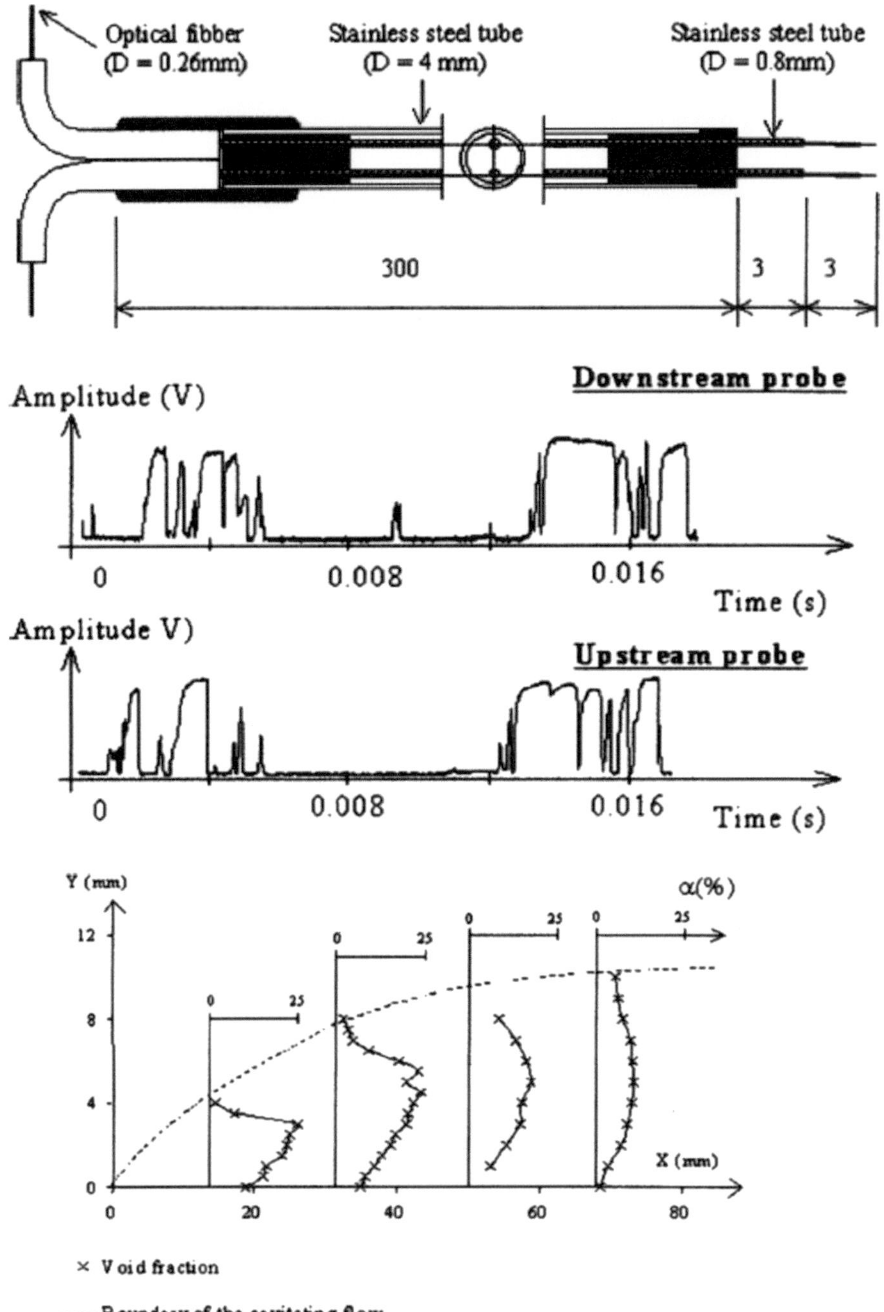

Fig. 12 The fiber optic probe used by Stutz and Reboud (2000) to measure flow within a partial cavity (**a**), example the time traces from two probes that can be used to measure volume fraction and phase speed (**b**), and a plot showing the average volume fraction within the cavity (**c**)

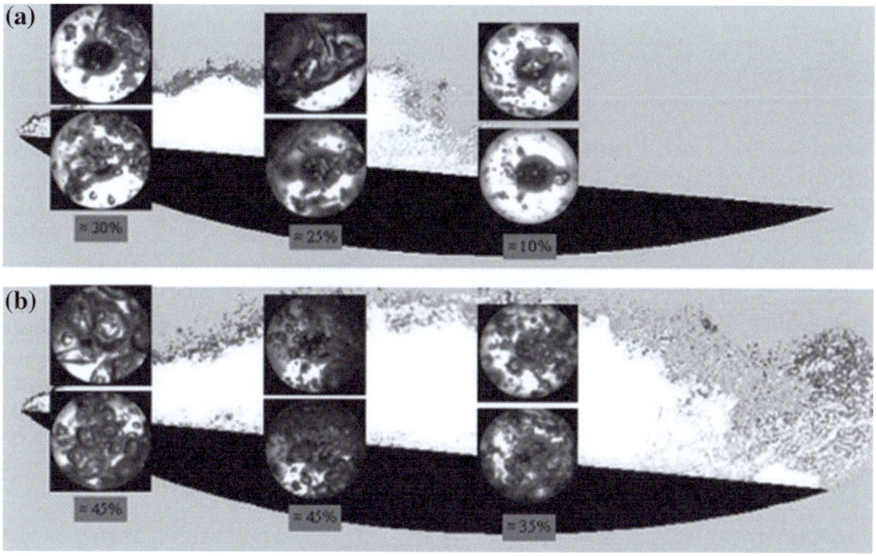

Fig. 13 Images of the bubbly flow within a partial cavity obtained by traversing an endoscope into the bubbly mixture (Coutier-Delgosha et al. 2006)

Beer–Lambert law provides a relationship between the transmitted, I, and incident, I_O, beam intensity:

$$\frac{I}{I_O} = e^{-\sum_{n=1}^{N} x_n \mu_n / \rho_n}$$

The attenuation coefficient is a function of material properties and photon energy, and is a known property for most common materials. Given this relationship, we can determine the void fraction, α, of a two-phase mixture, M, of liquid, L, and gas, G, for a monochromatic beam of photons:

$$\alpha = \ln\left(\frac{I_M}{I_L}\right) / \ln\left(\frac{I_G}{I_L}\right)$$

The transmitted beam intensities I_L, I_G, and I_M are values obtained for the pure liquid, pure gas, and mixture, respectively. For densitometry, average mixture void fraction is determined along a known linear beam path, while tomography involves the reconstruction of the two- or three-dimensional spatial distribution of attenuation through many measurements of linear path-averaged attenuation.

Stutz and Legoupil (2003) used X-ray densitometry to nonintrusively measure void fraction in a partial cavity. The setup consisted of a single row of 24 detectors that could acquire void fraction profiles along a line at a rate of 1000 samples per second. The measurements were compared with optical probes, and it was found that the maximum void fraction for the case of periodic shedding was about 25%.

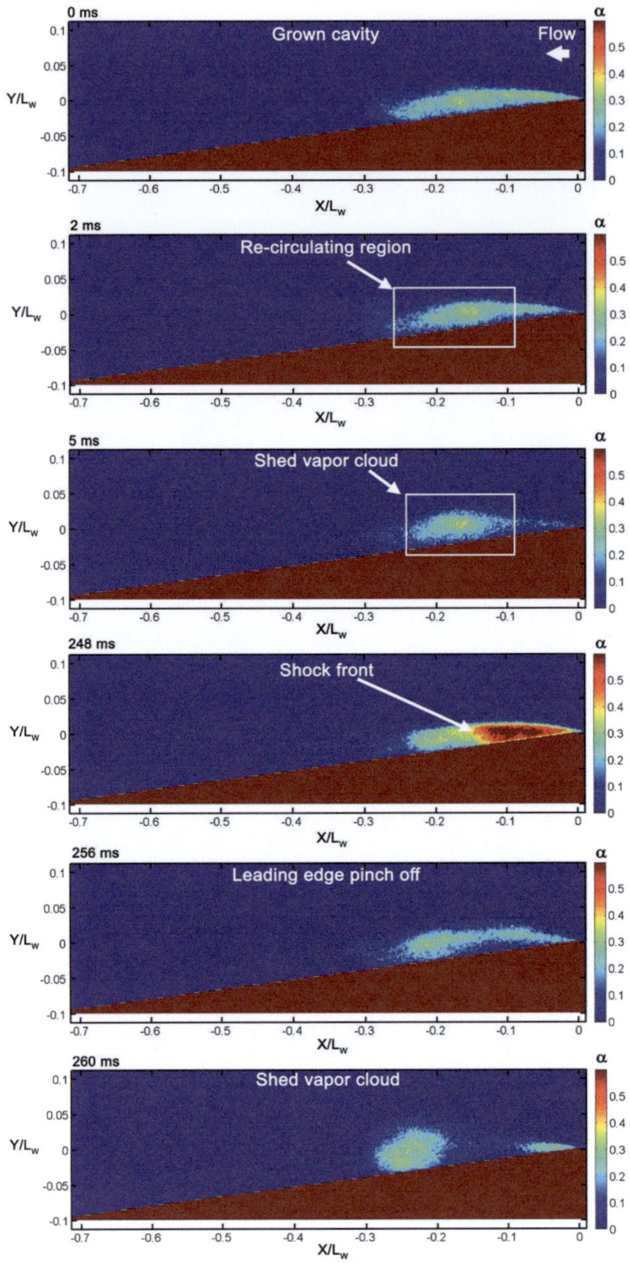

Fig. 14 A time series of X-ray densitometry-based images of a shedding partial cavity illustrating the presence of a condensation shock (Ganesh et al. 2016)

Coutier-Delgosha et al. (2007) used the same diagnostic setup to measure void fraction profiles on a plano-convex foil. They reported a maximum averaged void fraction values of close to 60% with instantaneous values exceeding 85%. These systems provided the time- and phase-averaged void fraction averaged across the span of the cavity. Mäkiharju et al. (2013) recently developed an cinemagraphic X-ray densitometry system that measures two-dimensional void fraction flow fields of gas–liquid flows, and this system was used to examine the dynamic void fraction within shedding partial cavities with frame rates up to 1000 per second Ganesh et al. (2016). Figure 14 shows X-ray densitometry images of the partial cavity forming at the apex of a wedge revealing the presence of a condensation shock. While the spatial and temporal resolution of these radiation-based techniques is presently nonideal for many cavitating flows, rapid technological advances (e.g., in X-ray detectors) will make these nonintrusive techniques increasingly useful.

Acknowledgements The first author is grateful to Prof. Salvetti and Prof. d'Agostino for their invitation to the Udine summer course on cavitating flows. The authors wish to acknowledge the support of the U.S. Office of Naval Research and the U.S. Naval Sea Systems Command for their ongoing support of their research.

References

Acosta, A. J., & Parkin, B. R. (1975). Cavitation inception-a selective review. *Journal of Ship Research, 19*(4), 193–205.

Adrian, R. J., & Westerweel, J. (2011). *Particle image velocimetry*. Cambridge University Press.

Arndt, R. E. (1981). Cavitation in fluid machinery and hydraulic structures. *Annual Review of Fluid Mechanics, 13*(1), 273–326.

Arndt, R. E. (2002). Cavitation in vortical flows. *Annual Review of Fluid Mechanics, 34*(1), 143–175.

Atlar, M. (2002). Final report of the specialist committee on water quality and cavitation. In *Proceedings of the 23rd ITTC*.

Avellan, F., Henry, P., & Ryhming, I. L. (1987). A new high speed cavitation tunnel for cavitation studies in hydraulic machinery. *Proceedings of international cavitation research facilities and techniques* (Vol. 57, pp. 49–60). Boston: ASME Winter Annual Meeting.

Balachandar, S., & Eaton, J. K. (2010). Turbulent dispersed multiphase flow. *Annual Review of Fluid Mechanics, 42*, 111–133.

Benjamin, T. B., & Ellis, A. T. (1966). The collapse of cavitation bubbles and the pressures thereby produced against solid boundaries. *Philosophical Transactions of the Royal Society of London A: Mathematical, Physical and Engineering Sciences, 260*(1110), 221–240.

Billet, M. L. (1985). *Cavitation nuclei measurements—a review*. Forum: Proceedings of ASME Cavitation and Multiphase Flow.

Boyer, C., Duquenne, A. M., & Wild, G. (2002). Measuring techniques in gas-liquid and gas-liquid-solid reactors. *Chemical Engineering Science, 57*(16), 3185–3215.

Brandner, P. A., Lecoffre, Y., & Walker, G. J. (2007). Design considerations in the development of a modern cavitation tunnel. *Proceedings of the 16th Australasian fluid mechanics conference* (pp. 630–637).

Brennen, C. E. (1995). *Cavitation and bubble dynamics*. Cambridge University Press.

Callenaere, M., Franc, J. P., Michel, J., & Riondet, M. (2001). The cavitation instability induced by the development of a re-entrant jet. *Journal of Fluid Mechanics, 444*, 223–256.

Cartellier, A. (1992). Simultaneous void fraction measurement, bubble velocity, and size estimate using a single optical probe in gas-liquid two-phase flows. *Review of Scientific Instruments, 63*(11), 5442–5453.
Ceccio, S. L. (2010). Friction drag reduction of external flows with bubble and gas injection. *Annual Review of Fluid Mechanics, 42*, 183–203.
Ceccio, S. L., & Brennen, C. E. (1991). Observations of the dynamics and acoustics of travelling bubble cavitation. *Journal of Fluid Mechanics, 233*, 633–660.
Ceccio, S. L., & Brennen, C. E. (1992). Dynamics of attached cavities on bodies of revolution. *Journal of Fluids Engineering, 114*(1), 93–99.
Ceccio, S. L., & George, D. L. (1996). A review of electrical impedance techniques for the measurement of multiphase flows. *Journal of Fluids Engineering, 118*(2), 391–399.
Ceccio, S. L., Gowing, S., & Gindroz, B. (1995). A comparison of csm bubble detection methods. In Proceedings of A.S.M.E. symposium on cavitation and gas-liquid flows in fluid machinery. FED (Vol. 226, pp. 43–50).
Chahine, G. L., & Kalumuck, K. M. (2003). Development of a near real-time instrument for nuclei measurement: The abs acoustic bubble spectrometer. *Proceedings of A.S.M.E./J.S.M.E. 4th joint fluids summer engineering conference* (pp. 183–191).
Chahine, G. L., & Shen, Y. T. (1986). Bubble dynamics and cavitation inception in cavitation susceptibility meters. *Journal of Fluids Engineering, 108*(4), 444–452.
Chambers, S. D., Bartlett, R. H., & Ceccio, S. L. (2000). Hemolytic potential of hydrodynamic cavitation. *Journal of Biomechanical Engineering, 122*(4), 321–326.
Chang, K. A., Lim, H. J., & Su, C. B. (2003). Fiber optic reflectometer for velocity and fraction ratio measurements in multiphase flows. *Review of Scientific Instruments, 74*(7), 3559–3565.
Chang, N., Ganesh, H., Yakushiji, R., & Ceccio, S. L. (2011). Tip vortex cavitation suppression by active mass injection. *Journal of Fluids Engineering, 133*(11301), 1–11.
Chang, N. A., & Ceccio, S. L. (2011). The acoustic emissions of cavitation bubbles in stretched vortices. *The Journal of the Acoustical Society of America, 130*(5), 3209–3219.
Chang, N. A., & Dowling, D. R. (2009). Ray-based acoustic localization of cavitation in a highly reverberant environment. *Journal of the Acoustical Society of America, 125*(5), 3088–3100.
Choi, J., & Ceccio, S. L. (2007). Dynamics and noise emission of vortex cavitation bubbles. *Journal of Fluid Mechanics, 575*, 1–26.
Coutier-Delgosha, O., Devillers, J. F., Pichon, T., Vabre, A., Woo, R., & Legoupil, S. (2006). Internal structure and dynamics of sheet cavitation. *Physics of Fluids (1994-present), 18*(1), 017103.
Coutier-Delgosha, O., Stutz, B., Vabre, A., & Legoupil, S. (2007). Analysis of cavitating flow structure by experimental and numerical investigations. *Journal of Fluid Mechanics, 578*, 171–222.
d'Agostino, L., & Acosta, A. J. (1991). A cavitation susceptibility meter with optical cavitation monitoring–part one: Design concepts. *Journal of Fluids Engineering, 113*(2), 261–269.
Dular, M., Bachert, R., Stoffel, B., & Širok, B. (2005). Experimental evaluation of numerical simulation of cavitating flow around hydrofoil. *European Journal of Mechanics-B/Fluids, 24*(4), 522–538.
Duraiswami, R., Prabhukumar, S., & Chahine, G. L. (1998). Bubble counting using an inverse acoustic scattering method. *Journal of the Acoustical Society of America, 104*(5), 2699–2717.
Durst, F. (1982). Review–combined measurements of particle velocities, size distributions, and concentrations. *Journal of Fluids Engineering, 104*(3), 284–296.
Elbing, B. R., Mäkiharju, S. A., Wiggins, A., Perlin, M., Dowling, D. R., & Ceccio, S. L. (2013). On the scaling of air layer drag reduction. *Journal of Fluid Mechanics, 717*, 484–513.
Escaler, X., Egusquiza, E., Farhat, M., Avellan, F., & Coussirat, M. (2006). Detection of cavitation in hydraulic turbines. *Mechanical Systems and Signal Processing, 20*(4), 983–1007.
Etter, R. J., Cutbirth, J. M., Ceccio, S. L., Dowling, D. R., & Perlin, M. (2005). High reynolds number experimentation in the U.S. Navy's William B. Morgan large cavitation channel. *Measurement Science and Technology, 16*(9), 1701–1709.
Foeth, E. J., Van Doorne, C. W. H., Van Terwisga, T., & Wieneke, B. (2006). Time resolved PIV and flow visualization of 3D sheet cavitation. *Experiments in Fluids, 40*(4), 503–513.

Ganesh, H., Mäkiharju, S. A., & Ceccio, S. L. (2016). Bubbly shock propagation as a mechanism for sheet-to-cloud transition of partial cavities. *Journal of Fluid Mechanics*, *802*, 37–78.
George, D. L., Iyer, C. O., & Ceccio, S. L. (2000a). Measurement of the bubbly flow beneath partial attached cavities using electrical impedance probes. *Journal of Fluids Engineering*, *122*(1), 151–155.
George, D. L., Torczynski, J. R., Shollenberger, K. A., O'Hern, T. J., & Ceccio, S. L. (2000b). Validation of electrical-impedance tomography for measurements of material distribution in two-phase flows. *International Journal of Multiphase Flow*, *26*(4), 549–581.
Gindroz, B. (1998). Cavitation nuclei and cavitation inception of marine propellers: State of the art at the dawn of the 21st century. *J.S.M.E. International Journal Series B*, *41*(2), 464–471.
Gindroz, B., & Billet, M. L. (1998). Influence of the nuclei on the cavitation inception for different types of cavitation on ship propellers. *Journal of Fluids Engineering*, *120*(1), 171–178.
Goldstein, R. (1996). *Fluid mechanics measurements*. CRC Press.
Gopalan, S., & Katz, J. (2000). Flow structure and modeling issues in the closure region of attached cavitation. *Physics of Fluids*, *12*(4), 895–911.
Gopalan, S., Katz, J., & Knio, O. (1999). The flow structure in the near field of jets and its effect on cavitation inception. *Journal of Fluid Mechanics*, *398*, 1–43.
Gowing, S., Briançon-Marjollet, L., Frechou, D., & Godeffroy, V. (1995). Dissolved gas and nuclei effects on tip vortex cavitation inception and cavitating core size. In *Proceedings of 5th international symposium on cavitation* (pp. 173–180).
Hassan, Y. A., Blanchat, T. K., Seeley, C. H., & Canaan, R. E. (1992). Simultaneous velocity measurements of both components of a two-phase flow using particle image velocimetry. *International Journal of Multiphase Flow*, *18*(3), 371–395.
Heindel, T. J. (2011). A review of X-ray flow visualization with applications to multiphase flows. *Journal of Fluids Engineering*, *133*(074001), 1–16.
Hewitt, G. F. (1978). *Measurement of two phase flow parameters*. Academic Press.
Iyer, C. O., & Ceccio, S. L. (2002). The influence of developed cavitation on the flow of a turbulent shear layer. *Physics of Fluids*, *14*(10), 3414–3431.
Katz, J., & Acosta, A. (1981). Observations of nuclei in cavitating flows. *Proceedings of I.U.T.A.M. symposium on mechanics and physics of bubbles in liquids* (pp. 123–132).
Katz, J., & Sheng, J. (2010). Applications of holography in fluid mechanics and particle dynamics. *Annual Review of Fluid Mechanics*, *42*, 531–555.
Katz, J., Gowing, S., O'Hern, T., & Acosta, A. (1984). A comparative study between holographic and light-scattering techniques of microbubble detection. *Proceedings of I.U.T.A.M. symposium on measuring techniques in gas-liquid two-phase flows* (pp. 41–66).
Kawanami, Y., Kato, H., Yamaguchi, H., Maeda, M., & Nakasumi, S. (2002). Inner structure of cloud cavity on a foil section. *J.S.M.E. International Journal Series B Fluids and Thermal Engineering*, *45*(3), 655–661.
Keller, A. P. (1972). The influence of the cavitation nucleus spectrum on cavitation inception, investigated with a scattered light counting method. *Journal of Fluids Engineering*, *94*(4), 917–925.
Keller, A. P. (1987). A vortex-nozzle cavitation susceptibility meter in routine application in cavitation inception measurements. In *Proceedings of euromech colloquium 222-unsteady cavitation and its effects*.
Keller, A. P. (2001). Cavitation scale effects—empirically found relations and the correlation of cavitation number and hydrodynamic coefficients. In *Proceedings of fourth international symposium on cavitation-CAV2001*.
Kjeldsen, M., Arndt, R. E. A., & Effertz, M. (2000). Spectral characteristics of sheet/cloud cavitation. *Journal of Fluids Engineering*, *122*(3), 481–487.
Kling, C. L., & Hammitt, F. G. (1972). A photographic study of spark-induced cavitation bubble collapse. *Journal of Fluids Engineering*, *94*(4), 825–832.
Koivula, T. (2000). On cavitation in fluid power. In *Proceedings of 1st FPNI-PhD symposium, Hamburg* (pp. 371–382).

Kubota, A., Kato, H., Yamaguchi, H., & Maeda, M. (1989). Unsteady structure measurement of cloud cavitation on a foil section using conditional sampling technique. *Journal of Fluids Engineering*, *111*(2), 204–210.

Kuhn de Chizelle, Y., Ceccio, S. L., & Brennen, C. E. (1995). Observations and scaling of travelling bubble cavitation. *Journal of Fluid Mechanics*, *293*, 99–126.

Kuiper, G. (1985). Reflections on cavitation inception. In *Proceedings of A.S.M.E. cavitaiton and multiphase flow forum*, FED-23.

Kumar, S. B., Dudukovic, M. P., Chaouki, J., Larachi, F., & Dudukovic, M. P. (1997). Computer assisted gamma and x-ray tomography: Applications to multiphase flow systems. *Non-invasive monitoring of multiphase flows* (pp. 47–103).

Laberteaux, K. R., & Ceccio, S. L. (2001a). Partial cavity flows. part 1. cavities forming on models without spanwise variation. *Journal of Fluid Mechanics*, *431*, 1–41.

Laberteaux, K. R., & Ceccio, S. L. (2001b). Partial cavity flows. part 2. cavities forming on test objects with spanwise variation. *Journal of Fluid Mechanics*, *431*, 43–63.

Lauterborn, W., & Bolle, H. (1975). Experimental investigations of cavitation-bubble collapse in the neighbourhood of a solid boundary. *Journal of Fluid Mechanics*, *72*(2), 391–399.

Lauterborn, W., & Hentschel, W. (1985). Cavitation bubble dynamics studied by high speed photography and holography: Part one. *Ultrasonics*, *23*(6), 260–268.

Lavigne, S. (1991). Le venturi analyseur de germes. *Journees DRET Cavitation*.

Le, Q., Franc, J. P., & Michel, J. M. (1993). Partial cavities: Pressure pulse distribution around cavity closure. *Journal of Fluids Engineering*, *115*(2), 249–254.

Le Corre, J.-M., Hervieu, E., Ishii, M., & Delhaye, J.-M. (2003). Benchmarking and improvements of measurement techniques for local-time-averaged two-phase flow parameters. *Experiments in fluids*, *35*(5), 448–458.

Lecoffre, Y., & Bonnin, J. (1979). Cavitation tests and nucleation control. In *Proceedings of A.S.M.E. international symposium on cavitation inception* (pp. 141–145).

Lecoffre, Y., Chantrel, P., & Teiller, J. (1987). Le grand tunnel hydrodynamique (GTH): France's new large cavitation tunnel for naval hydrodynamics research. In *Proceedings of A.S.M.E. international symposium on cavitation research facilities and techniques* (pp. 13–18).

Lee, I.-H., Mäkiharju, S., Ganesh, H., & Ceccio, S. L. (2016). Scaling of gas diffusion into limited partial cavities. *Journal of Fluids Engineering*.

Leger, A. T., & Ceccio, S. L. (1998). Examination of the flow near the leading edge of attached cavitation. part 1. detachment of two-dimensional and axisymmetric cavities. *Journal of Fluid Mechanics*, *376*, 61–90.

Leger, A. T., Bernal, L. P., & Ceccio, S. L. (1998). Examination of the flow near the leading edge of attached cavitation. part 2. incipient breakdown of two-dimensional and axisymmetric cavities. *Journal of Fluid Mechanics*, *376*, 91–113.

Li, C. Y., & Ceccio, S. L. (1996). Interaction of single travelling bubbles with the boundary layer and attached cavitation. *Journal of Fluid Mechanics*, *322*, 329–353.

Lindken, R., & Merzkirch, W. (2002). A novel piv technique for measurements in multiphase flows and its application to two-phase bubbly flows. *Experiments in Fluids*, *33*(6), 814–825.

Lucas, G. P., & Mishra, R. (2005). Measurement of bubble velocity components in a swirling gas-liquid pipe flow using a local four-sensor conductance probe. *Measurement Science and Technology*, *16*(3), 749–758.

Mäkiharju, S. A., Gabillet, C., Paik, B.-G., Chang, N. A., Perlin, M., & Ceccio, S. L. (2013). Time-resolved two-dimensional x-ray densitometry of a two-phase flow downstream of a ventilated cavity. *Experiments in Fluids*, *54*(7), 1–21.

McNulty, P. J., & Pearsall, I. S. (1982). Cavitation inception in pumps. *Journal of Fluids Engineering*, *104*(1), 99–104.

Mørch, K. A. (2007). Reflections on cavitation nuclei in water. *Physics of Fluids*, *19*(072104), 1–7.

Obreschkow, D., Kobel, P., Dorsaz, N., De Bosset, A., Nicollier, C., & Farhat, M. (2006). Cavitation bubble dynamics inside liquid drops in microgravity. *Physical Review Letters*, *97*(094502), 1–4.

O'Hern, T. J., Katz, J., & Acosta, A. J. (1985). *Holographic measurements of cavitation nuclei in the sea*. In Proceedings of A: S.M.E. cavitation and multiphase flow forum.
Ohl, C. D., Philipp, A., & Lauterborn, W. (1995). Cavitation bubble collapse studied at 20 million frames per second. *Annalen der Physik*, *507*(1), 26–34.
Ohl, C. D., Lindau, O., & Lauterborn, W. (1998). Luminescence from spherically and aspherically collapsing laser induced bubbles. *Physical Review Letters*, *80*(2), 393–396.
Oldenziel, D. M. (1982). A new instrument in cavitation research: the cavitation susceptibility meter. *Journal of Fluids Engineering*, *104*(2), 136–141.
Oweis, G. F., Choi, J., & Ceccio, S. L. (2004). Dynamics and noise emission of laser induced cavitation bubbles in a vortical flow field. *Journal of the Acoustical Society of America*, *115*(3), 1049–1058.
Pham, T. M., Larrarte, F., & Fruman, D. H. (1999). Investigation of unsteady sheet cavitation and cloud cavitation mechanisms. *Journal of Fluids Engineering*, *121*(2), 289–296.
Rae, W. H., & Pope, A. (1984). *Low-speed wind tunnel testing*, 2nd edition. Wiley.
Ran, B., & Katz, J. (1994). Pressure fluctuations and their effect on cavitation inception within water jets. *Journal of Fluid Mechanics*, *262*, 223–263.
Rood, E. P. (1991). Review–mechanisms of cavitation inception. *Journal of Fluids Engineering*, *113*(2), 163–175.
Rooze, J., Rebrov, E. V., Schouten, J. C., & Keurentjes, J. T. (2013). Dissolved gas and ultrasonic cavitation-a review. *Ultrasonics Sonochemistry*, *20*(1), 1–11.
Roth, H., Gavaises, M., & Arcoumanis, C. (2002). *Cavitation initiation, its development and link with flow turbulence in diesel injector nozzles*. S.A.E. Technical Paper, (2002-01-0214).
Sato, R., Mori, T., Yakushiji, R., Naganuma, K., Nishimura, M., Nakagawa, K., & Sasajima, T. (2003). Conceptual design of the flow noise simulator. In *Proceedings of joint A.S.M.E./J.S.M.E. 4th joint fluids summer engineering conference* (pp. 129–133).
Schiebe, F. R. (1972). Measurement of the cavitation susceptibility of water using standard bodies. *St. Anthony Falls Laboratory Project Report 118*, University of Minnesota.
Shen, Y. T., & Dimotakis, P. E. (1989). Viscous and nuclei effects on hydrodynamic loadings and cavitation of a naca 66 (mod) foil section. *Journal of Fluids Engineering*, *111*(3), 306–316.
Simoni, R. D., Hill, R. L., & Vaughan, M. (2002). The measurement of blood gases and the manometric techniques developed by donald dexter van slyke. *Journal of Biological Chemistry*, *277*(27), e16.
Straka, W. A., Meyer, R. S., Fontaine, A. A., & Welz, J. P. (2010). Cavitation inception in quiescent and co-flow nozzle jets. *Journal of Hydrodynamics, Series B*, *22*(5), 813–819.
Stutz, B. (2003). Influence of roughness on the two-phase flow structure of sheet cavitation. *Journal of Fluids Engineering*, *125*(4), 652–659.
Stutz, B., & Legoupil, S. (2003). X-ray measurements within unsteady cavitation. *Experiments in Fluids*, *35*(2), 130–138.
Stutz, B., & Reboud, J. L. (2000). Measurements within unsteady cavitation. *Experiments in Fluids*, *29*(6), 522–545.
Tanger, H., & Weitendorf, E. A. (1992). Applicability tests for the phase doppler anemometer for cavitation nuclei measurements. *Journal of Fluids Engineering*, *114*(3), 443–449.
Thoroddsen, S. T., Etoh, T. G., & Takehara, K. (2008). High-speed imaging of drops and bubbles. *Annual Review of Fluid Mechanics*, *40*, 257–285.
Tropea, C., Yarin, A. L., & Foss, J. F. (Eds.). (2007). *Springer handbook of experimental fluid mechanics*. Springer Science and Business Media.
Van der Kooij, J., & De Bruijn, A. (1984). Acoustic measurements in the NSMB depressurized towing tank. *International Shipbuilding Progress*, *31*(353), 13–25.
Vogel, A., & Lauterborn, W. (1988). Time-resolved particle image velocimetry used in the investigation of cavitation bubble dynamics. *Applied Optics*, *27*(9), 1869–1876.
Wetzel, J. M., & Arndt, R. E. A. (1994a). Hydrodynamic design considerations for hydroacoustic facilities: Part I- flow quality. *Journal of Fluids Engineering*, *116*(2), 324–331.

Wetzel, J. M., & Arndt, R. E. A. (1994b). Hydrodynamic design considerations for hydroacoustic facilities: Part II- pump design factors. *Journal of Fluids Engineering*, *116*(2), 332–337.

Wosnik, M., Schauer, T. J., & Arndt, R. E. (2003). Experimental study of a ventilated supercavitating vehicle. In *Proceedings of fifth international symposium on cavitation* (pp. 1–4).

Wu, Q., & Ishii, M. (1999). Sensitivity study on double-sensor conductivity probe for the measurement of interfacial area concentration in bubbly flow. *International Journal of Multiphase Flow*, *25*(1), 155–173.

Yu, P. W., & Ceccio, S. L. (1997). Diffusion induced bubble populations downstream of a partial cavity. *Journal of Fluids Engineering*, *119*(4), 782–787.

An Introduction to Flow-Induced Instabilities in Rocket Engine Inducers and Turbopumps

Luca d'Agostino, Angelo Cervone, Lucio Torre, Giovanni Pace, Dario Valentini and Angelo Pasini

Abstract The article reviews the main forms of flow-induced instabilities detected in the liquid propellant turbopumps systems of modern rocket engines, with special reference to rotating stall, rotating cavitation, cavitation surge and higher order surge modes, illustrating their characteristics, origin and damage potential.

1 Introduction

Propellant feed turbopumps are a crucial component of all liquid propellant rocket engines because of the severe limitations associated with the design of high power density, dynamically stable machines capable of meeting the extremely demanding suction, pumping and reliability requirements of modern space transportation systems. Figure 1 illustrates the power/weight ratio of the main propellant feed turbopumps used by US space rocket engines powering the Mercury, Gemini, Apollo, and Space Shuttle programs in the two decades from the early 60s to the 80s. The tenfold increase of the power density clearly indicates that today's rocket propellant feed turbopumps belong to a very special family of machines with respect to earlier more conventional land- or sea-based machines.

Such high power/weight ratios are obtained by running the impeller at the maximum allowable rotational speed, in order to reduce the size—and therefore the weight—of the machine. However, the resulting higher velocity of the flow also reduces the pressure on the suction sides of the blades, creating the conditions for the development of cavitation where the flow pressure drops below the vapor

L. d'Agostino (✉) · G. Pace · D. Valentini · A. Pasini
Dipartimento di Ingegneria Civile ed Industriale, Università di Pisa, Pisa, Italy
e-mail: luca.dagostino@ing.unipi.it

A. Cervone
T.U. Delft, Delft, The Netherlands

L. Torre
Alta S.p.A., Ospedaletto (Pisa), Italy

L. d'Agostino and M.V. Salvetti (eds.), *Cavitation Instabilities and Rotordynamic Effects in Turbopumps and Hydroturbines*, CISM International Centre for Mechanical Sciences 575, DOI 10.1007/978-3-319-49719-8_3

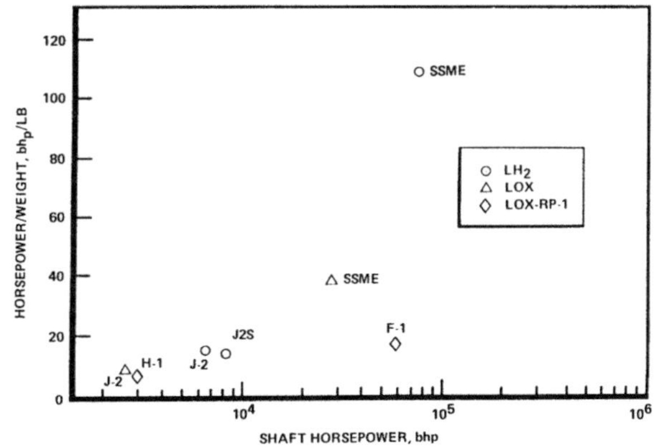

Fig. 1 Power/weight evolution of propellant feed turbopumps in the US space rocket engines (Rockwell International, personal communication)

pressure of the liquid. Indeed, operation under limited cavitation, and sometimes at supercritical speeds, is usually tolerated in rocket propellant feed turbopumps in order to maximize their performance, exposing the machine to the risk of developing dangerous cavitation-induced fluid dynamic and rotordynamic instabilities.

Cavitation usually represents the most stringent fluid mechanical constraint to the increase of the power density of hydraulic machinery and in particular the major source of degradation of the suction performance, efficiency, useful life, and reliability of turbopumps (Stripling and Acosta 1962; Stripling 1962). Its occurrence is, in fact, connected to the insurgence of undesirable phenomena ranging from efficiency deterioration to operational irregularities, mechanical erosion, noise, vibrations, as well as fluid mechanical and/or dynamical instabilities of the machinery exposed to the cavitating flow. This is particularly true in feed systems of cryogenic liquid propellant rocket engines, where propellant storage under saturation conditions greatly facilitates the occurrence of cavitation. More specifically, in space propulsion applications cavitation can provide the flow compliance and excitation for triggering dangerous fluid mechanic and/or rotordynamic instabilities (Sack and Nottage 1965; Natanzon et al. 1974; Brennen and Acosta 1973; Brennen and Acosta 1976; Ng and Brennen 1978; Braisted 1979; d'Agostino et al. 1998; Cervone et al. 2005; Cervone et al. 2006; d'Agostino and Salvetti 2007; d'Agostino 2013; Pace et al. 2015) or even, through the coupling with thrust generation, of the entire space vehicle (POGO auto-oscillations, Rubin 1966).

The most critical rotordynamic instability in high-performance turbopumps is the development of self-sustained lateral motions (whirl) of the impeller under the action of destabilizing forces of mechanical and fluid dynamic origin (Ehrich and Childs 1984). The role of rotordynamic fluid forces in promoting rotordynamic instabilities by significantly modifying, in conjunction with cavitation, the dynamic properties of the impeller, and therefore the critical speeds of the whole machine, has long been recognized (Rosenmann 1965). However, because of their complexity and the experimental difficulties in their measurement, so far rotordynamic fluid forces in cavitating inducers and turbopumps have received relatively little

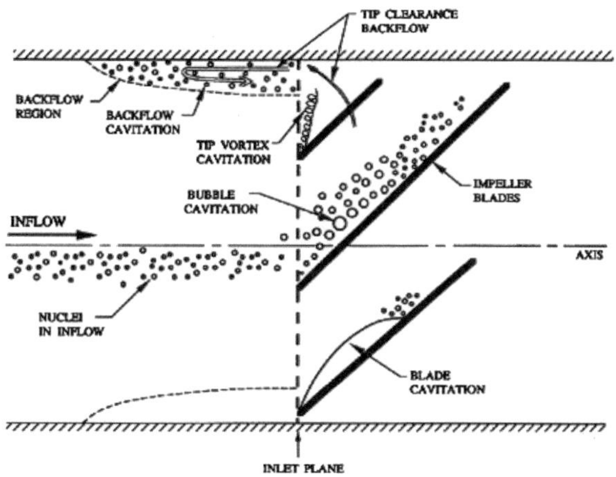

Fig. 2 Main form of cavitation in turbopumps (from Brennen 1994)

attention in the open literature (Franz 1989; Bhattacharyya 1994; Bhattacharyya et al. 1997; d'Agostino et al. 1998; Torre et al. 2010; Pasini et al. 2010; Pasini et al. 2011a, b; Torre et al. 2011a, b; Valentini et al. 2015).

Typically cavitation of the liquid propellant is confined to the turbopump inducer, an initial axial-flow or mixed-flow impeller routinely used to sufficiently raise the fluid pressure for suppressing cavitation in the main centrifugal impeller (s). The main forms of cavitation developing in turbopump inducers are schematically illustrated in Fig. 2.

Cavitation first appears in the core of the vortex generated by the blade tips when the Euler (or cavitation) number:

$$\sigma = \frac{p_{ref} - p_V}{\frac{1}{2}\rho_L \Omega^2 r_T^2}$$

(based on the reference pressure p_{ref}, the vapor pressure p_V, and the tip-speed dynamic pressure $\frac{1}{2}\rho_L \Omega^2 r_T^2$) drops below the inception value, which essentially depends on the geometry and operational conditions of the machine. Next, more extensive cavitation develops, starting in the leakage flow at the blade tips and progressively extending on the suction sides of the blades toward the hub. The bubbly, cloudy, or attached nature of blade cavitation depends on the dominant location and susceptibility of the nuclei responsible for the inception of cavitation, as well as on the thermodynamic state of the liquid and the nature of the flow (laminar/turbulent, separated/unseparated, etc.). Finally, especially at high blade loads, cavitation can also appear in the separated, recirculating and slowly co-rotating backflow that develops at the inlet of the inducer near the side wall of the casing as a consequence of the combined effects of tip blade and leading edge flow leakage. Figure 3 illustrates the typical progression of cavitation in turbopump inducers as a function of the NPSH (Net Positive Suction Head) or, in nondimensional form, of the inlet cavitation number.

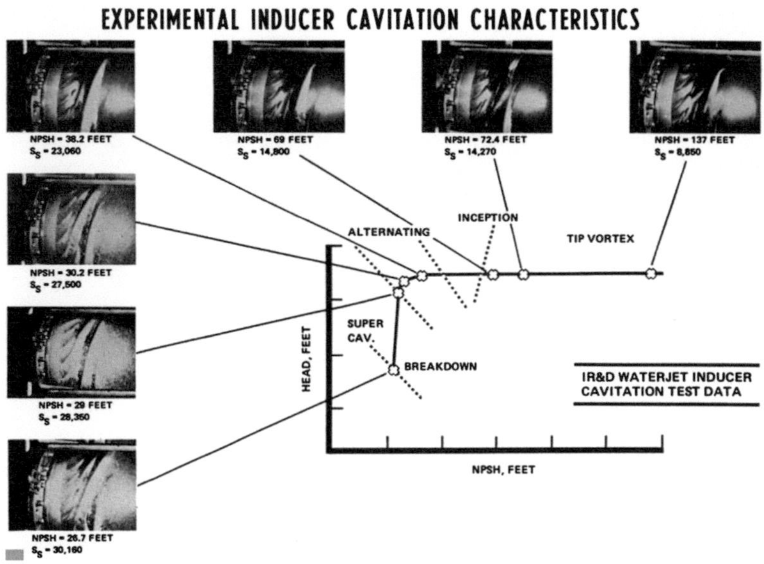

Fig. 3 Evolution of cavitation in the SSME liquid hydrogen turbopump inducer as a function of the NPSH (Rockwell International, personal communication)

With reference to Fig. 4, flow-induced instabilities of turbopumps can be conventionally classified as proposed by Brennen (1994) in: global oscillations (category A, including surge, cavitation auto-oscillations, rotating stall, rotating cavitation, unsteady partial-blade, and vortex cavitation, higher order cavitation surge), which affect the whole machine or even the entire pumping system on a large scale, local oscillations (category B, like blade flutter and rotor–stator interactions), of more limited impact, and rotordynamic instabilities (category C), caused by fluid forces on the impellers.

Another useful classification of cavitation-induced turbopump instabilities has been proposed by Franc (2001), who distinguishes between system instabilities, also involving the other components of the pumping system (inlet and outlet lines, tanks, valves), and intrinsic instabilities, whose features are dependent on the characteristics of the pump.

The cavitating behavior of a rotating machine can often be usefully interpreted with reference to the equivalent features of the flow through static cascades of hydrofoils. In this order of approximation, cavitation surge, sometimes observed in supercavitating hydrofoils (Wade and Acosta 1966) as a result of the interactions of long cavities with the dynamics the other components of the circuit, is probably the most well-known system instability. On the other hand, unsteady cloud cavitation, characterized by the periodic break-up of the cavity developing on highly loaded lifting surfaces and by the subsequent shedding of a large-scale vortex containing a cluster of many small bubbles (Kubota et al. 1989), is a typical example of intrinsic instability. Experiments carried in out facilities with different characteristics or

VIBRATION CATEGORY		FREQUENCY RANGE
A2	Surge	System dependent, $3-10\ Hz$ in compressors
A2	Auto-oscillation	System dependent, $0.1-0.4\Omega$
A1	Rotor rotating stall	$0.5\Omega-0.7\Omega$
A1	Vaneless diffuser stall	$0.05\Omega-0.25\Omega$
A1	Rotating cavitation	$1.1\Omega-1.2\Omega$
A3	Partial cavitation oscillation	$<\Omega$
C1	Excessive radial force	Some fraction of Ω
C2	Rotordynamic vibration	Fraction of Ω when critical speed is approached.
A4	Blade passing excitation (or B2)	$Z_R\Omega/Z_{CF}, Z_R\Omega, mZ_R\Omega$ (in stator frame) $Z_S\Omega/Z_{CF}, Z_S\Omega, mZ_S\Omega$ (in rotor frame)
B1	Blade flutter	Natural frequencies of blade in liquid
B3	Vortex shedding	Frequency of vortex shedding
A6	Cavitation noise	$1\ kHz-20\ kHz$

Fig. 4 Original classification of flow instabilities by Brennen (1994)

adjustable configurations (Tsujimoto et al. 1998) led to very similar Strouhal numbers for the cloud oscillations and proved the correlation between cloud cavitation and the re-entrant jet generated at the cavity closure for cavity lengths greater than about 50% of chord (Callenaere et al. 1998; Kawanami et al. 1997; Sakoda et al. 2001).

2 Rotating Stall

Rotating stall typically occur in compressors operating at high incidence, where it has been extensively investigated because of its critical impact on the performance and stability of the machine (Greitzer 1981). However, it has also been observed in noncavitating turbopumps at low flow rates operating near the maximum of the pumping characteristic. As a consequence of the distortions generated in the flow field, partial stall cells tend to rotate in the opposite direction of the impeller with an absolute speed equal to 50–70% of the rotor speed. Rotating stall is most frequently observed in compressors with a large number of blades, but it has also been detected in pumps with a just 3 or 4 blades. Actually, in Murai (1968), was the first to detect and investigate rotating stall in an axial-flow pump with 18 blades.

He measured stall propagation velocities between 0.45 and 0.6 times the impeller rotational speed, and observed that is was somewhat affected by the occurrence of a limited amount of cavitation.

A useful approximate criterion in compressors predicts the onset of rotating stall when the maximum of the total head rise curve is approached as the flow coefficient decreases. As a consequence, a number of authors postulated that also rotating stall in inducers could be associated to areas of positive slope of the pumping performance curve (see for example Kamijo et al. 1977). This is, however, no more than an approximation (Greitzer 1981). More sophisticated stall criteria typically take into consideration the evolution of the displacement thickness on the blade suction side. In particular, following this approach, Leiblein (1965) has been able to demonstrate the correlation of stall with a simple integral parameter of the flow, the diffusion factor, predicting the occurrence of stall on typical blade profiles when its value exceeds about 0.6.

3 Rotating Cavitation

At low Euler numbers, cavitating inducers and pump impellers can exhibit a phenomenon similar to rotating stall, known as rotating cavitation. Even the two phenomena look similar, significant differences exist. Rotating stall occurs at locations of the head-flow characteristic where the slope of the curve is positive and therefore unstable. Rotating cavitation, on the other hand, is normally observed in the zone of negative slope at cavitation numbers between 2 and 3 times larger than the breakdown value. Even if subsynchronous (or "backward rotating") rotating cavitation has sometimes been observed, usually it is supersynchronous (forward rotating with respect to the impeller blades) with frequencies between 1.1 and 1.2 times the pump's rotational speed, as indicated in Fig. 5 (from Brennen 1994) with reference to the inducer tested by Kamijo et al. (1977).

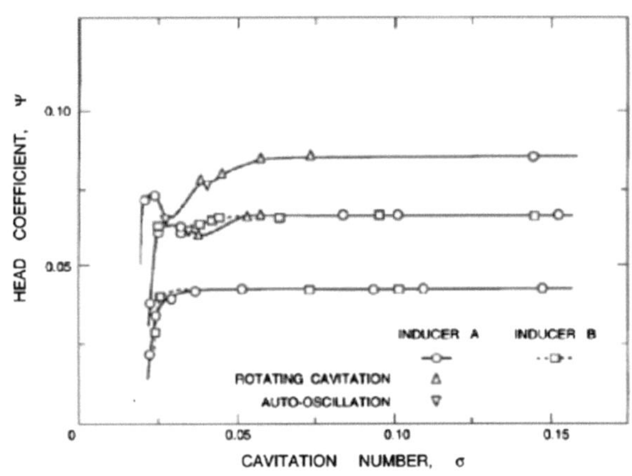

Fig. 5 Influence of rotating cavitation and auto-oscillation on the performance of a cavitating inducer (Brennen 1994)

The first observation of rotating cavitation was probably documented by Rosenmann (1965), who reported a vibration with similar characteristics in some of his experiments on inducers operating at cavitation numbers between 2 and 3 times higher than the breakdown value. However, Rosenmann did not recognize this phenomenon as rotating cavitation. It has first been identified explicitly in by Kamijo et al. (1977) and associated to the portion of the inducer performance curve with negative slope, where the head is starting to be affected by cavitation.

A detailed theoretical analysis of rotating cavitation was proposed by Tsujimoto et al. (1993). They used a simplified analytical model based on the actuator disk equations, able of reproducing fairly well the experimental results available at that time (in particular, those obtained by Kamijo et al. 1977). The most interesting results of their model were:

- Rotating cavitation is a completely different phenomenon from rotating stall, which is not significantly affected by the occurrence of cavitation. In particular, rotating cavitation is caused by a positive mass flow gain factor, while rotating stall is caused by the positive slope of the performance curve.
- The two driving parameters for the onset of rotating cavitation are the mass flow gain factor M and the cavitation compliance C. Other parameters, like flow coefficient and cavitation number, do not affect rotating cavitation directly.
- Two modes of rotating cavitation are predicted by the model: not only a "forward" rotating mode (i.e., rotating faster than the impeller speed), but also a "backward" rotating mode (slower than the impeller speed).

Similar results were obtained with a more refined model based on annular cascade equations, and presented by Watanabe et al. (1999). Several instability modes were obtained by the model as solutions of the characteristic equation, including rotating stall, forward and backward rotating cavitation, and various secondary radial modes.

Hashimoto et al. (1997a) first investigated the behavior of rotating cavitation on a 3-bladed inducer, observing that it was almost suppressed using a casing with a forward-facing step. In a second paper (Hashimoto et al. 1997b) they analyzed the influence of the rotational speed of the inducer, showing that the operational conditions for the occurrence of rotating cavitation, asymmetric cavitation and low-cycle oscillations (or cavitation surge) can be different depending on the geometry of the inducer casing. Rotating cavitation can eventually disappear at higher rotating speeds. Backward rotating cavitation, predicted by Tsujimoto et al. (1993), was first detected and recognized by analyzing the cross-correlation phase of pressure signals from transducers flush-mounted on the casing at different angular stations near the leading edges of the inducer. These results were also confirmed by Tsujimoto et al. (1997), who observed backward rotating cavitation with a propagation speed of about 0.9 times the inducer rotational speed.

Rotating cavitation has also been detected by Zoladz (2000) in both water and LOX tests of the Fastrac engine turbopump, where a connection with cavitation surge has also been observed. The frequency of rotating cavitation decreased (tending to synchronous frequency) as the cavitation number approached breakdown conditions. This result is consistent with the predictions of the linear 2D cascade model with closed cavities proposed by Tsujimoto 2001, who also found that 4-bladed inducers are expected to show higher rotating cavitation frequencies than 3-bladed ones and, furthermore, that the onset point of cavitation corresponds to a cavity length equal to about 70% of the blade spacing.

Recent 3D simulations (see for example Kimura et al. 2006; Kang et al. 2007) indicate that the interaction between the tip leakage vortex and the next blade plays a central role in the onset of rotating cavitation. The design of inducer leading edge shape is therefore of great importance for the suppression of the rotating cavitation (Cervone et al. 2012; Pace et al. 2015). Furthermore, it has been recognized that rotating instability modes can be controlled by changing the shape of the inducer casing. In particular, Shimiya et al. (2006), have shown that the use of J-Grooves on the inducer casing can effectively reduce or suppress rotating cavitation, but also promotes the occurrence of cavitation surge modes generated by the interference of the tip leakage vortex cavitation and the leading edge of the next blade. At higher cavitation numbers, these modes can be reduced by extending the J-Groove further upstream of the inducer leading edges, but cannot be suppressed in this way at lower cavitation numbers. Furthermore, the use of J-Groove seems not to be effective for the control of the rotating cavitation at higher flow coefficients near design conditions.

More recently, rotating cavitation has been observed at Pratt-Whitney Rocketdyne laboratories in a 3-bladed LH2 inducer (Sergeant et al. 2008). The 2D simulations reported by Ohta and Kajishima (2008), also predicted a forward-running rotating cavitation mode in a 3-bladed inducer and a backward-running mode with similar characteristics in a 4-bladed inducer.

4 Other Rotating Instabilities

Attached uneven cavitation (also called "steady asymmetric cavitation" or "synchronous rotating cavitation") and alternate blade cavitation are synchronous rotating instabilities frequently occurring in cavitating inducers (see Fig. 6). Attached uneven cavitation is characterized by nonuniform length of the cavities developing on the inducer blades. It can be considered as a single-cell rotating cavitation moving with the same angular velocity of the inducer and generates a peak in the pressure spectra at the synchronous frequency. Alternate blade cavitation is frequently observed in inducers with an even number of blades when alternate blades experience different intensities of cavitation. It can be considered as a synchronous rotating cavitation mode with a number of cells equal to one half of the number of blades and generates a peak in the pressure spectra at a frequency equal to the synchronous frequency multiplied by the number of cells.

Experimental observations of synchronous rotating instabilities were presented by Hashimoto et al. (1997a). Attached uneven cavitation was observed in both a

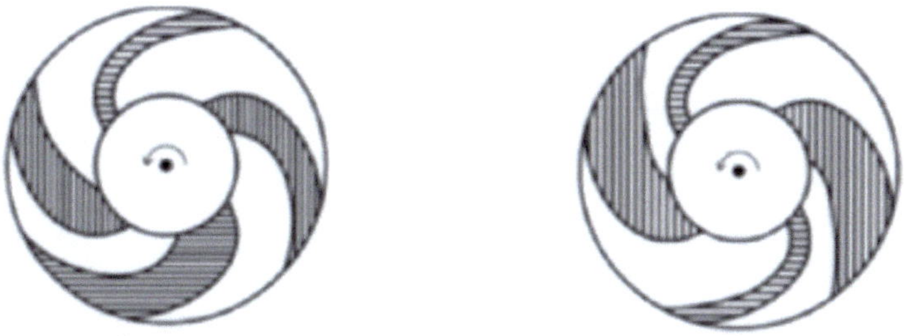

Fig. 6 Schematic representation of the asymmetric cavitation (*left*) and the alternate blade cavitation (*right*)

3-bladed and a 4-bladed inducer operating at lower cavitation numbers near breakdown conditions. Alternate blade cavitation was observed on the 4-bladed inducer at higher cavitation numbers, around 2–3 times the breakdown value. Both instabilities occurred at high flow coefficients close to the design point. Not surprisingly, alternate blade cavitation, unlike attached uneven cavitation, did not induce significant shaft vibrations because of the polar symmetry of the perturbed flow field.

A detailed analysis of the influence of the blade number, angle, and solidity on synchronous rotating instabilities is presented in by Furukawa et al. (2002). It was found that the range of cavitation numbers for which it is possible to detect both attached uneven cavitation and alternate blade cavitation tends to be narrower for inducers with higher number of blades, longer blade length, or higher blade angle.

Using a quasi-3D analytical model of steady cavitation in a 2-bladed inducer, Horiguchi et al. (2003b) predicted that alternate blade cavitation starts to occur when the cavity length on the inducer blades reaches about 65% of the blade spacing, confirming the results obtained for rotating cavitation from simpler 2D analyses of blade cascades.

An analysis of the characteristics of synchronous uneven cavitation on a 3-bladed inducer is presented by Shimagaki et al. (2006). Figure 7 shows the waterfall plot of mid-chord pressure fluctuations and the suction characteristic of the inducer at its design flow coefficient. A significant decrease of the head is present when synchronous asymmetric cavitation occurs, and the authors report that the head loss can be different depending on which inducer blade experiences the asymmetry in cavitation.

Finally, Yoshida et al. (2008) conducted a detailed experimental and numerical investigation to elucidate the mechanism of generation and development of asymmetric cavitation by means of measurements taken by pressure transducers installed along the blade channels. They reported that the synchronous blade vibration generated by uneven cavitation appeared to be a self-excited vibration, hydrodynamically coupled with the rotordynamic characteristics of the impeller.

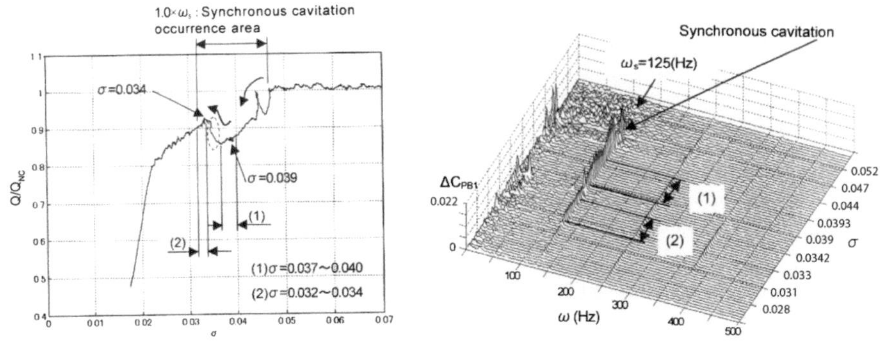

Fig. 7 *Left* suction performance curve of a 3-bladed inducer at the design flow coefficient. *Right* waterfall plot of the power spectra of mid-chord pressure fluctuations, for the same flow conditions (Shimagaki et al. 2006)

Backflow vortex cavitation was first detected and identified by Tsujimoto et al. (1997) in inducers operating at flow coefficients lower than design conditions and cavitation numbers either close to breakdown or significantly higher (3–4 times the breakdown value). As illustrated in Fig. 8, it is a rotating instability of the inducer backflow with the formation of multiple cells of axial vorticity (5 cells in Tsujimoto et al. 1997), which slowly rotate at an angular speed less than 0.2 times the inducer rotating speed.

The occurrence of a backflow vortex instability with 8 cells rotating at about 32% of the impeller speed has been documented by Zoladz (2000) with reference to the tests of the Fastrac turbopump. Imamura et al. (2003) showed that backflow

Fig. 8 Schematic representation of the "backflow vortex cavitation" instability (Tsujimoto et al. 1997)

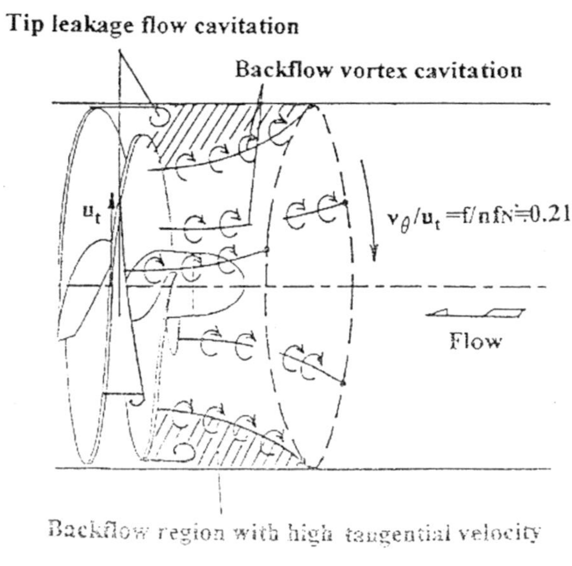

vortex cavitation can be suppressed by means of J-grooves in the inducer casing, similar to those used for the control of rotating cavitation.

Rotating choke is another form of cavitation instability first identified by Tsujimoto and Semenov (2002), Semenov et al. (2004), with reference to the phenomenon observed by Shimura et al. (2002), Uchiumi et al. (2003) in the LH2 inducer of the Japanese H-II engine and indicated as "rotating stall cavitation." It occurred near the inducer operational conditions (with a negative slope of the performance curve) at rotating speeds close to the second critical speed of the rotor, and was eventually eliminated by redesigning the inducer casing with slightly conical tip blade shape. Tsujimoto and Semenov (2002) indicated that rotating choke is not related to blade stall as originally postulated, but rather to the transition of pumping performance from negative to positive slope when breakdown are approached, as illustrated in Fig. 9.

They proposed to name this instability "rotating choke" in order to highlight its dependence on the "choking" cavitation conditions that develop near inducer breakdown. As the Euler number decreases the cavities on the suction side of the blades increase, eventually "choking" the flow and rapidly increasing the head losses as soon as they extend into the throat between adjacent blades. If this head loss is sufficiently large the slope of the pumping characteristic becomes positive. Under these conditions choking in a blade channel propagates in the opposite direction of the inducer rotation with the same mechanism as for rotating stall.

A similar kind of "rotating stall" was observed by the author and his collaborators in a series of tests conducted in the Cavitating Pump Rotordynamic Test Facility at Alta, Pisa, Italy, on a 3-bladed commercial inducer (Cervone et al. 2005) and, later, on a 2-bladed inducer with the same blade geometry of the 3-bladed LOX inducer for the Vinci engine (Cervone et al. 2006). In these tests, however, the phenomenon was detected at high blade loads (flow coefficients lower than the design value) over a wide range of cavitation conditions, even if it was more intense

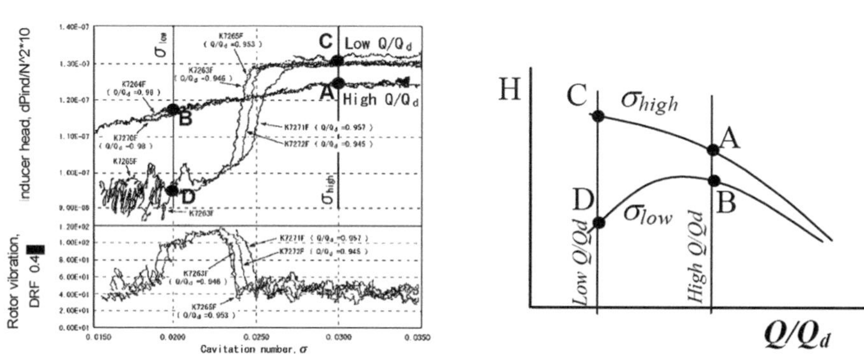

Fig. 9 *Left* Suction performance and rotor vibration as a function of the cavitation number, at two different flow coefficients (95% and 98% of the operating conditions), for the liquid hydrogen inducer of the Japanese LE-7 engine. *Right* schematic representation of the change of slope in the performance curve (Tsujimoto and Semenov 2002)

near breakdown. The rotating nature of the phenomenon was later confirmed by flow visualization by means of a high-speed camera (Cervone et al. 2007).

In a series of experiments on several inducers intended for use in liquid hydrogen Uchiumi and Kamijo (2008) showed that "rotating stall cavitation" occurred when the flow incidence on the leading edges of the inducer was larger than about 43% of the tip blade angle measured from the tangential direction.

5 Cavitation Surge

Surge and cavitation surge (or cavitation auto-oscillation) are instabilities of pumping systems resulting in pressure and flow rate oscillations that can often generate excessive vibration and performance degradation of the machine and, in the worst case, even affect the integrity of the entire system.

With reference to Fig. 10, consideration of the effects of small perturbations shows that the operational point O in the left diagram, at the intersection between the head characteristic and the load curve of the system, is quasi-statically stable if the slope of the load characteristic is larger than the slope of the pumping characteristic (Greitzer 1981). Hence, for instance, point A in the right diagram is stable, B is neutrally stable, and C is unstable. Operation of the machine under unstable conditions leads to the development of potentially dangerous large-amplitude surge oscillations both in compressors and pumps. As frequently observed in nonlinear dynamic oscillation, also surge instabilities of pumping systems usually display an hysteresis cycle (Greitzer 1981).

Clearly the rapid increase of the slope of the performance characteristic of cavitating turbopumps as breakdown is approached suggests that in these machines cavitation surge becomes increasingly more likely at low cavitation numbers.

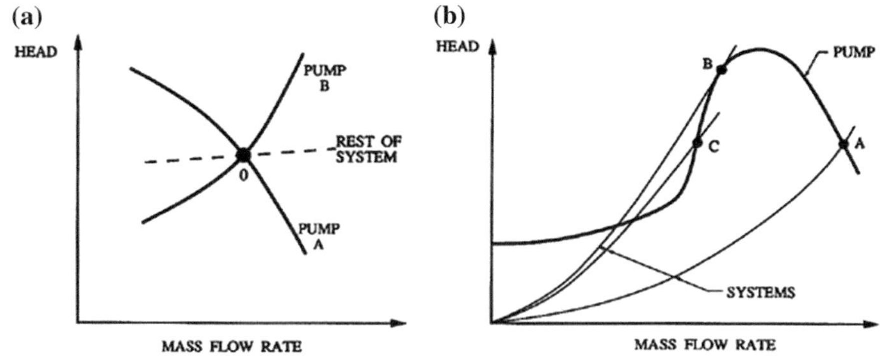

Fig. 10 Schematic of stable and unstable characteristic curves of a pumping system (Brennen 1994)

Indeed in cavitating pumps violent oscillations of the pressure and flow rate of the entire pumping system have routinely been observed when the cavitation number is decreased at values for which the head rise begins to be affected, as documented by Rosenman (1965), Sack and Nottage (1965), Natanzon et al. 1974, Brennen and Braisted (1980), among others. This surge phenomenon, named "cavitation auto-oscillation" in the original classification by Brennen (1994), has been later indicated as "cavitation surge" by other researchers. It can lead to very large fluctuations across the pumping system and can also cause substantial axial forces on the shaft, on the order of 20% of the steady axial thrust on the impeller. Brennen and Braisted (1980), and later Yamamoto (1991), showed that the frequency of cavitation surge oscillations is roughly proportional to the pump's rotating speed and the square root of the cavitation number.

With reference to space propulsion applications, the onset of cavitation surge is especially dangerous in liquid propellant rocket feed systems because the resulting flow rate fluctuations, by modulating the engine thrust, can trigger self-sustained POGO oscillations of the entire space vehicle, a global instability characterized by extremely dangerous large-amplitude fluctuations of the thrust (NASA 1970).

A violent cavitation surge instability at very low frequency (<10 Hz) was observed by Hashimoto et al. (1997a) on a 3-bladed model of the liquid oxygen inducer of the Japanese LE-7 engine operating near breakdown conditions and at flow coefficients close to the nominal design value. The authors called this instability "low cycle oscillations" and reported a significant interaction with steady uneven cavitation and backflow vortices. A similar surge mode, but at higher frequency (about 18 Hz), was detected on a similar 3-bladed inducer by Tsujimoto et al. (1997), who interpreted it as the result of a resonance between the vibration mode of the system at 18 Hz and the oscillations of the blade cavities due to the occurrence of rotating cavitation.

A surge mode instability at about 2 Hz was detected by Furukawa et al. (2001) in water tests on a 2-bladed inducer with a relatively high tip blade angle (about 11 degrees) operating near breakdown conditions at flow coefficients significantly lower than the nominal design value (Fig. 11). The onset pressure of this instability mode is reported to coincide with the value for which the cavities on the blades reach the inlet of the blade channels.

The dependence of the frequency of cavitation surge on the Euler number is probably the consequence of a strong connection between this form of instability and rotating cavitation, as documented by Zoladz (2000) in connection with the tests carried out on the LOX turbopump of the Fastrac engine. If Ω is the pump rotational speed, f_{rc} is the frequency of rotating cavitation instability and Z is the number of blades of the inducer, the frequency f of the cavitation surge mode is reported to be equal to:

$$f = Z(f_{rc} - \Omega)$$

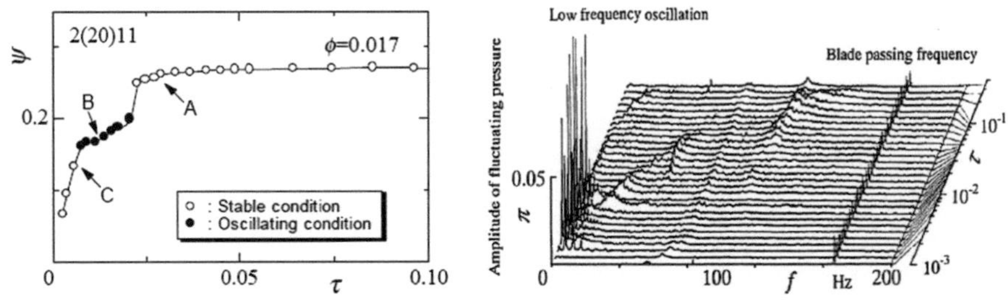

Fig. 11 *Left* suction performance curve of a 2-bladed inducer at $\varphi = 0.017$, including the field of existence of the surge mode instability. *Right* waterfall plot of the power spectrum of outlet pressure fluctuations on the same inducer, under the same test conditions (Furukawa et al. 2001)

Several observations under different test conditions confirmed this relationship, which is also compatible with the cavitation surge frequencies reported by other research groups on different pumps.

The influence of number of blades, blade angle and blade solidity on the cavitation surge instability was investigated by Furukawa et al. (2002) by means of experimental observations on several test inducers. It was found that:

- when the blade number is increased (and, consequently, the head rise is decreased), the range of cavitation numbers for which the cavitation surge is detected tends to become wider;
- when blade solidity is increased (by increasing the axial length of the blades) and head rise is therefore increased, the range of cavitation numbers for the existence of cavitation surge becomes narrower;
- finally, when the blade angle is decreased, the field of existence of the cavitation surge in terms of s tends to be wider.

6 Higher Order Cavitation Instabilities

Free oscillation modes with frequencies significantly higher than the inducer rotational speed are generically indicated as higher order instabilities and can involve unstable flow motion in the streamwise (higher order surge modes) and azimuthal (higher order rotating modes) directions. They have been detected and investigated in detail only in recent years and are universally recognized as extremely important in the design and characterization of space rocket inducers because of their potential for coupling with the lower critical speeds of the turbopump or with the flutter frequencies of the inducer blades, possibly leading to catastrophic failures of the machine. As an example, the failure of the Japanese H-II rocket in 1999 is thought to have been caused by the fatigue rupture of the inducer blades under the resonant excitation of an unstable high-order cavitation instability

(NASDA 2000a, b). Furthermore, several recent flight tests on important components of the Space Shuttle main engine have suffered from vibration problems at frequencies compatible with the characteristic frequencies of high-order rotating cavitation instabilities observed in the propellant feed turbopumps.

High-order streamwise and azimuthal instabilities were first predicted on theoretical grounds by Tsujimoto (2001) as unstable modes resulting from the application of his 2D linear cavity model originally developed for the investigation of rotating cavitation. He anticipated the occurrence of these modes with system-independent frequencies between 4.2 and 4.7 times the inducer rotating speed and at Euler numbers equal or larger than about four times the breakdown value, just above those characteristic of conventional rotating cavitation. The author also suggested that the high-order azimuthal mode might be caused by the modulation of the inlet backflow structure, also in consideration that it was predicted to occur even without cavitation on the blades.

Experimental evidence of the occurrence of higher order instabilities was first provided by Fujii et al. (2002) in a test campaign on a 3-bladed inducer similar to the LOX inducer of the LE-7 turbopump, whose main results are illustrated in Fig. 12. The spectral analysis of the pressure fluctuations, recorded at the inlet and outlet cross-sections of the inducer when operating at nearly nominal flow and variable inlet pressure, indicated the presence, together with conventional rotating cavitation denoted by 1 in the figure, of two additional higher order instabilities:

- A higher order surge (denoted by 2 in the figure) with a frequency about 5 times the rotational speed of the inducer at cavitation numbers higher than for conventional rotating cavitation.
- A higher order rotating cavitation (denoted by 3 in the figure) at frequencies and cavitation numbers close to those of the higher order surge oscillations. It was

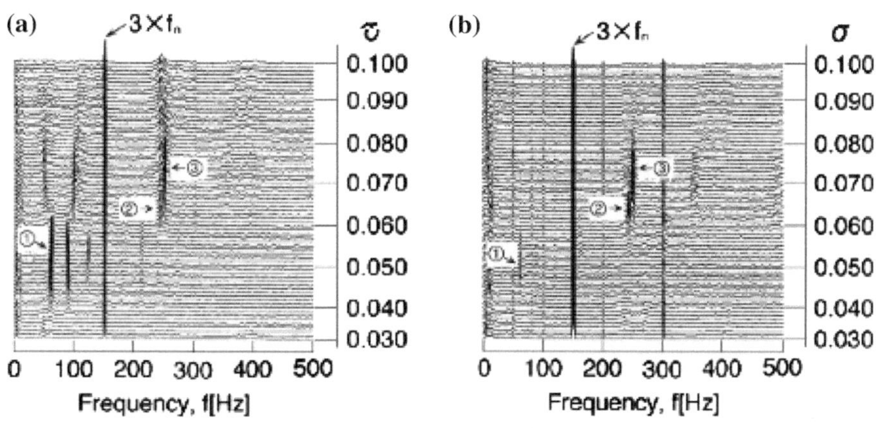

Fig. 12 Waterfall plots of the power spectra of inlet and outlet pressure fluctuations in a 3-bladed inducer at flow coefficient $\varphi = 0.080$ ([28] Fujii et al. 2002)

> also found that the cavity size fluctuation for this instability mode is smaller than for conventional rotating cavitation. Furthermore, while in conventional rotating cavitation the outlet pressure fluctuations are significantly smaller than those at the inlet, for the higher order rotating cavitation mode the situation is reversed.

In the same paper the authors also indicate that the above instabilities could be eliminated almost completely by means of the introduction of a suitable step in the design of the inducer casing. Finally, the detection of a higher order surge instability with the same features observed by Fujii et al. (2002) was confirmed by Tsujimoto and Semenov (2002) also in full-scale tests on the liquid hydrogen turbopump of the LE-7 engine.

Figure 13 summarizes the results of similar tests conducted in the U.S. by Subbaraman and Patton (2006) on a 4-bladed inducer for the characterization of higher order instabilities at frequencies up to ten times the rotational speed of the impeller.

The plot in the figure clearly shows the occurrence of higher order surge cavitation (HOSC) and higher order rotating cavitation (HORC) at nearly nominal flow conditions and cavitation numbers equal to or higher than about four times the breakdown value. The HORC and HOSC frequencies only differ by about one tenth of the inducer rotational speed and, not surprisingly, the HORC component displays both forward and backward running modes. These higher order instabilities also produced many nonlinear modulations and harmonics over the whole range of experimental observation, including a component at very low frequency indicated as "Low Freq Cav-Induced" in the above plot. The most intense harmonic

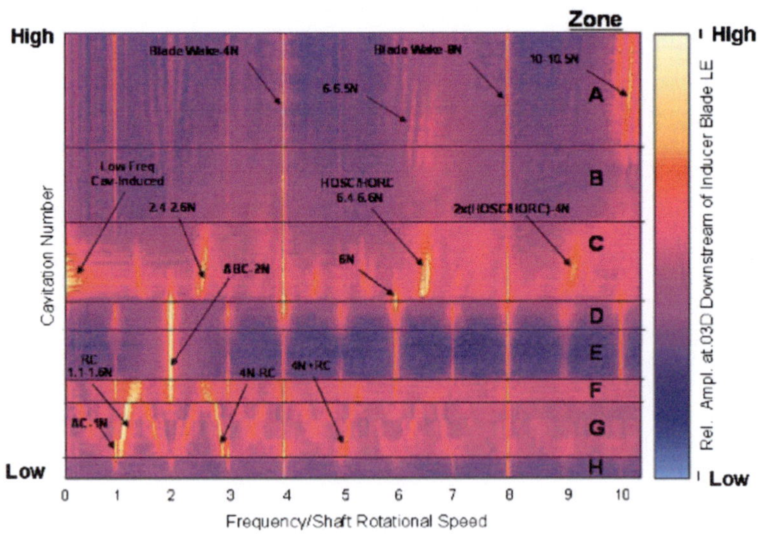

Fig. 13 2D Color plot of the power spectra of the pressure fluctuations versus the cavitation number in a 4-bladed inducer, at a flow coefficient equal to 1.02 times the design value (Subbaraman and Patton 2006)

component occurred at a frequency equal to 6.4–6.6 times the rotational speed, which was therefore considered as the "nominal" frequency of the instability.

The author present in the paper a "Tip Vortex Suppressor", consisting of some devices able to provide a re-circulate flow from the inducer discharge to the inducer inlet. This device showed to be quite effective, shifting the higher order phenomena to low values of the flow coefficient (significantly lower than the operating point) and eventually suppressing most of them when an additional through flow was provided by the device. However, in the latter case the detected oscillations due to alternate blade cavitation were significantly stronger.

A higher order surge mode was detected by Cervone et al. (2006) on a 2-bladed inducer with the same blade geometry of the 3-bladed LOX inducer for the Vinci engine. Two harmonics of the phenomenon have been identified, at 4.4 and 6.6 times the inducer rotational speed. The phenomenon was observed at all values of the flow coefficient and cavitation number, but became more evident at higher flow rates near the design point and tended to disappear as breakdown conditions were approached.

7 Conclusions and Perspectives

In the past generation the technology and knowledge base of liquid propellant feed turbopumps has been completely revolutionized. The development of the propulsion systems for the Space Shuttle in the 1970s and of the Japanese launchers in the 1980s and early 1990s have represented corner points in the quest for higher performance and more detailed insight on the hydrodynamics of the flow field through the machine. As a consequence of these development efforts the design of turbopumps for space propulsion applications has rapidly evolved and their power density has raised by more than one order of magnitude with the successful introduction of faster, lighter and often supercritical machines. A large number of new types of fluid dynamic instabilities, mostly induced by the occurrence of cavitation, have been identified both theoretically and experimentally, and field experience has confirmed their potentially dangerous nature. In addition, operation at supercritical speeds with non-negligible cavitation has significantly increased the importance of rotordynamic instabilities and the need for more accurate prediction of the critical speeds and their dependence on the properties of the flow.

Today cavitation phenomena in hydraulically operating turbomachinery cannot be safely predicted by theoretical or numerical methods because of the complexity of the physical phenomena involved, the imperfections and approximations of the physical models, and the insufficient temporal and spatial resolution of current numerical simulation tools. Hence experimentation still plays an essential role for technology progress in this field. Operational and economic limitations also clearly indicate that detailed experiments can only be effectively carried out on turbopumps under fluid dynamic and thermal cavitation similarity.

In recognition of these aspects, the research group at Pisa University has designed and realized the Cavitating Pump Rotordynamic Test Facility (CPRTF), the first openly documented facility in Europe—and one of the few in the world—capable of carrying out the direct measurement of the unsteady rotordynamic fluid forces on scaled turbopumps in fluid dynamic and thermal cavitation similarity.

The final objective of the research in this field consists in giving the designers of high-performance turbopumps a deeper insight on the development of dangerous flow instabilities of crucial importance in rocket propulsion applications. In more practical terms, the experimental results will shed some light on the dependence of these instabilities on the scaling parameters describing the operational conditions of the machine and the intensity of thermal effects, which commonly occur in most cryogenic propellants. By proper extrapolation based on well-established similarity laws, this information will allow for realistic predictions of these phenomena in full-scale turbomachines to be made, thereby effectively contributing to identify the operational regimes that are likely to develop into dangerous and potentially destructive fluid dynamic instabilities.

Together with the numerical simulations and theoretical analyses, the results of similarity experiments represent an integrated approach to the analysis of cavitation-induced instabilities in liquid propellant fuel feed systems and are expected to effectively contribute to the advancement of technology in an important sector of space engineering.

References

Bhattacharyya, A. (1994). *Internal Flows and Force Matrices in Axial Flow Inducers*. Report No. E249.18, California Institute of Technology, Pasadena, USA.

Bhattacharyya, A., Acosta, A. J., Brennen, C. E., & Caughey, T. K. (1997). Rotordynamic forces in cavitating inducers. *ASME Journal of Fluids Engineering, 199*(4), 768–774.

Braisted, D. M. (1979). *Cavitation Induced Instabilities Associated with Turbomachines*. In Report No. E184.2, California Institute of Technology, Pasadena, CA, USA.

Brennen, C. E. (1994). *Hydrodynamics of pumps*. Oxford University Press, Oxford Engineering Science Series 44.

Brennen, C. E., & Acosta, A. J. (1973). Theoretical, quasi-static analysis of cavitation compliance in turbopumps. *Journal of Spacecraft, 10*(3), 175–179.

Brennen, C. E., & Acosta, A. J. (1976). The dynamic transfer function for a cavitating inducer. *ASME Journal of Fluids Engineering, 98*, 182–191.

Brennen, C. E., & Braisted, D. M. (1980). Stability of hydraulic systems with focus on cavitating pumps. In *Proceedings of IAHR Conference*, Tokyo, Japan.

Callenaere, M., Franc, J. P., & Michel, J. M. (1998, April 7–10). Influence of cavity thickness and pressure gradient on the unsteady behaviour of partial cavities. In *3rd International Symposium on Cavitation*, Grenoble, France.

Cervone, A., Testa, R., Bramanti, C., Rapposelli, E., & d'Agostino, L. (2005). Thermal Effects on Cavitation Instabilities in Helical Inducers. *AIAA Journal of Propulsion and Power, 21*(5), 893–899.

Cervone, A., Torre, L., Bramanti, C., Rapposelli, E., & d'Agostino, L. (2006). Experimental characterization of cavitation instabilities in a two-bladed axial inducer. *AIAA Journal of Propulsion and Power, 22*(6), 1389–1395.

Cervone, A., Bramanti, C., Torre, L., Fotino, D., & d'Agostino, L. (2007). Setup of a high-speed optical system for the characterization of flow instabilities generated by cavitation. *ASME Journal of Fluids Engineering, 129*(7), 877–885.

Cervone, A., Pace, G., Torre, L., Pasini, A., Bartolini, S., Agnesi, L., & d'Agostino, L. (2012, February 27–March 2). Effects of the leading edge shape on the performance of an axial three bladed inducer. In *14th International Symposium on Transport Phenomena and Dynamics of Rotating Machinery, ISROMAC-14*, Honolulu, HI, USA.

d'Agostino, L. (2013, September 19–22). On the hydrodynamics of rocket propellant engine inducers and turbopumps. In *6th International Conference on Pumps and Fans with Compressors and Wind Turbines (IPCF 2013)*, Beijing, China, IOP Publishing, IOP Conference Series: Materials Science and Engineering 52 (2013) 012004.

d'Agostino, L., d'Auria, F., & Brennen, C. E. (1998). A three-dimensional analysis of rotordynamic forces on whirling and cavitating helical inducers. *ASME Journal of Fluids Engineering, 120*, 698–704.

d'Agostino, L., & Salvetti, M. V. (Eds.). (2007). Fluid dynamics of cavitation and cavitating turbopumps. In *CISM Courses and Lectures No. 496, International Centre for Mechanical Sciences*. Vien and New York: Springer.

Ehrich, F., & Childs, S. D. (1984, May). Self-excited vibrations in high performance turbomachinery. In *Mechanical Engineering* (pp. 66–79).

Franc, J. P. (2001, June 20–23). Partial cavity instabilities and re-entrant jet. In *CAV2001, International Symposium on Cavitation*, Pasadena, CA, USA.

Franz, R. J. (1989). *Experimental Investigation of the Effect of Cavitation on the Rotordynamic Forces on a Whirling Centrifugal Pump Impeller*. Ph.D. Thesis, California Institute of Technology, Pasadena, USA.

Fuji, A., Azuma, S., Yoshida, Y., & Tsujimoto, Y. (2002, February 10–14). Higher order rotating cavitation in an inducer. In *Proceedings of ISROMAC-9—The 9th International Symposium on Transport Phenomena and Dynamics of Rotating Machinery*, Honolulu, Hawaii, USA.

Furukawa, A., Ishizaka, K., & Watanabe, S. (2001, June 20–23). Experimental estimate of helical inducer blade forces in cavitation surge condition. In *CAV2001, International Symposium on Cavitation*, Pasadena, California USA.

Furukawa, A., Ishizaka, K., & Watanabe, S. (2002). Experimental study of cavitation induced oscillation in two bladed inducers. In *Space Launcher Liquid Propulsion: 4th International Conference on Space Launcher Technology*, Liege, Belgium.

Greitzer, E. M. (1981). The stability of pumping systems. ASME Journal of Fluids Engineering, *103*, 193–242.

Hashimoto, T., Yoshida, H., Funatsu, S., Hishimoto, J., Kamijo K., & Tsujimoto Y. (1997a). Rotating cavitation in three and four-bladed inducers. In *33th AIAA/ASME7SAE/ASEE Joint Propulsion Conference and Exhibit*, Seattle, USA.

Hashimoto, T., Yoshida, H., Watanabe, M., Kamijo, K., & Tsujimoto, Y. (1997b). Experimental study on rotating cavitation of rocket propellant pump inducers. *Journal of Propulsion and Power,* 13(4).

Horiguchi, H., Arai, S., Fukutomi, J., Nakase, Y., & Tsujimoto, Y. (2003b, July 6–10). Quasi three-dimensional analysis of cavitation in an inducer. In *4th ASME-JSME Joint Fluids Engineering Conference*, Honolulu, HIa, USA.

Imamura, H., Kurokawa, J., Matsui, J., & Kikuchi, M. (2003, November 1–4). Suppression of cavitating flow in inducer by J-Groove. In *CAV2003, International Symposium on Cavitation*, Osaka, Japan.

Kamijo, K., Shimura, T., & Watanabe, M. (1977). *Experimental investigation of cavitating inducer instability*. ASME Paper 77/WA-FE-14.

Kang, D., Cervone, A., Yonezawa, K., Horiguchi, H., Kawata, Y., & Tsujimoto, Y. (2007). Effect of blade geometry on tip leakage vortex of inducer. In *The 9th Asian International Conference on Fluid Machinery*, Jeju, South Korea.

Kawanami, Y., Kato, H., Yamaguchi, H., Tagaya, Y., & Tanimura, M. (1997). Mechanism and control of cloud cavitation. *Journal of Fluids Engineering, 119*, 788–795.

Kimura, T., Yoshida, Y., Hashimoto, T., & Shimagaki, M. (2006). Numerical simulation for unsteady cavitating flow in a turbopump inducer. In *CAV2006, 6th International Symposium on Cavitation*, Wageningen, The Netherlands.

Kubota, A., Kato, H., Yamaguchi, H., Maeda, M. (1989). Unsteady structure measurement of cloud cavitation on a foil section using conditional sampling technique. Journal of Fluids Engineering, *111*, 204–210.

Lieblein, S. (1965). Experimental flow in two-dimensional cascades. In *Aerodynamic Design of Axial Flow Compressors, NASA SP-36*, 101–149.

Murai, H. (1968). Observations of cavitation and flow patterns in an axial flow pump at low flow rates. *Memoirs of the Institute of High Speed Mechanics, 24*(246), Tohoku University.

NASA. (1970). *Prevention of Coupled Structure-Propulsion Instability*. NASA SP-8055, Space Vehicle Design Criteria Manuals.

NASDA. (2000a). Report No. 94, May 2000.

NASDA. (2000b). Report No. 96, June 2000.

Natanzon, M. S., et al. (1974). Experimental investigation of cavitation induced oscillations of helical inducers. *Fluid Mechanics Soviet Research, 3*(1), 38–45.

Ng, S. L., & Brennen, C. E., (1978). Experiments on the dynamic behavior of cavitating pumps. *ASME Journal of Fluids Engineering, 100*, 166–176.

Ohta, T., & Kajishima, T. (2008, February 17–22). Transition of different unsteady cavitating flows in 2D cascade with flat blades. In *Proceedings of ISROMAC-12—The 12th International Symposium on Transport Phenomena and Dynamics of Rotating Machinery*, Honolulu, HI, USA.

Pace, G., Valentini, D., Pasini, A., Torre, L., Fu, X., d'Agostino, L. (2015). Geometry effects on flow instabilities of different three-bladed inducers. ASME J. Fluids Engineering, *137*(4)/011102-1, 041304.

Pasini, A., Torre, L., Cervone, A., & d'Agostino L. (2010). Rotordynamic forces on a four-bladed inducer. In *Proceedings of 46th AIAA/ASME/SAE/ASEE Joint Propulsion Conference*, Nashville, USA.

Pasini, A., Torre, L., Cervone, A., & d'Agostino, L. (2011a). Characterization of the rotordynamic forces on tapered axial inducers by means of a rotating dynamometer and high-speed movies. In *Proceedings of WIMRC 3rd International Cavitation Forum 2011*, University of Warwick, United Kingdom.

Pasini, A., Torre, L., Cervone, A., & d'Agostino, L. (2011b). Continuous spectrum of the rotordynamic forces on a four-bladed inducer. *ASME J. Fluids Engineering, 133*(12).

Rosenmann, W. (1965). Experimental investigations of hydrodynamically induced shaft forces with a three bladed inducer. In *Proceedings of ASME Symposium on Cavitation in Fluid Machinery*.

Rubin, S. (1966). Longitudinal *Instability of Liquid Rockets Due to Propulsion Feedback (POGO). Journal of Spacecraft and Rockets, 3*(8), 1188–1195.

Sack, L. E., & Nottage, H. B. (1965). System oscillations associated to cavitating inducers. ASME Journal of Basic Engineering, *87*, 917–924.

Sakoda, M., Yakushiji, R., Maeda, M., & Yamaguchi, H. (2001, June 20–23). Mechanism of cloud cavitation generation on a 2-D hydrofoil. In *CAV2001, International Symposium on Cavitation*, Pasadena, California, USA.

Sergeant, S. R., Sorensen, K. P., & McGlynn, R. D. (2008, February 17–22). Measurement of cavitation induced strain on a 3 bladed inducer. In *Proceedings of ISROMAC-12—The 12th International Symposium on Transport Phenomena and Dynamics of Rotating Machinery*, Honolulu, Hawaii, USA.

Semenov, Y. A., Fujii, A., & Tsujimoto, Y. (2004). Rotating choke in cavitating turbopump inducer. *ASME Journal of Fluids Engineering, 126*, 87–93.

Shimagaki, M., Hashimoto, T., Watanabe, M., Hasegawa, S., Nakamura, N., & Shimura, T. (2006). Unsteady pressure fluctuations in an inducer. JSME International Journal of Fluids and Thermal Engineering, Series B, *49*(3), 806–811.

Shimiya, N., Fujii, A., Horiguchi, H., Uchiumi, M., Kurokawa, J., & Tsujimoto, Y. (2006). Suppression of cavitation instabilities in an inducer by J-Groove. In *CAV2006, 6th International Symposium on Cavitation*, Wageningen, The Netherlands.

Shimura, T., Yoshida, M., Kamijo, K., Uchiumi, M., & Yasutomi, Y. (2002, February 10–14). Cavitation induced vibration caused by rotating-stall-type phenomenon in LH2 turbopump. In *Proceedings of ISROMAC-9—The 9th International Symposium on Transport Phenomena and Dynamics of Rotating Machinery*, Honolulu, Hawaii, USA.

Stripling, L. B. (1962). Cavitation in turbopumps, part 2. *ASME Journal of Basic Engineering*, *84* (3), p. 329".

Stripling, L. B., & Acosta, A. J. (1962). Cavitation in turbopumps—part 1. *ASME Journal of Basic Engineering*, *84*, 326–338.

Subbaraman, M., & Patton, M. (2006). Suppressing higher-order cavitation phenomena in axial inducers. In *CAV2006, 6th International Symposium on Cavitation*, Wageningen, The Netherlands.

Torre, L., Pasini, A., Cervone, A., Pecorari, L., Milani, A., & d'Agostino, L. (2010). Rotordynamic forces on a three bladed inducer. In *Proceedings of Space Propulsion Conference*, San Sebastian, Spain.

Torre, L., Pasini, A., Cervone, A., & d'Agostino, L. (2011a). Continuous spectrum of the rotordynamic forces on a four-bladed inducer. In *ASME/JSME/KSME Joint Fluids Engineering Conference*, Hamamatsu, Japan.

Torre, L., Pasini, A., Cervone, A., & d'Agostino, L. (2011). Experimental characterization of the rotordynamic forces on space rocket axial inducers. ASME Journal of Fluids Engineering, *133* (10).

Tsujimoto, Y. (2001, June 20–23). Simple rules for cavitation instabilities in turbomachinery. In *CAV2001, International Symposium on Cavitation*, Pasadena, California, USA.

Tsujimoto, Y., Kamijo, K., & Yoshida, Y. (1993). A theoretical analysis of rotating cavitation in inducers. *Journal of Fluids Engineering*, *115*.

Tsujimoto, Y., Yoshida, Y., Maekawa, Y., Watanabe, S., & Hashimoto, T. (1997). Observations of oscillating cavitation of an inducer. *ASME Journal of Fluids Engineering, 119,* 775–781.

Tsujimoto, Y., Watanabe, S., & Horiguchi H. (1998, November 1–6). Linear analyses of cavitation instabilities of hydrofoils and cascades. In *Proceedings of US-Japan Seminar: Abnormal Flow Phenomena in Turbomachinery*, Osaka, Japan.

Tsujimoto, Y., & Semenov, Y. A. (2002). New types of cavitation instabilities in inducers. In *Space Launcher Liquid Propulsion: 4th International Conference on Space Launcher Technology*, Liege, Belgium.

Uchiumi, M., Kamijo, K., Hirata, K., Konno, A., Hashimoto, T., & Kobayasi, S. (2003). Improvement of inlet flow characteristics of LE-7A liquid hydrogen pump. Journal of Propulsion and Power,*19*(3).

Uchiumi, M., & Kamijo, K. (2008, February 17–22). Occurrence range of a rotating-stall-type phenomenon in a high head liquid hydrogen inducer. In *Proceedings of ISROMAC-12—The 12th International Symposium on Transport Phenomena and Dynamics of Rotating Machinery*, Honolulu, HI, USA.

Valentini, D., Pace, G., Pasini, A., Torre, L., & d'Agostino, L. (2015). Influences of the operating conditions on the rotordynamic forces acting on a three-bladed inducer under forced whirl motion. Journal of Fluids Engineering, *137*(7), 071304.

Wade, R. B., & Acosta, A. J. (1966). Experimental observations on the flow past a plano-convex hydrofoil. Journal of Basic Engineering, *87*, 273–283.

Watanabe, S., Yokota, K., Tsujimoto, Y., & Kamijo, K. (1999). Three-dimensional linear analysis of rotating cavitation in inducers using an annular cascade model. *ASME Journal of Fluids Engineering, 121,* 866–871.

Yamamoto, K. (1991). Instability in a cavitating centrifugal pump. JSME International Journal, Series II, *34*, 9–17.

Yoshida, Y., Kazami, Y., Nagaura, K., Shimagaki, M., Iga, Y., & Ikohagi, T. (2008). Interaction between uneven cavity length and shaft vibration at the inception of synchronous rotating cavitation. In *Proceedings of ISROMAC-12—The 12th International Symposium on Transport Phenomena and Dynamics of Rotating Machinery,* Honolulu, HI, USA, February 17–22.

Zoladz, T. (2000). Observations on rotating cavitation and cavitation surge from the development of the fastrac engine turbopump. In *36th AIAA/ASME/SAE/ASEE Joint Propulsion Conference*, Huntsville, AL, USA.

Three-dimensional Simulation of Cavitation Instabilities

Yoshinobu Tsujimoto

1 Three-Dimensional Simulation of Cavitation Instabilities

In real three-dimensional inducers, most of the cavitation instabilities appear when the steady cavity length at the tip becomes about 65% of the blade spacing. This agrees with the result of two-dimensional stability analysis (Watanabe et al. 1999; Horiguchi et al. 2000). However, the two-dimensional stability analysis is based on the blade surface cavitation while the cavitation occurs most significantly at the tip in real inducers. So, we need to clarify the cause of cavitation instabilities under the conditions with real three-dimensional tip cavitation. This generally requires unsteady, three-dimensional calculation of cavitating flow.

1.1 Simulation of Alternate Blade Cavitation (Kang et al. 2009a)

Since the flow under alternate blade cavitation is steady, the cause of alternate blade cavitation can be found from steady calculations. So, steady flow calculations were made on two pitches of a four-bladed inducer. The commercial software, ANSYS-CFX11.0, was used for the simulation of the cavitating flow through the inducer. Three-dimensional Reynolds averaged Navier–Stokes (RANS) equations were solved by finite volume method with hybrid unstructured grids. The k-ω turbulence model was used. The cavitation model is based on the homogenous multiphase flow framework of the CFD solver taking into the dynamics of cavi-

Y. Tsujimoto (✉)
Osaka University, Osaka, Japan
e-mail: tujimoto@me.es.osaka-u.ac.jp

L. d'Agostino and M.V. Salvetti (eds.), *Cavitation Instabilities and Rotordynamic Effects in Turbopumps and Hydroturbines*, CISM International Centre for Mechanical Sciences 575, DOI 10.1007/978-3-319-49719-8_4

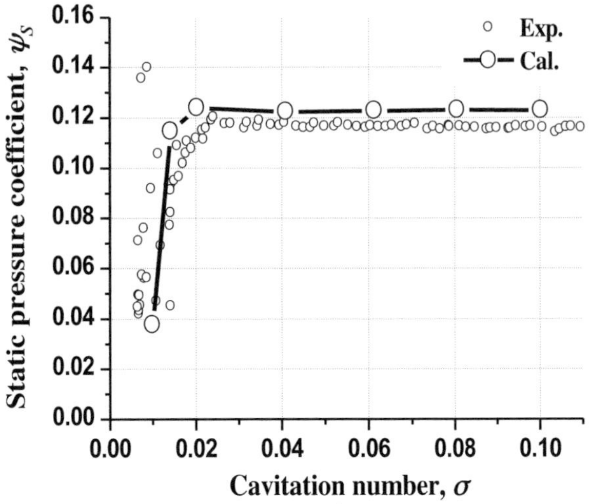

Fig. 1 Suction performance curve

tation bubbles by solving a simplified Reyleigh–Plesset equation. The effect of cavitation is taken into account by adding a special source term into the continuity equation. The working fluids are water and its vapor. The total pressure and no circumferential velocity were specified at the inlet and the mass flow rate was specified at the outlet. The rotational speed was fixed at 3000 rpm, which is the same as the experiments.

Figure 1 shows the suction performance and Fig. 2 the cavity length as compared with experimental results (Kang, Yonezawa et al. 2009a). As shown in Fig. 2, alternate blade cavitation starts to occur when the cavity length becomes about 65% of the blade spacing, *h*, at the tip.

Figure 3 shows the shape of the cavity shown by void fraction and the velocity field in a cylindrical plane with $r/R_t = 0.98$. In the present calculation with a bubbly flow cavity model, the velocity vector is not parallel to the cavity surface but penetrates the surface, unlike the two-dimensional blade surface cavitation.

In order to find the cause of alternate blade cavitation, the velocity disturbance caused by the cavitation was evaluated by subtracting the non-cavitating flow vector from cavitating flow vector and is shown in Fig. 4. Near the leading edge where the cavitation bubbles grow, a source like outward flow is found. Towards the cavity trailing edge where the cavity bubbles collapse, a jet-like flow occurs. This flow has an axial flow component directed downstream. This component has an effect to reduce the incidence angle to the next blade. At $\sigma = 0.06$, the flow towards the cavity trailing edge is about to interact with the leading edge of the next blade.

Figure 5 shows the velocity disturbance in meridional planes. Near the cavity leading edge shown in (a) and (c), where the bubbles are growing, we observe an outward flow. Near the cavity trailing edge shown in (d), a strong axial flow disturbance can be observed. The leading edge of the upstream blade (Blade 2) is

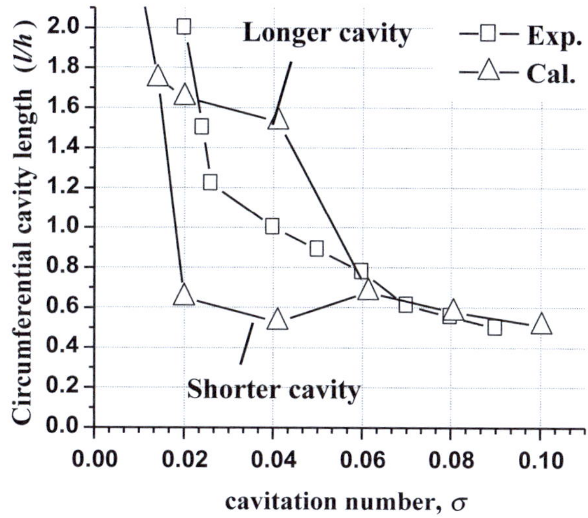

Fig. 2 Cavity length at tip

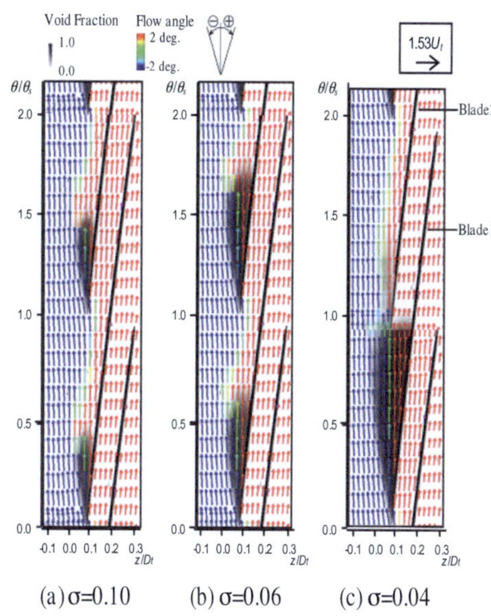

Fig. 3 Relative velocity vector, void fraction, and flow angle

exposed to the strong axial flow and the cavity on this blade is suppressed, as shown in Fig. 4c, due to the decreased incidence angle to the blade. The radial component of the disturbance flow in the meridional plane is much smaller than the axial component. This is one of the reasons why the two-dimensional analysis can predict the cavitation instabilities correctly.

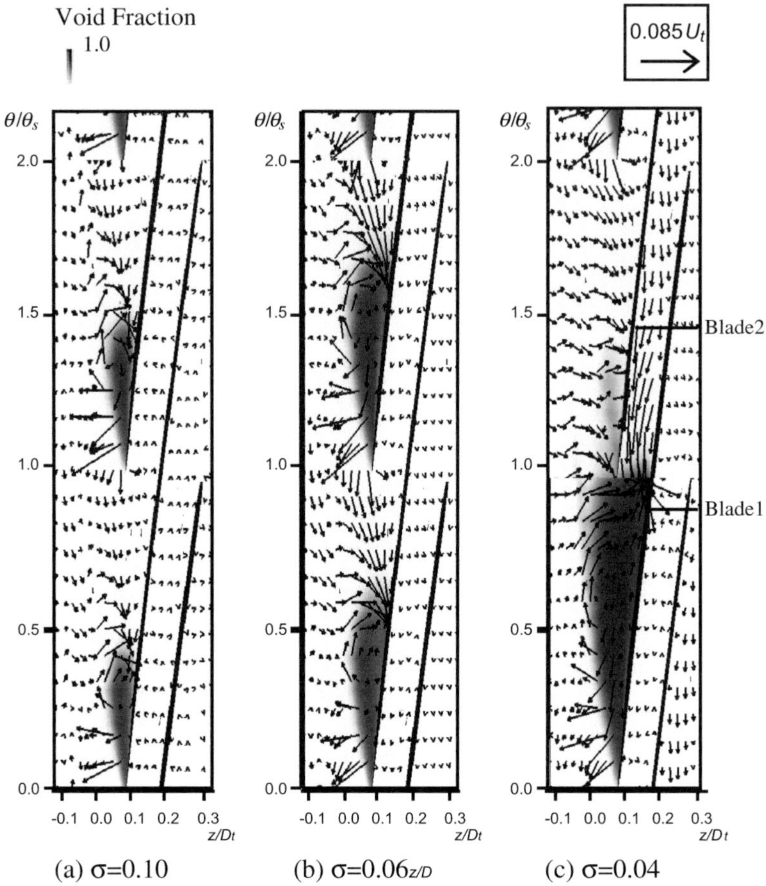

Fig. 4 Disturbance velocity vector and void fraction

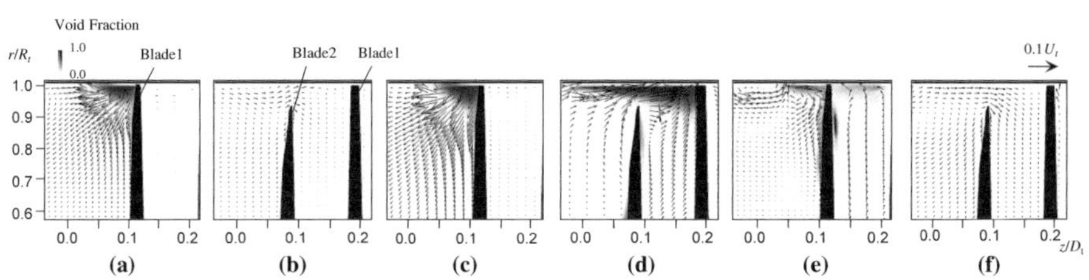

Fig. 5 Disturbance velocity and void fraction in meridional plane. $\sigma = 0.06$: (a) $\theta/\theta_s = 0.2$, (b) $\theta/\theta_s = 0.9$ $\sigma = 0.04$: (c) $\theta/\theta_s = 0.2$, (d) $\theta/\theta_s = 0.9$, (e) $\theta/\theta_s = 1.2$, (f) $\theta/\theta_s = 1.9$

1.2 Simulation of Rotating Cavitation

Encouraged by the success of the simulation of alternate blade cavitation, unsteady calculations were made on a three-bladed inducer to simulate rotating cavitation. The time pitch was 1/400 of one rotation of the inducer.

Figure 6 shows the rotating cavitation simulated in a three-bladed inducer at the design flow coefficient $\phi = 0.078$ and the cavitation number $\sigma = 0.04$. The propagation velocity ratio $\omega/\Omega = 5/4 = 1.25$ is close to experimental results. Figure 7 shows the velocity disturbance in the meridional plane at $\theta/\theta_s = 0.9$ under rotating cavitation. At 16.65–19.15 Rev., a strong axial flow component directed downstream can be observed. The leading edge of upstream blade, Blade 2, is exposed to this downward flow and the cavity on this blade is smaller as shown in Fig. 6.

Figure 8 shows the pressure distribution under rotating cavitation. As typically shown at 16.05 Rev. on the blade whose leading edge is at $\theta/\theta_s = 0$ (Blade 1), a high pressure region can be seen at the trailing edge of the cavity. This is caused by the collapse of the cavity or the flow deceleration at the cavity trailing edge. With the growth of the cavity, this higher pressure region extends to the suction side of the next blade, as shown at 17.3 Rev. This also has the effect to decrease the cavity length on the next blade.

Although not shown here, cavitation surge occurred at $\sigma = 0.050$, 0.045, 0.035. It was found that, when the cavity becomes longer, the leading edge of the next blade is exposed to the axial flow disturbance at the trailing edge of the cavity and hence the cavity is decreased. Thus, the cavitation surge can also be explained by the

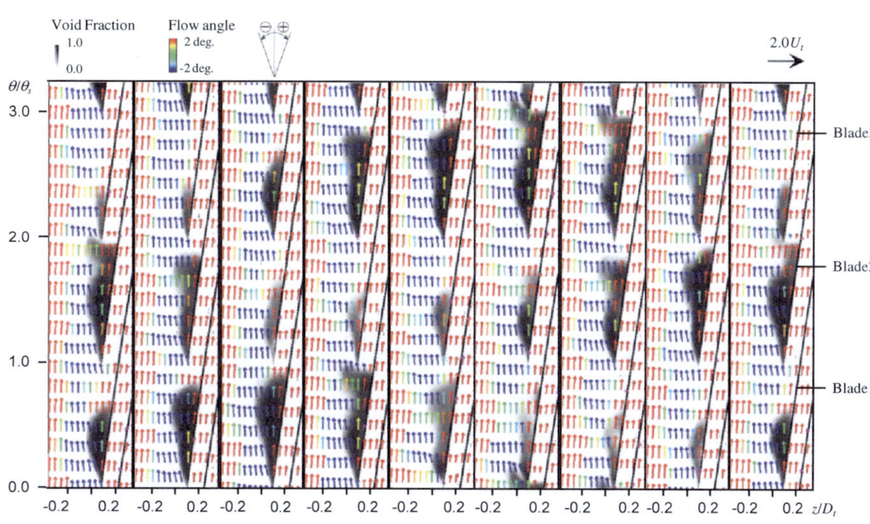

Fig. 6 Velocity field in a cylindrical plane at $r/R_t = 0.98$ under rotating cavitation, $\phi = 0.078$, $\sigma = 0.04$

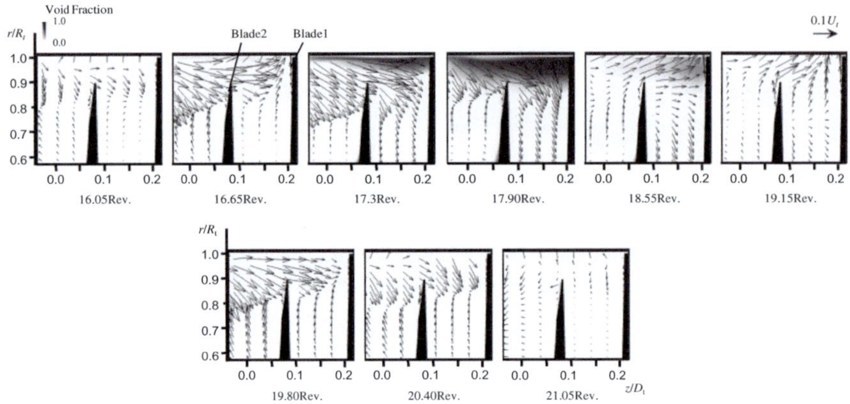

Fig. 7 Velocity disturbance in a meridional plane at $\theta/\theta_s = 0.9$ under rotating cavitation

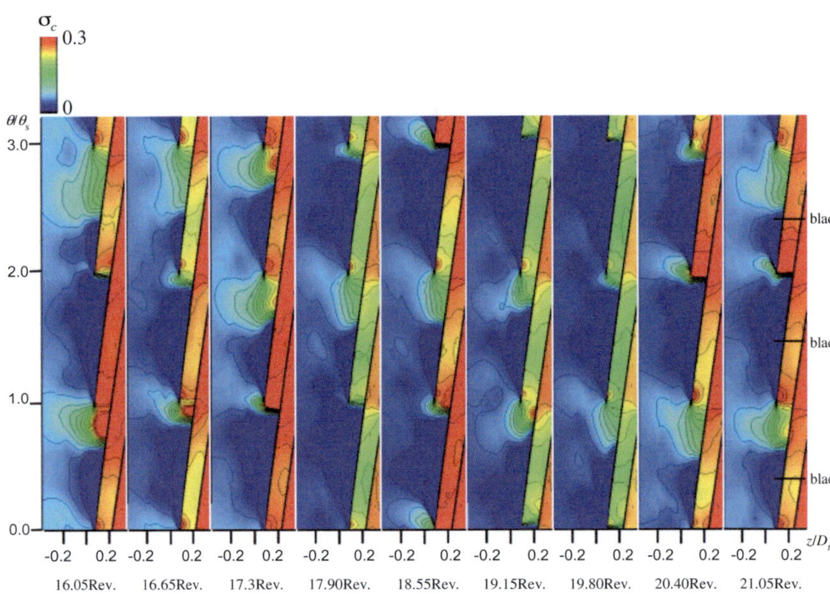

Fig. 8 Pressure distribution in a cylindrical plane at $r/R_t = 0.98$ under rotating cavitation

interaction of the flow towards the cavity trailing edge with the leading edge of the next blade.

As mentioned above, in real three-dimensional flow with significant tip cavitation, there exists a local flow near the cavity trailing edge in which the incidence angle to the next blade is decreased. Cavitation instabilities such as alternate blade cavitation, rotating cavitation, and cavitation surge start to occur when this local flow starts to interact with the leading edge of the next blade. Figure 9 shows the

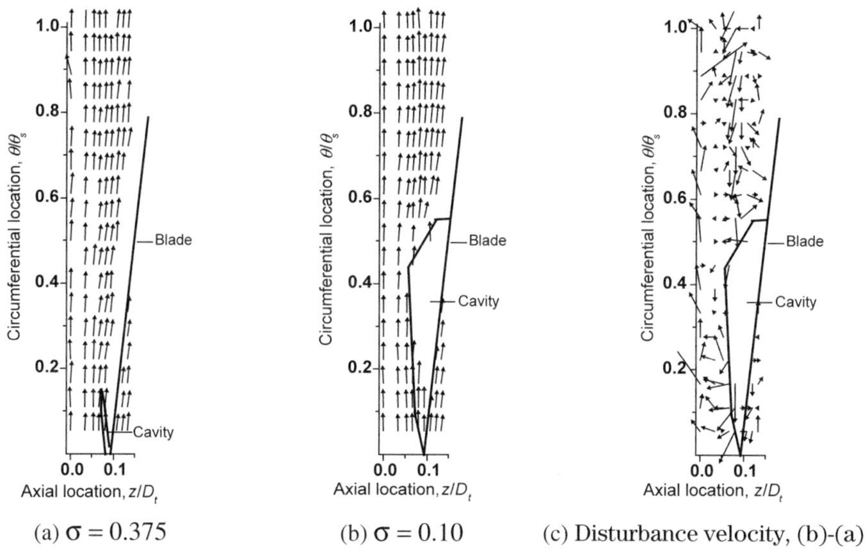

Fig. 9 Velocity field measured by LDV, $r/R_t = 0.98$, $\phi = 0.078$

relative velocity field around the tip cavity, obtained by experiments. As shown in (b), the flow penetrates into the cavity and does not round about the cavity in the same way as for the calculated result shown in Fig. 3. By subtracting the non-cavitating velocity vector shown in (a) from the cavitating flow vector shown in (b), the disturbance flow vector is obtained and shown in (c). The disturbance field shown in (c) is similar to that in Fig. 4 with outward flow near the cavity leading edge and the flow towards the cavity trailing edge. Thus, the existence of disturbance flow field predicted by CFD was confirmed by experiments.

In many cases, the tip cavitation is more significant than blade surface cavitation. This means that cavitation instabilities can be avoided by designing the inducers so that the interaction of the tip cavity with the next blade can be avoided. Actually several inducers were designed based on this guideline and it was confirmed by experiments that cavitation instabilities are successfully suppressed (Kang et al. 2009b).

2 Suppression of Rotating Cavitation Using a Circumferential Groove on the Casing

Since it was found in the last section that cavitation instabilities occur due to the interaction of the flow near the cavity trailing edge with the leading edge of the next blade, it might be possible to avoid cavitation instabilities by avoiding the interaction by trapping the tip leakage flow by a circumferential groove on the casing.

This method has been used for suppressing rotating stall for compressors (Hah et al. 2008). This section describes about the application of this method for inducers to illustrate the usefulness of CFD for developments (Kang et al. 2010).

2.1 Design of Circumferential Grooves

First, we numerically investigated the effect of groove geometries on the tip leakage vortex and found the optimum geometries to "pack" the tip leakage vortex in the circumferential groove.

Figure 10 shows the geometry of circumferential groove tested. The width of groove is 19.5 mm and the upstream edge is located 3.5 mm upstream of the leading edge of blade tip. The depth of groove is 10 mm for the impeller with diameter 149.8 mm. Table 1 shows the static pressure coefficient ψ_s obtained from the calculation. With the grooved casing, the pressure coefficient is about 5% lower than that with the straight casing.

2.2 Flow Characteristics

Figure 11 shows the cavity geometries at $\sigma = 0.04$. The cavitation number σ is defined as $(p_1 - p_v)/(\rho U_t^2/2)$, where p_v is the vapor pressure and p_1 is the inlet pressure. The cavity is shown by the surface with the void fraction $\alpha = 0.01$. With the grooved casing, the cavity volume is remarkably reduced as compared with the

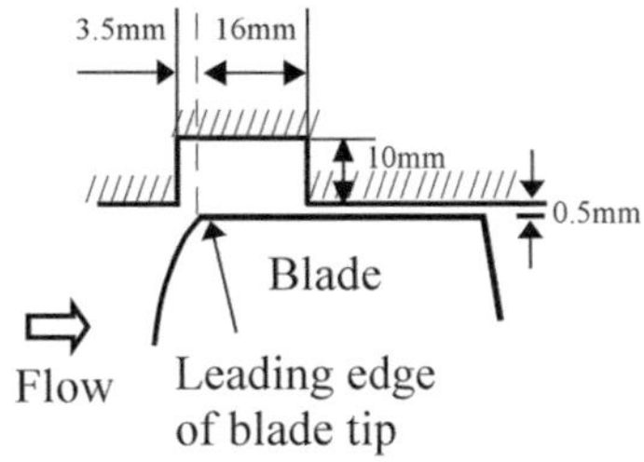

Fig. 10 Geometry of circumferential groove

Table 1 Static pressure coefficient at $\phi/\phi_d = 1$

	Straight casing	Grooved casing
ψ_s $(\sigma = \infty)$	0.1263	0.1213
ψ_s $(\sigma = 0.1)$	0.1256	0.1190
ψ_s $(\sigma = 0.04)$	0.1267	0.1214

straight casing and the separating point of tip leakage vortex cavity is moved downstream.

To clarify the effect of cavity, the velocity disturbance caused by the cavity was evaluated by subtracting the velocity of non-cavitating flow from that of cavitating flow. The disturbance velocity vectors and void fraction distributions in z, θ and r, z-planes are shown in Fig. 12. With the straight casing, the disturbance flow near the cavity trailing edge has an axial component which reduces the incidence angle to the next blade. It has been shown in the last section that the axial disturbance flow causes alternate blade cavitation, cavitation surge, and rotating cavitation. With the grooved casing, the disturbance flow is weaker and no positive axial flow is found near the leading edge of the blade. This is preferable to suppress cavitation instabilities.

Figures 13 and 14 show the velocity vectors and void fraction distributions at $\phi/\phi_d = 1.0$ and $\sigma = 0.04$ in r-z planes. With the straight casing, the tip leakage vortex

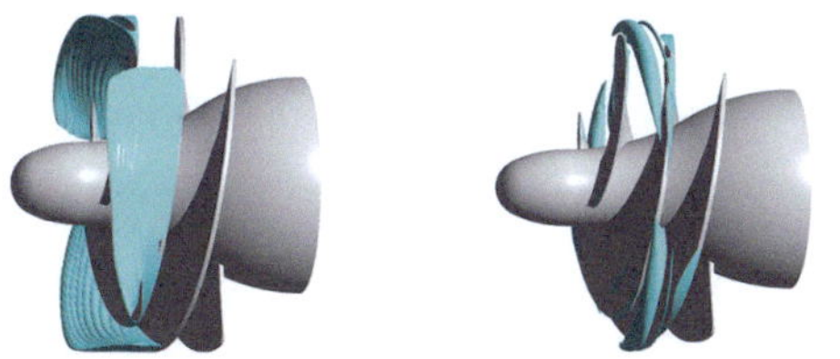

(a) Straight casing **(b)** Grooved casing

Fig. 11 Cavitation geometry, $\phi/\phi_d = 1$, $\sigma = 0.04$, $\alpha = 0.01$

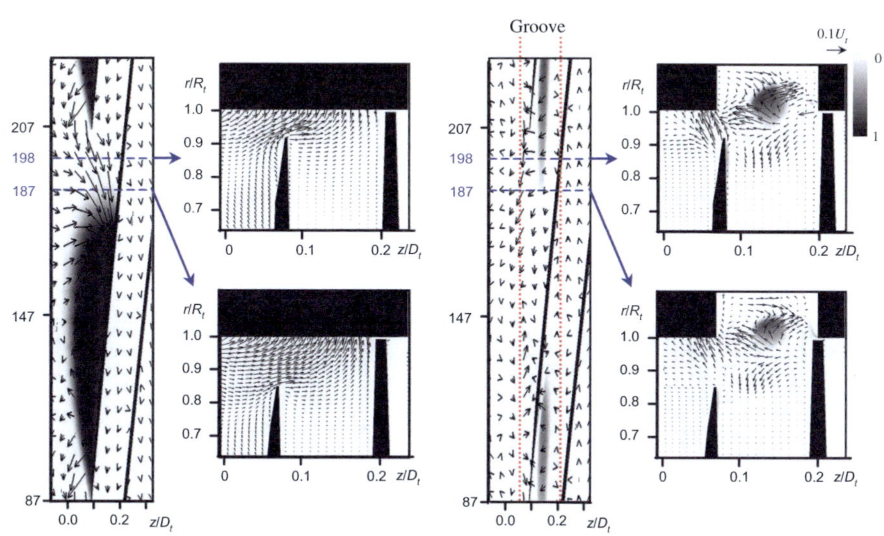

(a) Straight casing **(b)** Grooved casing

Fig. 12 Disturbance velocity and void fraction, $r/R_t = 0.98$, $\phi/\phi_d = 1$, $\sigma = 0.04$

core appears near the leading edge at blade tip shown in Fig. 13a, and then it extends upstream as θ is increased. The cavity appears not in the tip leakage vortex but in the lower pressure region around the junction of blade suction surface and the casing.

With the grooved casing, the tip leakage vortex cavity develops only in the circumferential groove. The vortex is enhanced by the jet at $\theta = 170°$ from the clearance between the pressure surface of the blade and the rear wall of the groove. At $\theta = 110°$, a weak cavity is observed on the pressure surface. This is caused by the interaction with the tip leakage vortex cavity. However, the head $\psi_s = 0.1214$ at $\phi = 0.04$ is not significantly lower than the head $\psi_s = 0.121$ under non-cavitating condition. Inward radial velocity is observed in the upstream edge of the circumferential groove, suppressing the tip leakage vortex cavity on the suction side of blade tip.

Figure 15 shows the blade pressure distributions at $r/R_t = 0.98$. At $\sigma = 0.10$ a cavity appears on the suction surface in $90° < \theta < 135°$ with the straight casing but no cavity is found with the grooved casing. With the grooved casing, negative blade loading is found near the leading edge and the blade loading is moved downstream. Thus, we can control the blade loading by the groove. At $\sigma = 0.04$ the cavitating region is extended to $90° < \theta < 175°$ with the straight casing. With the grooved casing, the effect of cavitation number is smaller although a cavity appears in $100° < \theta < 120°$ on the pressure surface. The pressure on the pressure surface is

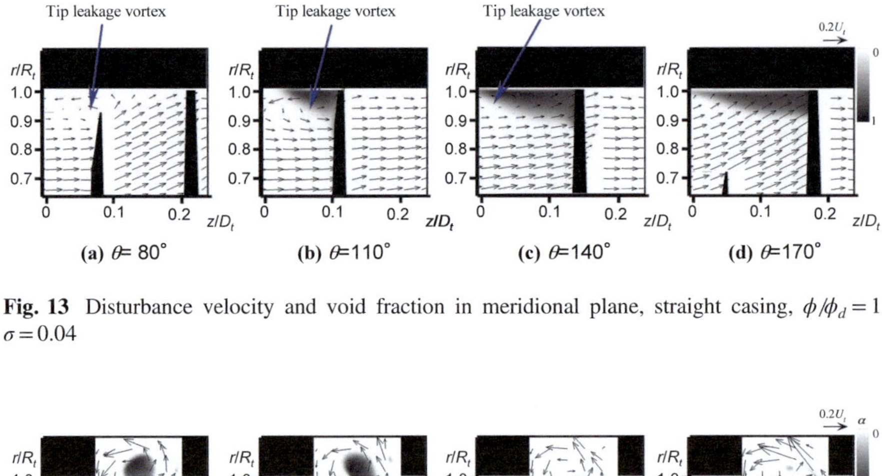

Fig. 13 Disturbance velocity and void fraction in meridional plane, straight casing, $\phi/\phi_d = 1$, $\sigma = 0.04$

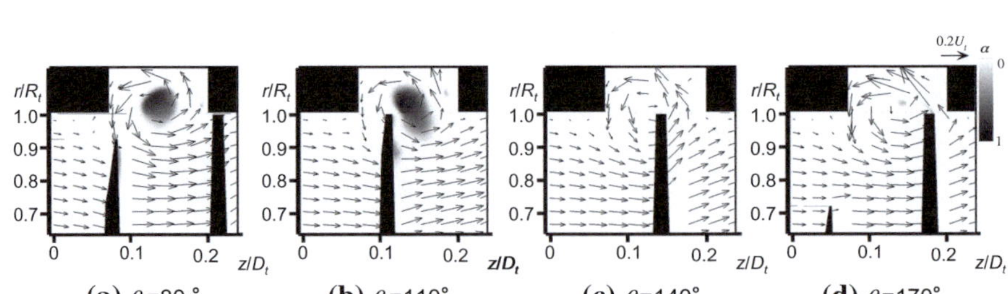

Fig. 14 Disturbance velocity and void fraction in meridional plane, grooved casing, $\phi/\phi_d = 1$, $\sigma = 0.04$

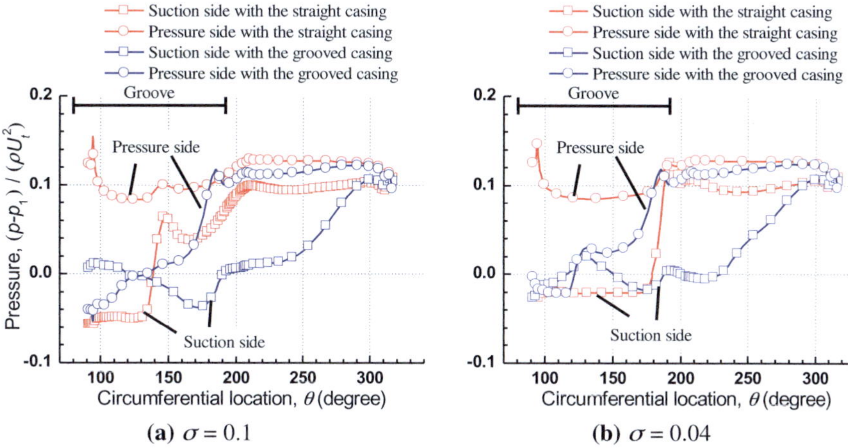

Fig. 15 Pressure distribution at $r/R_t = 0.98$, $\phi/\phi_d = 1$

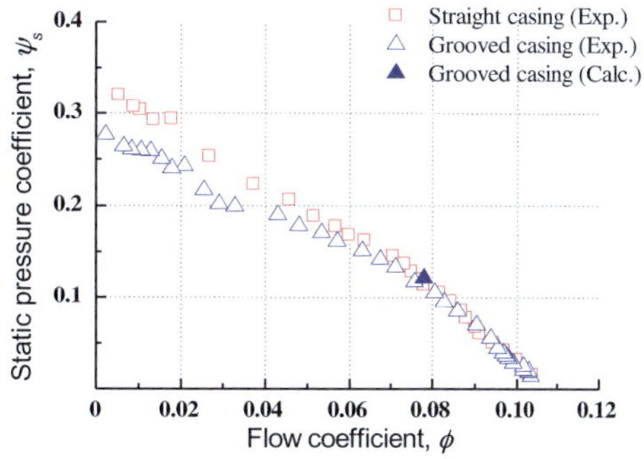

Fig. 16 Non-cavitating performance curve

lower and the cavity appears caused by the interaction with the tip leakage vortex from the preceding blade.

2.3 Experimental Results

2.3.1 Non-cavitating Performance

Figure 16 shows the non-cavitating performance curves obtained from the experiments and the computation. With the grooved casing, the static pressure coefficients at lower flow coefficients are lower than that with the straight casing. However, the decrease of static pressure coefficient is acceptable at the design flow coefficient, $\phi_d = 0.078$.

2.3.2 Suction Performance

Figure 17 compares the suction performance curve with and without the groove. With the straight casing (Tsujimoto et al. 1997), the test was conducted at 3000 rpm with orifice plates at the inlet, while the test was conducted at 4500 rpm without the orifice plates for the case with the grooved casing. The larger scatter with the straight casing is caused by cavitation instabilities with some effects of bubbles from the orifice plates. We find that the suction performance is somewhat improved by the groove.

2.3.3 Cavitation Instabilities

Figures 18, 19 and 20 show the spectra of inlet pressure fluctuations measured by the pressure transducers flush mounted 44 mm upstream of the blade leading edge at the tip. The horizontal axis shows the frequency, the vertical axis shows the magnitude of the pressure fluctuation $\Delta\psi$ defined as $\Delta p/\rho U_t^2)$ and the depth axis shows the cavitation number σ. N is the frequency 50 Hz of the impeller rotation. The phase difference of pressure fluctuations measured at two circumferential

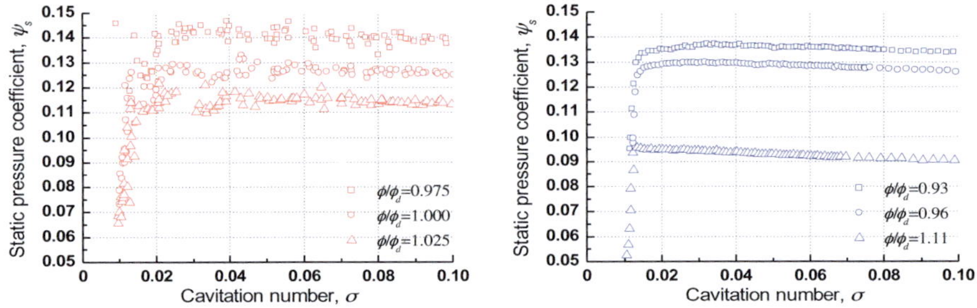

(a) Straight casing **(b)** Grooved casing

Fig. 17 Suction performance curve

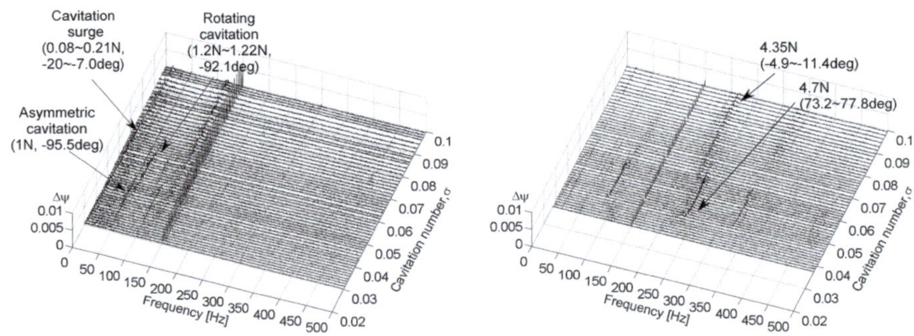

(a) Straight casing **(b)** Grooved casing

Fig. 18 Inlet pressure fluctuation, $\phi/\phi_d = 0.9$

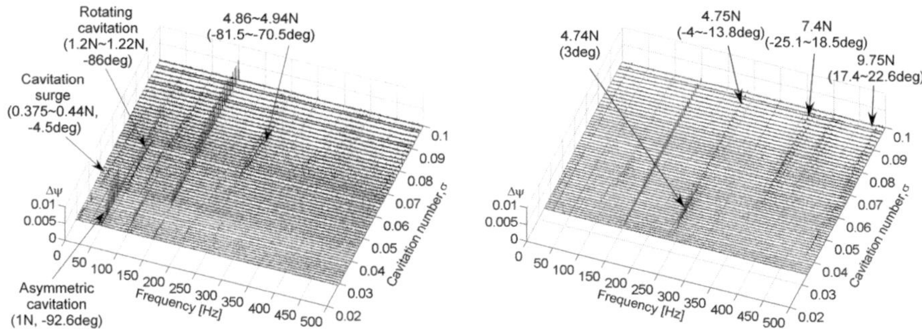

(a) Straight casing **(b)** Grooved casing

Fig. 19 Inlet pressure fluctuation, $\phi/\phi_d = 1$

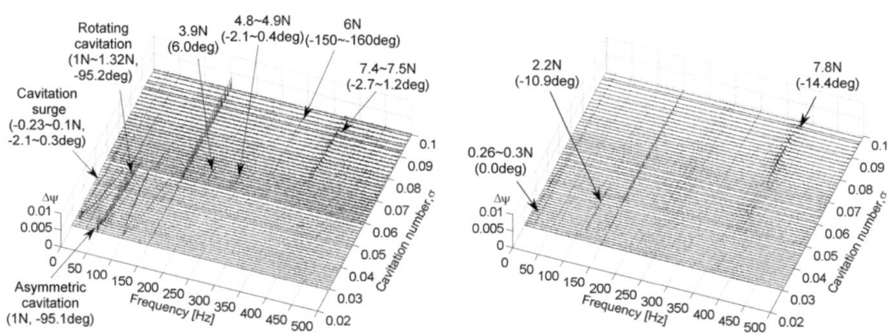

(a) Straight casing **(b)** Grooved casing

Fig. 20 Inlet pressure fluctuation, $\phi/\phi_d = 1.1$

locations 90° apart are shown for typical components. The negative/positive phase difference means that the pressure pattern rotates in the same/opposite direction of the impeller rotation.

In $0 < f < 150$ Hz, with the circumferential groove, rotating cavitation, cavitation surge, and asymmetric cavitation were successfully suppressed at all flow rates except for weak instabilities observed at $\phi/\phi_d = 1.1$.

In $150 < f < 500$ Hz, instabilities with higher frequency components are observed. A higher frequency component with $4.8N$ was observed in the initial engine firing tests and water tests on liquid hydrogen turbopump inducer for HII rocket (Tsujimoto and Semenov 2002). In this case, the phase difference was 0°. The higher frequency components are important since they can be closer to the natural frequencies of blade bending mode. For this reason, extensive studies have been made to avoid the higher frequency components. In the present tests, we find components around $5N$, $7 \sim 8N$, and $10N$. The phase difference is different from case to case but can be assorted into groups around zero (surge mode), $-90°$ (one cell rotating forward) and $+90°$ (one cell rotating backward). The cause of these higher frequency modes is discussed later.

2.3.4 Cavity Geometry

Figures 21 and 22 compare the pictures of cavities at the design flow coefficient at $\phi/\phi_d = 1.0$ with and without the groove. We observe the following differences.

1. With the straight casing, the cavity occupies the entire tip region while the cavity mostly occurs in the groove with the grooved casing. This agrees with the numerical results shown in Figs. 13 and 14.
2. With the straight casing, the backflow vortex cavitation extends more upstream than the case with the groove. With the grooved casing, the backflow vortex cavitation extends only slightly from the upstream edge of the groove.

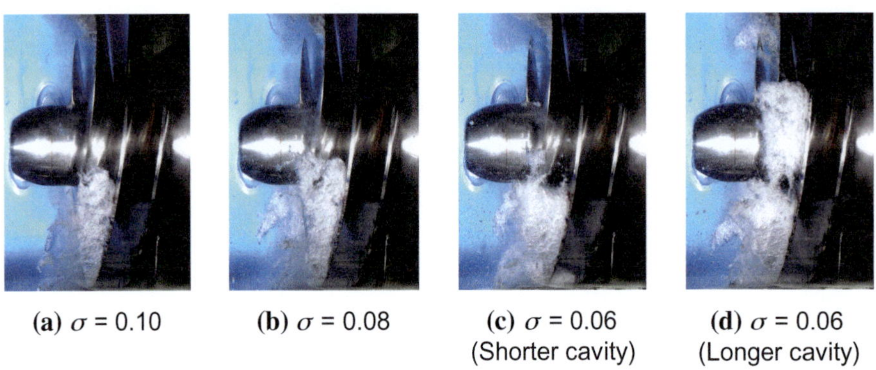

Fig. 21 Cavitation geometry, straight casing, $\phi/\phi_d = 1$

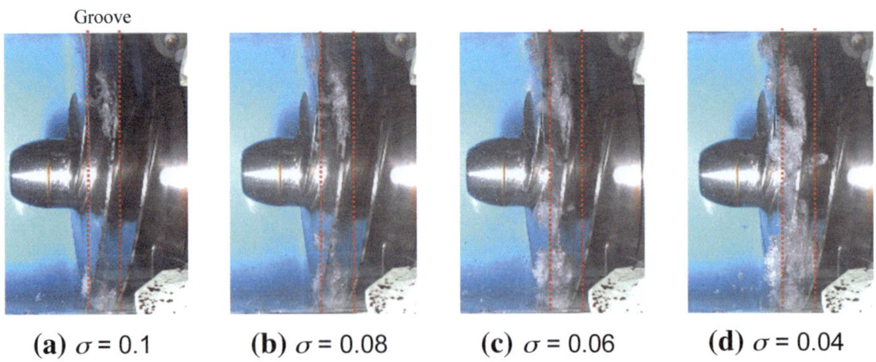

Fig. 22 Cavitation geometry, grooved casing, $\phi/\phi_d = 1$

Although not shown, it was found that the increase of cavity volume with the decrease of flow rate was much smaller with the grooved casing.

2.3.5 Propagation of Backflow Vortex Cavity

Figure 23 shows the pictures of cavities from high-speed video to examine the propagation of backflow vortex cavities for the case with the grooved casing, at $\phi/\phi_d = 1.0$ and $\sigma = 0.045$. The propagation speed Ω_v is measured from the pictures and number of vortices N_v was evaluated by counting the passage of vortices. The results are shown in Table 2 where Ω_n is the impeller speed.

For the case without groove, the backflow vortex structure has been studied extensively (Yokota et al. 1998) and it was found that the vortex structure is caused by the roll-up of the shear layer between the straight main flow and swirling backflow. It was found that the number of vortices is determined from the stability of vortices and the number decreases with the decrease of flow rates caused by the inward shift of the radial location of the vortices. It was also found that both the velocity and the number of vortices fluctuate largely with time. So, the values in Table 2 are only representative values.

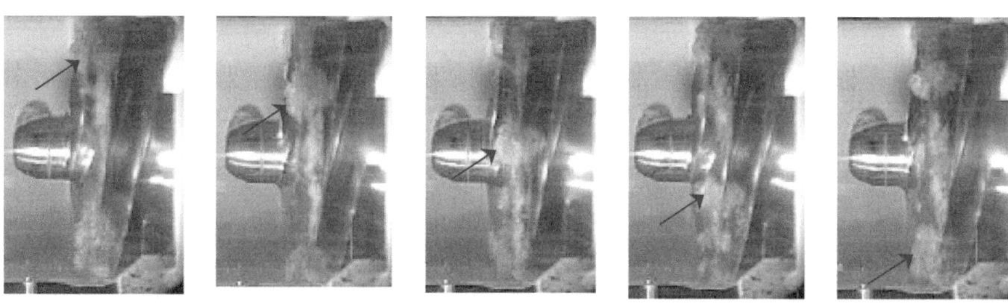

(a) 0 s **(b)** 0.0088s **(c)** 0.0177s **(d)** 0.0267s **(e)** 0.0355s

Fig. 23 Propagation of backflow vortex cavities, $\phi/\phi_d = 1$, $\sigma = 0.045$

Table 2 Propagation speed and number of vortices

ϕ/ϕ_d	σ	Ω_{vv}	N_v
0.9 (Grooved casing)	0.055	0.16 Ω_n	7.99
1.0 (Grooved casing)	0.045	0.13 Ω_n	8.84
1.0 (Grooved casing)	0.080	0.15 Ω_n	10.7
1.1 (Grooved casing)	0.080	0.11 Ω_n	16.4
1.0 (Straight casing)	0.050	0.12 Ω_n	8–16

2.4 Cause of Higher Frequency Oscillations

2.4.1 Unsteady Calculation

To confirm the suppression of rotating cavitation and to clarify the cause of higher frequency components, unsteady calculations were made on full passage with the grooved casing, at $\phi/\phi_d = 0.9$ and $\sigma = 0.05$.

Figure 24 shows the spectrum of inlet pressure fluctuation at the same location with the experiments. As expected, we do not have significant component around $N = 50$ Hz, suggesting that the rotating cavitation has been suppressed. In addition to the blade passing component and its modulations around $3N = 150$ Hz we observe a component with $5.44N$.

Figure 25 shows the three-dimensional representation of iso-pressure surface of $\psi_c = (p - p_v)/\rho U_t^2 = 0.083$ and the pressure distributions in the planes of $r/R_t = 0.87$

Fig. 24 Spectrum of inlet pressure oscillation, $\phi/\phi_d = 0.9$, $\sigma = 0.05$

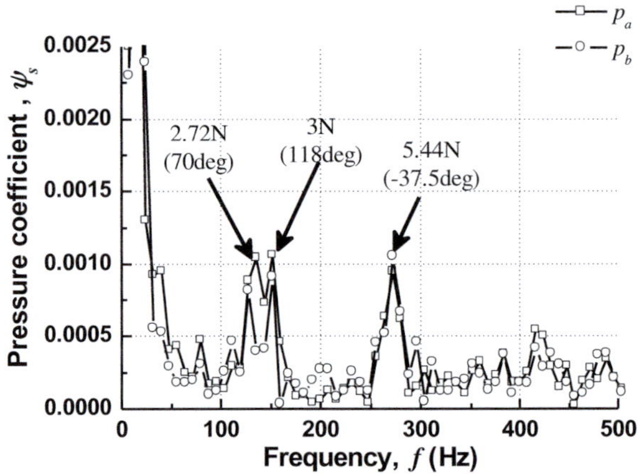

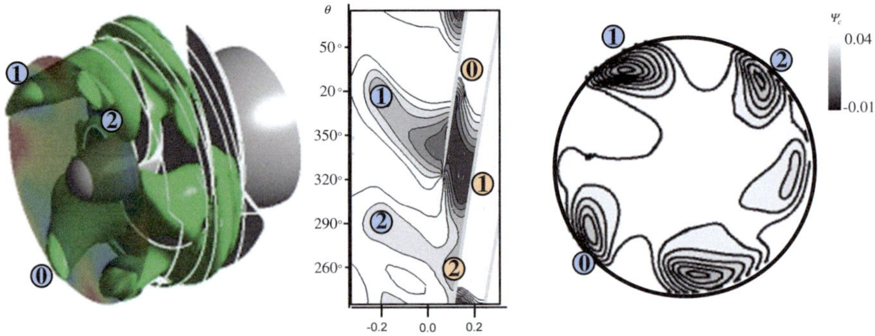

(a) Iso-pressuresurface with Ψ_c=0.083

(b) Pressure distribution in the plane of r/R_t=0.86

(c) Pressure distribution at z/D_t=-0.2

Fig. 25 Iso-pressure surface and pressure distribution, $\phi/\phi_d = 0.9$, $\sigma = 0.05$

and of $z/D_t = -0.2$. We observe five vortices interacting with the blades. It was found that the vortices rotate at an absolute speed of $0.122\Omega_n$. Then, the vortex passage frequency observed on a blade is $5\times(1-0.112)\Omega_n = 4.44\Omega_n$. This is different from the frequency $5.44N$ found in the upstream pressure fluctuation.

2.4.2 Rotating Modes Due to Interaction

It is known that rotating modes occur due to rotor–stator interaction (Tyler and Sofrin 1962). We first consider a rotor with R blades ($r = 1, \ldots R$) rotating with an angular speed of Ω and a stator with S blades ($s = 1, \ldots S$) as shown in Fig. 26.

We represent the pressure component caused by the interaction of a stator blade $s = 1$ with R rotor blades ($r = 1, \ldots R$) by

$$p_{m,n}^{s=1}(\theta, t) = a_{m,n} \cdot \cos(m\theta - nR\Omega t), \tag{1}$$

where m and n are the harmonic order in θ and t. Then the pressure pattern caused by the interaction of a stator blade $s = q$ with R rotor blades ($r = 1, \ldots R$) can be represented by

$$p_{m,n}^{s=q}(\theta, t) = a_{m,n} \cdot \cos\left\{m\left[\theta - \frac{2\pi}{S}(q-1)\right] - nR\Omega\left[t - \frac{2\pi}{S\Omega}(q-1)\right]\right\} \tag{2}$$

By adding up the effects of all stator blades ($q = 1 \sim S$), we obtain

$$\begin{aligned} p_{m,n}(\theta, t) &= a_{m,n} \cdot \sum_{q=1}^{S} \cos\left[m\theta - nR\Omega t - (m - nR)\frac{2\pi}{S}(q-1)\right] \\ &= \begin{cases} S \cdot a_{m,n} \cdot \cos(m\theta - nR\Omega t) & m = nR + kS \\ 0 & m \neq nR + kS \end{cases} \end{aligned} \tag{3}$$

where k is an arbitrary integer.

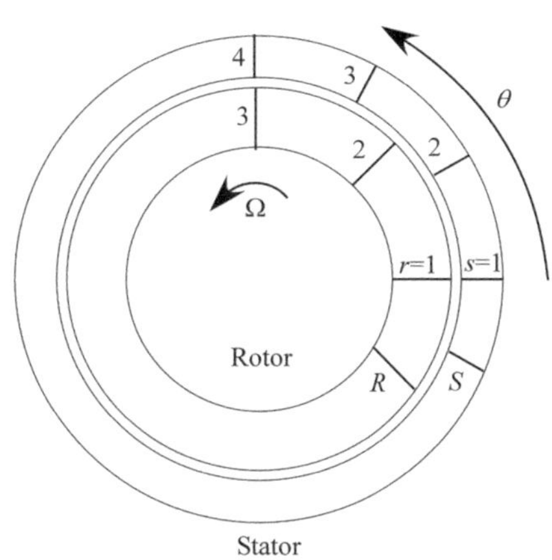

Fig. 26 Rotor–stator interaction model

We consider the case when the "stator" is also rotating and represent the stator speed by Ω_v and the rotor speed by Ω_n. We introduce an absolute circumferential coordinate $\theta^* = \theta + \Omega_v t$. By putting $\theta = \theta^* - \Omega_v t$ and $\Omega = \Omega_n - \Omega_v$, we obtain

$$p_{m,n}(\theta^*, t) = S \cdot a_{m,n} \cdot \cos\left\{m\theta^* - [nR(\Omega_n - \Omega_v) + m\Omega_v]t\right\} \tag{4}$$

for the case with $m = nR + ks$.

Now we examine the pressure pattern caused by the interaction of N_v inlet backflow vortices with the impeller blades. We replace the number of stator blades S with the number of vortices N_v.

Table 3 shows possible cases of

$$m = nR + kN_v \tag{5}$$

for the case of $R = 3$ and the frequency

$$\omega = nR(\Omega_n - \Omega_v) + m\Omega_v \tag{6}$$

obtained by assuming $\Omega_n = 0.15$. We find

1. The circumferential mode number m changes easily with the change of vortex number Nv.
2. The frequency is nearly constant in the groups of $N_v = (5, 6, 7)$, $N_v = (8, 9, 10)$, $N_v = (11, 12, 13)$ and changes largely between $N_v = 7$ and 8 or 10 and 11.

This corresponds to the experimental observations that

1. Various modes occur with nearly the same frequency.
2. The frequencies are around 5, 7, and 10 Ω_n.

Table 4 compares the components observed in the experiments and the frequency of possible modes obtained by Eq. (6). Experimental values are used for Ω_v. Although not perfect, general agreement is obtained. This suggests that the higher

Table 3 Mode analysis with $k = -1$, $R = 3$ and $\Omega_v = 0.15\Omega_n$

	$m = nR + kN_v$ ($n = 2$, $N_v = 5, 6, 7$)	$m = nR + kN_v$ ($n = 3$, $N_v = 8, 9, 10$)	$m = nR + kN_v$ ($n = 4$, $N_v = 11, 12, 13$)
$m = 1$	$1 = 2 \times 3 - 1 \times 5$ $\omega = 5.5\Omega_n$	$1 = 3 \times 3 - 1 \times 8$ $\omega = 7.8\Omega_n$	$1 = 4 \times 3 - 1 \times 11$ $\omega = 10.35\Omega_n$
$m = 0$	$0 = 2 \times 3 - 1 \times 6$ $\omega = 5.1\Omega_n$	$0 = 3 \times 3 - 1 \times 9$ $\omega = 7.65\Omega_n$	$0 = 4 \times 3 - 1 \times 12$ $\omega = 10.2\Omega_n$
$m = -1$	$-1 = 2 \times 3 - 1 \times 7$ $\omega = 4.95\Omega_n$	$-1 = 3 \times 3 - 1 \times 10$ $\omega = 7.5\Omega_n$	$-1 = 4 \times 3 - 1 \times 13$ $\omega = 10.1\Omega_n$

Table 4 Comparison of the experimental and CFD results with the analysis

	Experiment, CFD					Analysis			
ϕ/ϕ_d	σ	N_v	ω	Ω_{vv}	phase delay (deg)	m	n	N_v	ω
0.9 (Groove, Exp.)	0.055	7.99	$4.7\Omega_n$	$0.16\Omega_n$	73.2–77.8	–1	2	7	$4.88\Omega_n$
1.0 (Groove, Exp.)	0.045	8.84	$4.74\Omega_n$	$0.13\Omega_n$	3	0	2	6	$5.22\Omega_n$
1.0 (Groove, Exp.)	0.080	10.7	$4.75\Omega_n$	$0.15\Omega_n$	−4–13.8	0	2	6	$5.1\Omega_n$
			$7.4\Omega_n$		−25.1–18.5	0	3	9	$7.65\Omega_n$
			$9.75\Omega_n$		17.4–22.6	0	4	12	$10.2\Omega_n$
1.1 (Groove, Exp.)	0.080	16.4	$7.8\Omega_n$	$0.11\Omega_n$	−14.4	0	3	9	$8.01\Omega_n$
1.0 (Straight, Exp.)	0.080	8 ~ 16	$4.86 \sim 4.94\Omega_n$	$0.11\Omega_n$	−81.5– −70.5	1	2	5	$5.40\Omega_n$
0.9 (Groove, CFD)	0.050	5	$5.44\Omega_n$	$0.112\Omega_n$	−37.5	1	2	5	$5.44\Omega_n$

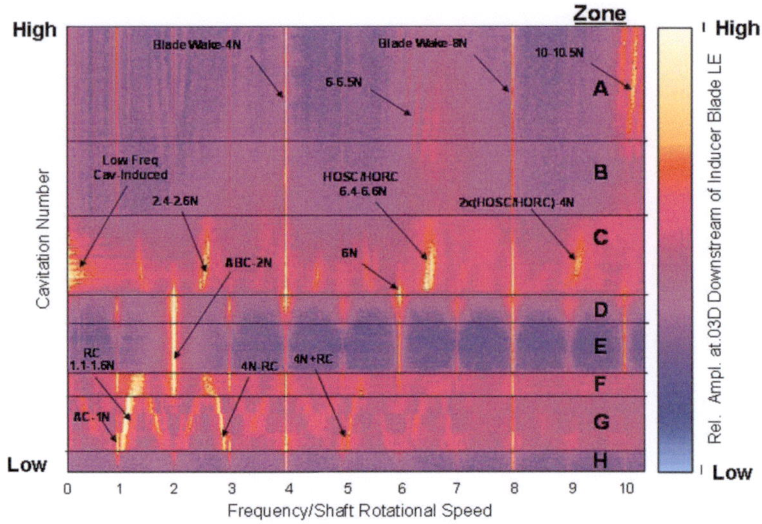

Fig. 27 Spectra of pressure fluctuation in a four-bladed inducer, from Subbaraman and Patton, 2006

frequency components are caused by the interaction of the backflow vortices with the blades. The discrepancy might be caused by the unsteady nature of the backflow vortices; the number of vortices and the propagation velocity changes irregularly with time.

Figure 27 shows the spectra of pressure fluctuation in a four-bladed inducer measured at $0.03D_t$ axial distance downstream from the blade leading edge at the tip (Subbaraman and Patton 2006). We have a case of $0 = 2 \times 4 - 1 \times 8$ for $m = nR + kN_v$. For this case, the estimated frequency with $\Omega_v = 0.15$ is $\omega = 2 \times 4 \times (1 - 0.15)\ \Omega_n = 6.8$. Another case is $0 = 3 \times 4 - 1 \times 12$ with the frequency $\omega = 3 \times 4 \times (1 - 0.15)\ \Omega_n = 10.2$. The frequencies are 4/3 times the frequencies for the three-bladed inducer shown in Table 4. Figure 27 includes the components close to these frequencies.

3 Conclusions

It has been shown that CFD has been developed to the level that three-dimensional unsteady cavitating code can be used to simulate cavitation instabilities such as rotating cavitation and cavitation surge. These cavitation instabilities are caused by the interaction of local flow near cavitation trailing edge with the leading edge with the next blade. This shows that cavitation instabilities can be avoided by avoiding the interaction. As an example, an avoidance method using circumferential groove on the casing was shown. This lead to the explanation of higher frequency oscillation which may cause structural resonance.

Qualitative simulation of cavitation instabilities is rather easy as compared with the simulation of instabilities caused by flow separation since the pressure around cavity is kept nearly constant at the saturated vapor pressure while the back pressure in separated flow region largely depends on the shear stress. The flow disturbance near the cavity trailing edge occurs through the continuity relation, and not through the momentum relation in which the shear stress plays an important role. However, the precise prediction of onset range of cavitation instabilities is still difficult since it depends also on the shear stress. So, further progress in CFD technique is needed in order to examine the occurrence of cavitation instabilities at the design stage of machines.

Acknowledgements The present manuscript is a summary of researches with many colleagues. Their contributions are acknowledged by referencing their works.

References

Hah, C., Mueller, M., & Schiffer, H. (2008). Aerodynamic study of circumferential grooves in a transonic axial compressor. *ASME Fluid Engineering Conference, 55232,* 1–8.

Horiguchi, H., Watanabe, S., & Tsujimoto, Y. (2000). A Linear stability analysis of cavitation in a finite blade count impeller. *ASME Journal of Fluids Engineering, 122*(4), 798–805.

Kang, D., Yonezawa, C., Horiguchi, H., Kawata, Y., & Tsujimoto, Y. (2009a). Cause of Cavitation instabilities in three-dimensional inducer. *International Journal of Fluid Machinery and Systems, 2*(3), 206–214.

Kang, D., Watanabe, T., Yonezawa, K., Horiguchi, H., Kawata, Y., & Tsujimoto, Y. (2009b). Inducer Design to Avoid Cavitation Instabilities. *International Journal of Fluid Machinery and Systems, 2*(4), 439–448.

Kang, D., Arimoto, Y., Yonezawa, K., Horiguchi, H., Kawata, Y., Hah C., & Tsujimoto, Y. (2010). Suppression of cavitation instabilities in an inducer by circumferential groove and explanation of higher frequency components. *International Journal of Fluid Machinery and Systems, 3*(2)

Subbaraman, M., & Patton, M. (2006). Suppressing higher order cavitation phenomena in axial inducers. *Sixth International Symposium on Cavitation* CAV2006 pp. 1–13.

Tsujimoto, Y., Yoshida, Y., Maekawa, M., Watanabe, S., & Hashimoto, T. (1997). Observations of oscillating cavitations of an inducer. *ASME Journal of Fluids Engineering, 119*(4), 775–781.

Tsujimoto, Y. and Semenov, Y. (2002). New types of cavitation instabilities in inducers. In *4th International Conference on Launcher Technology "Space Launcher Liquid Propulsion"* pp. 1–10.

Tyler, J. M. and Sofrin, T. G. (1962). Axial compressor noise studies. In Transactions Society of Automotive Engineers *70*, pp. 309–322.

Watanabe, S., Sato, K., Tsujimoto, T., & Kamijo, K. (1999). Analysis of rotating cavitation in a finite pitch cascade using a closed cavity mode and a singularity method. *ASME Journal of Fluids Engineering, 121*(4), 834–840.

Yokota, K., Kurahara, K., Kataoka, D., & Tsujimoto, Y. (1998). AS study of swirling backflow and vortex structure in the inlet of an inducer. *Journal of the Japan Society of Mechanical Engineers, 64–622,* 51–58.

Rotordynamics of Turbopumps and Hydroturbines

Yoshinobu Tsujimoto

1 Introduction

In modern turbomachinery, acceptability and reliability depend heavily on the degree of vibration and noise, and therefore, it is becoming a common practice to carry out shaft vibration analysis at the development stage. In such analysis, the estimation of the unsteady fluid dynamic forces acting on various parts of the shaft system is crucial for the successful simulation of shaft vibration.

The unsteady fluid dynamic forces on the impeller include

1. Unbalanced fluid dynamic forces due to manufacturing errors.
2. Forces due to interaction of impeller vanes with stationary structures such as stator vanes or volute.
3. Forces caused by fluid dynamic flow instabilities such as surge, rotating stall, precession of inlet or outlet vortex, and von Karman vortices from vanes and struts.

Although there might be some interaction of these forces with structural vibration as exemplified by the lock-in of the von Karman vortices, these forces can be basically treated as (given) external forces in the preliminary rotordynamic analysis. In most cases the frequency of these forces is proportional to the impeller speed. The shaft vibration amplitude increases when the excitation frequency is close to one of the natural frequencies of the shaft vibration, as shown in Fig. 1a.

There is another class of fluid forces, which occur in response to shaft vibration. This is herein called "rotordynamic forces." In many cases the fluid forces are treated as equivalent mass, damping, and stiffness and all of which alter the dynamic characteristics of the shaft system. The rotordynamic forces may cause

Y. Tsujimoto (✉)
Osaka University, Osaka, Japan
e-mail: tujimoto@me.es.osaka-u.ac.jp

L. d'Agostino and M.V. Salvetti (eds.), *Cavitation Instabilities and Rotordynamic Effects in Turbopumps and Hydroturbines*, CISM International Centre for Mechanical Sciences 575, DOI 10.1007/978-3-319-49719-8_5

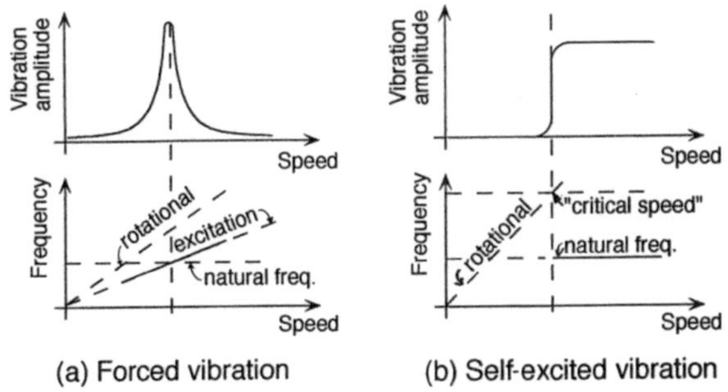

Fig. 1 Forced and self-excited vibrations

self-excited vibration of the rotor. Generally, this occurs when the shaft speed exceeds a certain "critical speed" and the amplitude does not decrease even if the speed becomes much larger than the critical speed. This is quite different from forced vibrations. The frequency is maintained as one of the natural frequencies of the shaft system. These circumstances are shown in Fig. 1b, which is partly modified from a figure in Ehrich and Child's article (1984).

Rotordynamic forces are found in various parts of turbomachinery such as hydrodynamic bearings, annular or labyrinth seals and impellers. In the first part of this note we focus on rotordynamic impeller forces simply because the author has not been involved in other forces. Textbooks such as Vance (1988), Ehrich (1992), and Childs (1993) cover other rotordynamic forces and rotordynamic analysis. Brennen's book (1994) focuses more on the fluid dynamic explanations of the forces. However, there are certain cases when self-excited vibration occurs even when the rotational speed is lower than the critical speed. These cases are treated in the second part of this note.

2 Example of Forced and Self-Excited Vibrations in Pumps

It might be appropriate to start with an example of forced and self-excited vibrations (Yoshida et al. 1992). Figure 2 shows an experimental apparatus designed to reduce the natural frequency of the shaft system using a pendulum structure.

Figure 3 shows the shaft vibration of a 2-D centrifugal impeller with an outlet blade angle $\beta_2 = 25°$ in a vaned diffuser with the vane angle $\beta = 20°$, at various values of flow coefficient ϕ (the design point is $\phi_d = 0.1$). The ratio of the natural frequency, Ω^*, to the rotational frequency, ω, is set at $\Omega^*/\omega = 0.206$. ε/d_2 is the vibration amplitude/impeller diameter ratio and Ω/ω is the vibration frequency/rotational frequency (400 rpm = 6.67 Hz) ratio. $\Delta p/\rho u_2^2$ is the normalized pressure fluctuation at the impeller outlet. We observe forced vibrations caused by the impeller (with $\Omega/\omega \approx 0.8$, near $\phi = 0.016$, denoted by IRS) and diffuser (with $\Omega/\omega \approx 0.05$, near $\phi = 0.04$, denoted by DRS) rotating stalls at smaller ϕ, and

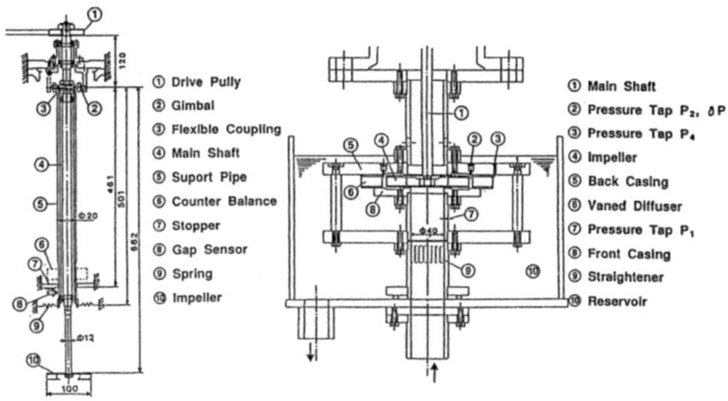

Fig. 2 Experimental apparatus for the measurements of shaft vibrations at supercritical speed

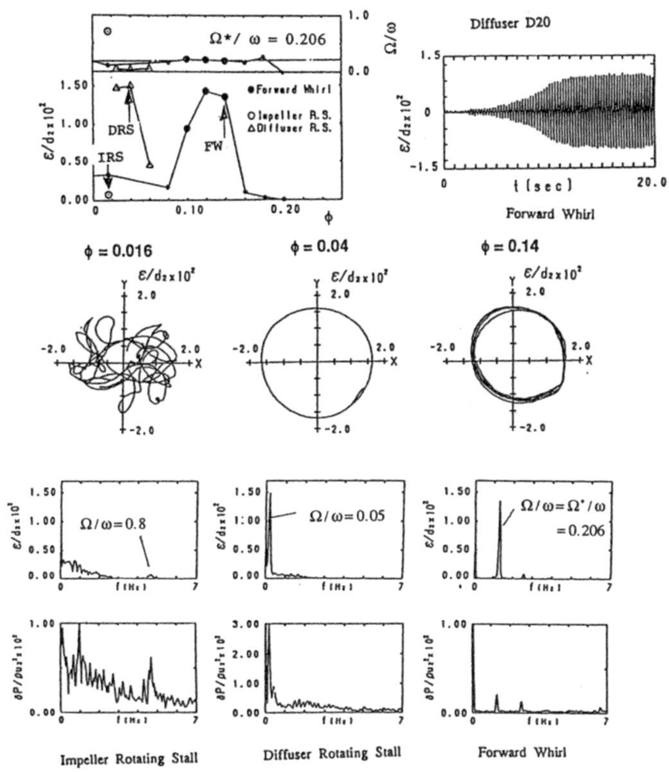

Fig. 3 Forced and self-excited shaft vibrations at supercritical speed ($\Omega^{*}/\omega = 0.206$)

self-excited vibration (with $\Omega/\omega = \Omega^{*}/\omega = 0.206$, near $\phi = 0.14$ denoted by "Forward Whirl", FW) at higher flow rate. Figure 4 shows the effects of shaft natural frequency, Ω/ω^{*}, on each component of the shaft vibration. We observe that

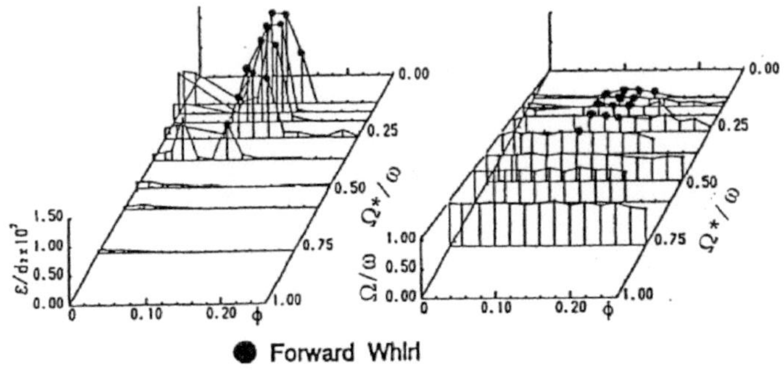

(a) Self-excited vibrations

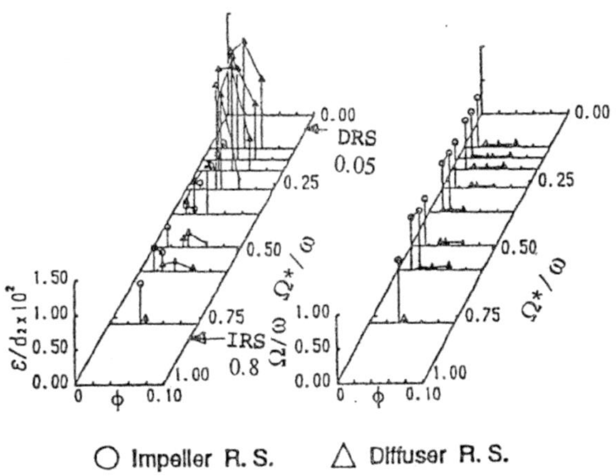

(b) Forced vibrations

Fig. 4 Effects of critical speed/shaft speed ratio, Ω^*/ω, on self-excited and forced vibrations

the frequencies of forced vibration caused by impeller and diffuser stalls are kept nearly constant, and that their magnitude increases as the natural frequency, Ω^*/ω, approaches their own frequencies, $\Omega/\omega \approx 0.08$ for impeller rotating stall, $\Omega/\omega \approx 0.05$ for diffuser rotating stall. The frequency of self-excited vibration is identical to the natural frequency Ω^*/ω and it occurs when $\Omega^*/\omega < 0.5$. In other words, self-excited vibration occurs when the rotational frequency, ω, exceeds twice the natural frequency, $2\Omega^*$. This is caused by the rotordynamic forces on the impeller, which is the main focus of the present note.

3 Effects of Rotordynamic Force and Moment on Rotordynamic Instability

We consider an overhung rotor rotating with an angular velocity ω and executing a whirling motion with an angular velocity Ω with a whirl radius $|\varepsilon|$ as shown in Fig. 5b. The fluid force F and moment M are represented as follows

$$F = |\varepsilon| f e^{i(\Omega t + \theta)} = \varepsilon f e^{i\theta} \tag{1}$$

$$M = |\varepsilon| m e^{i(\Omega t + \varphi)} = \varepsilon m e^{i\varphi} \tag{2}$$

The displacement of the shaft $\varepsilon = |\varepsilon| e^{i\Omega t}$ caused by the force F^* and the moment M^* applied on the shaft by the rotor can be represented as follows, by using the stiffness coefficients k and k_m of the shaft.

$$\varepsilon = F^*/k + i M^*/k_m \tag{3}$$

If we consider the inertia $-m_a \ddot{\varepsilon}$ and the damping force on the rotor $-c\dot{\varepsilon}$, the momentum equation of the rotor for the lateral vibration can be given by

$$F^* = F - m_a \ddot{\varepsilon} - c\dot{\varepsilon} \tag{4}$$

If we assume that the angular displacement of the rotor is so small that the inertia moment and damping can be neglected, we can write

$$M^* = M \tag{5}$$

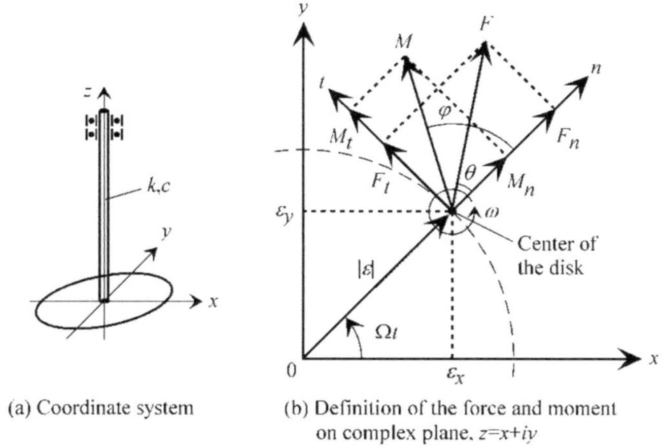

Fig. 5 Force and moment acting on the rotor in whirling motion

By putting Eqs. (3) and (5) into (4), we obtain

$$m_a\ddot{\varepsilon} + c\dot{\varepsilon} + k\varepsilon = F + i\frac{k}{k_m}M \equiv \varepsilon\left[\left(f_n - \frac{k}{k_m}m_t\right) + i\left(f_t + \frac{k}{k_m}m_n\right)\right] \equiv \varepsilon f_e e^{i\theta^*}, \tag{6}$$

where

$$f_n = f\,cos\theta, \quad f_t = f\,sin\theta, \quad m_n = m\,cos\varphi, \quad m_t = m\,sin\varphi$$

$$f_e = |f_{en} + if_{et}|, \quad f_{en} = f_n - (k/k_m)m_t, \quad f_{et} = f_t + (k/k_m)m_n$$

Then, Eq. (6) can be rewritten as

$$m_a\ddot{\varepsilon} + c\dot{\varepsilon} + (k - f_e e^{i\theta^*})\varepsilon = 0 \tag{7}$$

If we put $\varepsilon = e^{\lambda t}$ in Eq. (7), and assume that c and f_e are sufficiently small, we obtain

$$\lambda = -\frac{c}{2m_a} \mp \frac{k_I}{2\sqrt{m_a k_R}} \pm i\sqrt{\frac{k_R}{m_a}} \tag{8}$$

where $k_R = k - f_e\,cos\theta^*$ and $k_I = -f_e\,sin\theta^*$. Equation (8) shows that the whirl frequency Ω is given by

$$\Omega = \Omega_R = \pm\sqrt{\frac{k_R}{m_a}} \tag{9}$$

The forward whirl with $\Omega_R = \sqrt{k_R/m_a}$ grows when

$$-\frac{c}{2m_a} - \frac{k_I}{2\sqrt{m_a k_R}} > 0$$

That is,

$$f_{et}\frac{1}{2m_a\Omega_R} > \frac{c}{2m_a} \quad \text{or} \quad |\varepsilon|f_{et} > c|\varepsilon|\Omega_R \tag{10}$$

This means that the whirl radius $|\varepsilon|$ grows when the equivalent tangential force $|\varepsilon|f_{et} = |\varepsilon|\{f_t + (k/k_m)m_n\}$ becomes larger than the damping force $c|\varepsilon|\Omega_R$. Equation (10) also shows that the backward whirl with $\Omega_R = -\sqrt{k_R/m_a}$ occurs when $|\varepsilon|f_{et} < -c|\varepsilon|\Omega_R$. The contribution of the tangential force f_t was shown by Shoji and Ohashi et al. (1981). Present result shows that the whirl instability can occur also when the normal moment m_n has the same sign as the whirl angular velocity Ω.

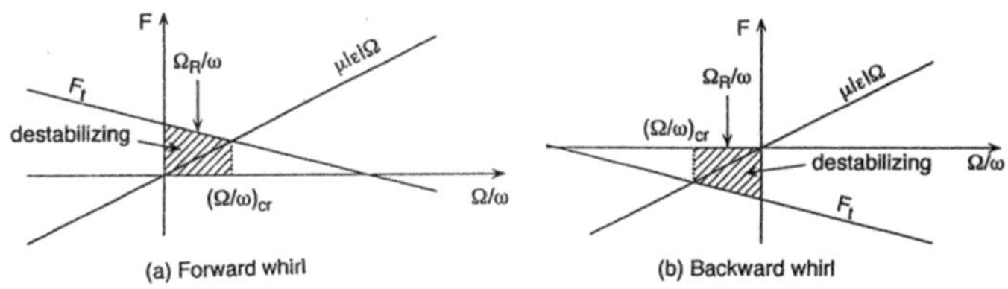

Fig. 6 Destabilizing regions for forward and backward whirl

We should stress here that the radial force, f_{en}, has the effect of simply modifying the shaft effective stiffness but that the tangential force, f_{et}, has the effect of promoting the whirling instability in the direction of f_{et}.

Generally, the rotordynamic forces (f_n, f_t) are functions of the whirl velocity ratio, Ω/ω, where Ω and ω are whirl and impeller rotational frequencies. For the case of Fig. 6a or b, forward or backward whirl instability will occur when

$$\Omega_R/\omega < \left|(\Omega/\omega)_{cr}\right| \tag{11}$$

where Ω_R is the shaft natural frequency of Eq. (9) and $(\Omega/\omega)_{cr}$ is the critical whirl frequency ratio at which

$$f_t = \mu\,|\varepsilon|\,\Omega_R \tag{12}$$

For the forward whirl, Eq. (12) shows that the instability occurs when the rotational frequency, ω, exceeds a critical speed, ω_{cr}:

$$\omega > \omega_{cr} = \Omega_R/(\Omega/\omega)_{cr} \tag{13}$$

As shown above, it is most appropriate to represent the rotordynamic forces (f_n, f_t) as functions of the whirl velocity ratio, Ω/ω. They are made dimensionless as

$$F_n = f_n/\left[\left(\rho\pi R^2 L\right)\omega^2\varepsilon\right] \tag{14}$$

$$F_t = f_t/\left[\left(\rho\pi R^2 L\right)\omega^2\varepsilon\right], \tag{15}$$

where R and L are representative radial and axial length dimensions of the device. ε is used to represent $|\varepsilon|$ hereafter. For centrifugal impellers, the impeller outlet radius, r_2, and breadth, b_2, are used for R and L.

4 Rotordynamic Forces on Centrifugal Impellers

In this section we consider only the whirling forces F. The fluid force moment will be discussed in the next section.

4.1 Two-Dimensional Impeller in a Vaneless Space

As the most fundamental case, Ohashi and Shoji (1987) measured the rotordynamic forces on a 2-D centrifugal impeller executing a circular motion with a constant radius ε at various whirl speed ratios, Ω/ω, in a vaneless space as shown in Fig. 7.

The forces on the outer surfaces of frond and back shrouds have been subtracted by using the data of a dummy impeller without flow channel. Figure 8 shows the dependences of F_n and F_t on the whirl speed ratio, Ω/ω, at five different flow rates. This figure shows that

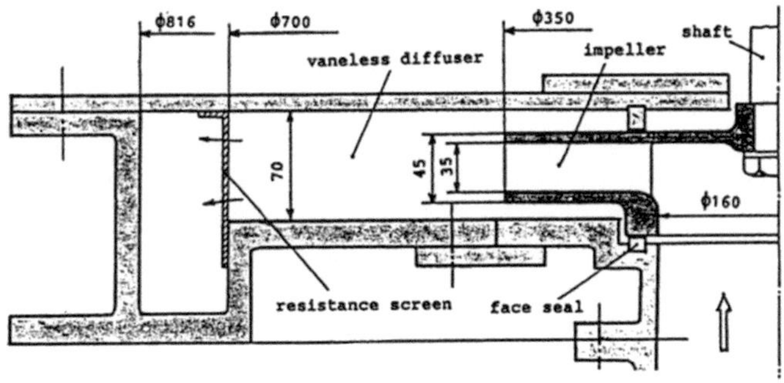

Fig. 7 2-D impeller in vaneless diffuser

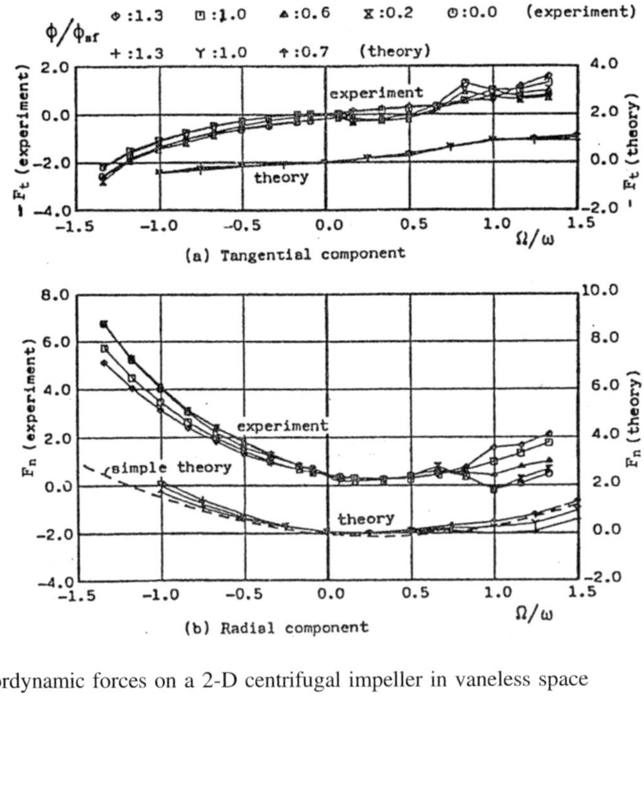

Fig. 8 Rotordynamic forces on a 2-D centrifugal impeller in vaneless space

1. Rotordynamic forces are basically stabilizing with $F_t < 0$ for $\Omega/\omega > 0$ and $F_t > 0$ for $\Omega/\omega < 0$ except for the cases of $\phi/\phi_{sf} < 0.6$, where ϕ_{sf} is the incidence free flow coefficient.
2. At smaller flow rates with $\varphi/\varphi_{sf} < 0.6$, the tangential force becomes positive for $0 < \Omega/\omega < 0.5$. Thus, the impeller forces become destabilizing for forward whirl at smaller flow rates.

The radial component is explained as the total of centrifugal force on the added mass, M_a, for lateral acceleration, and the Kutta–Joukovsky force on the impeller circulation due to the whirl velocity, $\varepsilon\Omega$. This is expressed by

$$F_n = M_a/\left(\rho\pi r_2^2 b_2\right)(\Omega/\omega)^2 - (\psi/\eta)(\Omega/\omega) \tag{16}$$

where ψ is the head coefficient and η is the hydraulic efficiency. This result is plotted in Fig. 8 as "simple theory".

Another type of 2-D flow analysis has been made by Tsujimoto and Acosta (1987). In this model the analysis is simplified by assuming an infinite number of impeller vanes. However, the effects of vaneless diffuser and of impeller loss are taken into account, instead. Figure 9 shows the results. At the shock-free flow coefficient $\varphi/\varphi^* = 1.0$ the rotordynamic force is stabilizing. However, at reduced flow rates, a destabilizing region appears at small positive whirl speed ratio. Large changes of forces occur when the flow coefficient and the whirl speed ratio approach the onset flow coefficients and the stall propagation velocity ratios of impeller and vaneless diffuser rotating stalls.

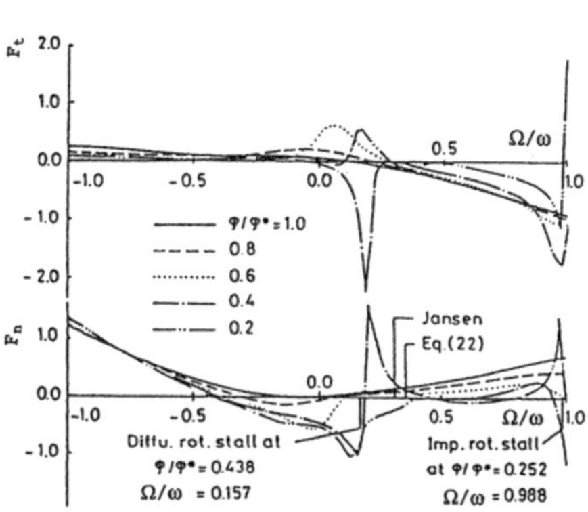

Fig. 9 Rotordynamic forces on a 2-D centrifugal impeller with $r_1/r_2 = 0.5$ and $\beta = 30°$ in a vaneless diffuser of radius ratio $r_3/r_2 = 1.5$

4.2 Interaction with Volutes and Vaned Diffusers

Experiments at Caltech (Chamieh et al. 1985: Jery et al. 1985) were started with a 3-D impeller (Impeller X) combined with a volute casing (Volute A). The measured rotordynamic forces were significantly larger than the results shown in Fig. 8 and a destabilizing region was found for $0 < \Omega/\omega < 0.5$ even at the design point. To find out the reason of the differences, a two-dimensional impeller (Impeller Z) was tested in Volute A (Franz and Arndt 1986). The results are shown in Fig. 10.

The magnitude of the forces is equivalent to that shown in Fig. 8 but a destabilizing region is found at $0 < \Omega/\omega < 0.3$ even near the design flow coefficient $\phi_d = 0.092$. This is caused by the interaction with the volute.

To confirm this, a 2-D linear inviscid flow analysis assuming an infinite number of impeller vanes was made by representing the effect of volute by a singularity method (Tsujimoto et al. 1988a). The results are shown in Fig. 11.

We observe reasonable agreements with the 2-D impeller experiments and the destabilizing region $0 < \Omega/\omega < 0.3$ is well simulated by the model. It was also confirmed by the analysis that the destabilizing region disappears when the volute is neglected or when the normal velocity disturbance on the volute is allowed. It was also found by the calculation that the radial location of the volute tongue r_t should

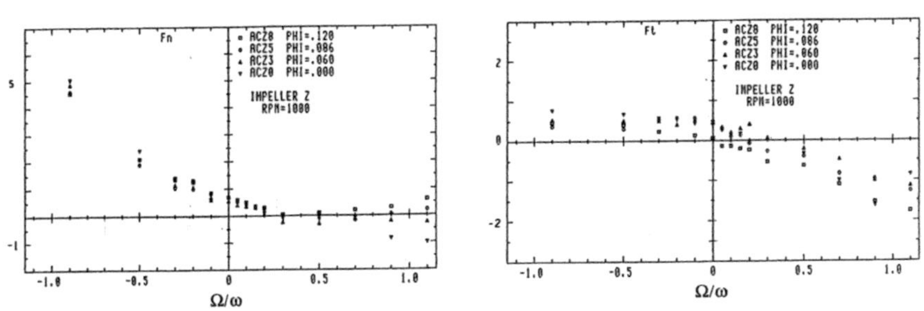

Fig. 10 Rotordynamic forces on a 2-D impeller (Impeller Z) in a volute (Volute A)

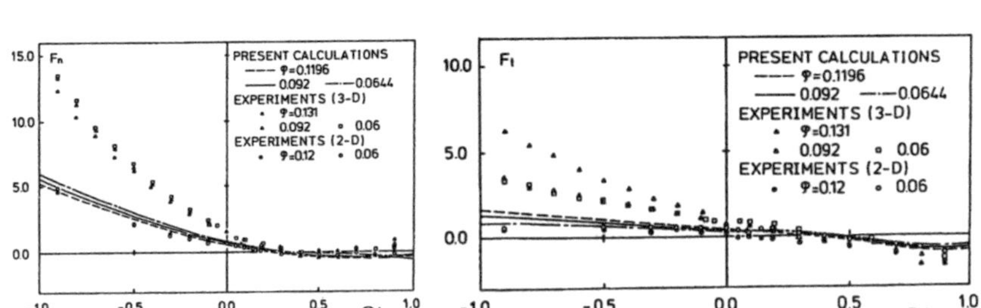

Fig. 11 Comparison of theoretical results with experiments on 2-D (Impeller Z) and 3-D (Impeller X) impellers in Volute A

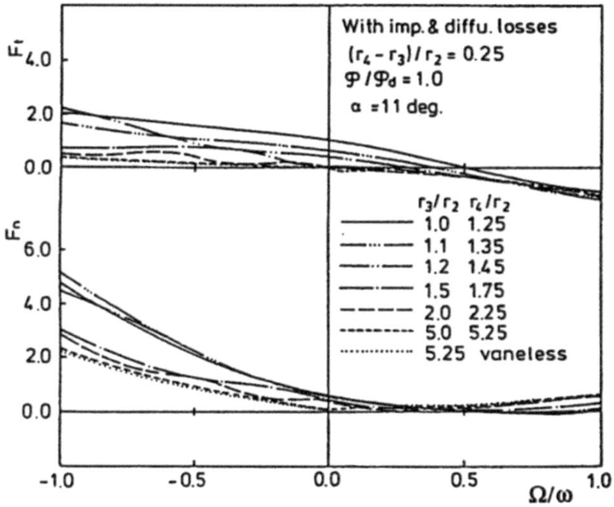

Fig. 12 Effects of the diffuser inlet/impeller outlet radius ratio r_3/r_2 at the design flow rate, diffuser vane angle $\alpha = 11°$

satisfy $r_t/r_2 > 1.6$ to substantially remove the destabilizing region. These facts show that the destabilizing region is caused by the interaction with the volute.

A similar 2-D inviscid flow analysis was made to examine the effects of the interaction with vaned diffusers (Tsujimoto et al. 1988). Figure 12 shows the effects of the diffuser inlet/impeller outlet radius ratio, r_3/r_2 at the design flow rate. A destabilizing tangential force occurs at $0 < \Omega/\omega < 0.5$ for the cases with realistic values of r_3/r_2. This is the cause of the self-excited vibration shown in Fig. 4.

The impeller forces can be calculated from the pressure distributions around the inlet and outlet of the impeller, momentum flux at the inlet and outlet of the impeller, and the change of the momentum of the fluid in the impeller. It was found that the component due to the pressure distribution around the impeller is the largest and determines the rotordynamic forces.

The impeller forces under the conditions of strong interaction with a volute or a vaned diffuser can be explained as follows. We consider the cases when the angle of the volute or the diffuser vanes measured from the circumferential direction is very small. In this case the fluctuation of radial absolute velocity is strongly regulated by the existence of the volute or the vane surfaces. The radial velocity fluctuation can be divided into the components caused by

(a) whirling motion,
(b) dislocation of the steady flow around the impeller represented by a source and a vortex,
(c) Nonuniform relative flow through the impeller flow passage.

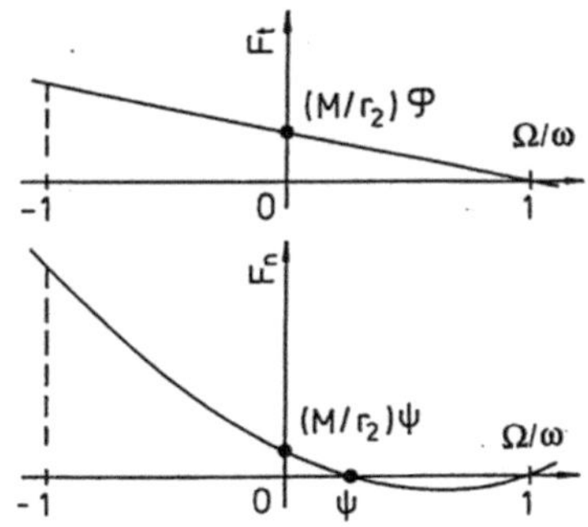

Fig. 13 Rotordynamic forces under strong interaction with a volute or a vaned diffuser

Under the condition with the strong interaction, the component (c), i.e., the nonuniform flow through the impeller, occurs so as to cancel the components (a) and (b). Then the pressure distribution around the impeller is established such as to balance the inertia due to the fluctuation of the flow through the impeller. By integrating the pressure distribution around the impeller, we obtain the following rotordynamic forces

$$F_t = (M/r_2)(1 - \Omega/\omega)\phi \tag{17}$$

$$F_n = (M/r_2)(1 - \Omega/\omega)(\psi - \Omega/\omega) \tag{18}$$

where ϕ and ψ are the flow and head coefficients of the impeller, and

$$M = \int_{r_1}^{r_2} \frac{r_2}{r\, sin^2\beta(r)} dr$$

is the inertial length of the impeller flow channel ($\beta(r)$ is the local vane angle from the tangent). The results of Eqs. (17) and (18) are shown in Fig. 13.

4.3 Rotordynamic Forces on the Shroud

We have seen that the rotordynamic forces on a 3-D impeller are significantly larger than those on a 2-D impeller (Fig. 11). In order to determine the rotordynamic forces on the shroud, experiments were made by Guinzburg et al. (1994), using the apparatus shown in Fig. 14.

The leakage flow was generated by using an external pump. The forces were normalized by Eqs. (14) and (15) using the axial length of the leakage path for L. The leakage flow rate, Q, was normalized as $\phi = Q/2\pi r_2^2 H\omega$, where H is the clearance between the rotor and casing. Figure 15 shows the effect of leakage flow rate. The dimensional magnitude of the rotordynamic forces is the same order as that of 3-D impeller and a destabilizing region is found at small positive whirl. The

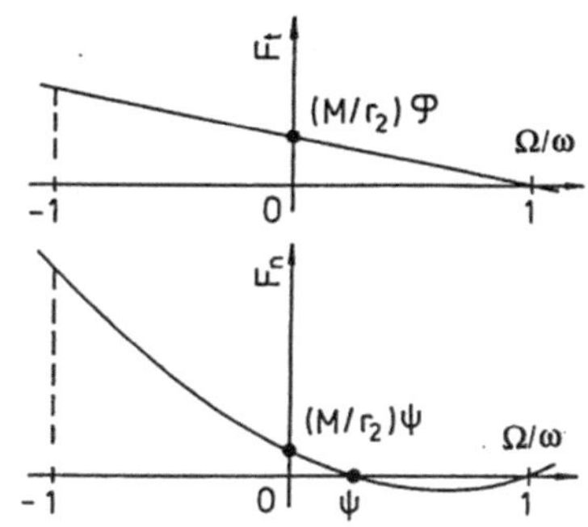

Fig. 14 Experimental apparatus for the measurement of shroud forces

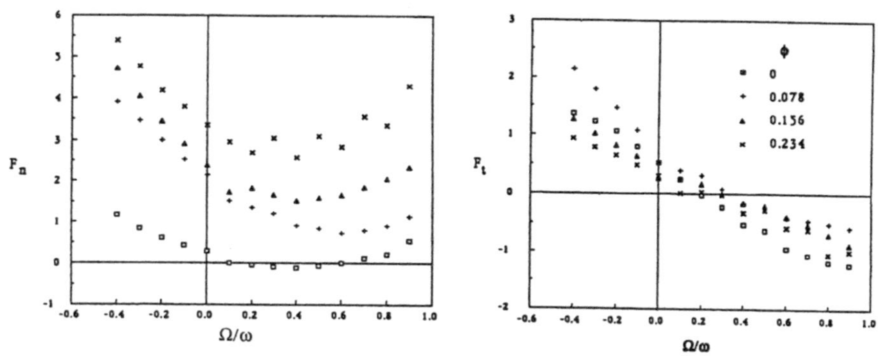

Fig. 15 Effects of leakage flow rate, $\varepsilon = 0.0254$ cm, $H = 0.14$ cm, at 1000 rpm

normal force significantly increases as the leakage flow increases. It was also found by giving inlet swirl, that the tangential force and the destabilizing region is significantly increased by adding a swirl at the inlet, in the direction of the rotor rotation (Fig. 16). It was also found that the magnitude of the force is inversely proportional to the clearance.

For the calculation of the shroud forces a bulk flow model proposed by Childs (1989) was examined. In this model, dependent variables are averaged over the clearance and the averaged 2-D flow in the variable clearance is calculated taking account of the empirical shear stress on the wall. Figure 16 shows an example of calculated results. Reasonable agreement is obtained for $\Omega/\omega < 0.5$ without inlet swirl ($\Gamma = 0$), but the calculated results are too sensitive to the inlet swirl. The large changes of the forces in $\Omega/\omega > 0.5$ are called "resonance" but this was not observed throughout the experiments.

Other examples showing the importance of the shroud forces are given in Figs. 17 and 18 (Ohashi et al. 1991). In Fig. 18, the results with and without the spacer are shown, together with the case of a spacer having radial grooves to reduce the swirl of leakage flow. The forces are larger for the case with the spacer, i.e., with smaller clearance. More important is the fact that a destabilizing region appears at

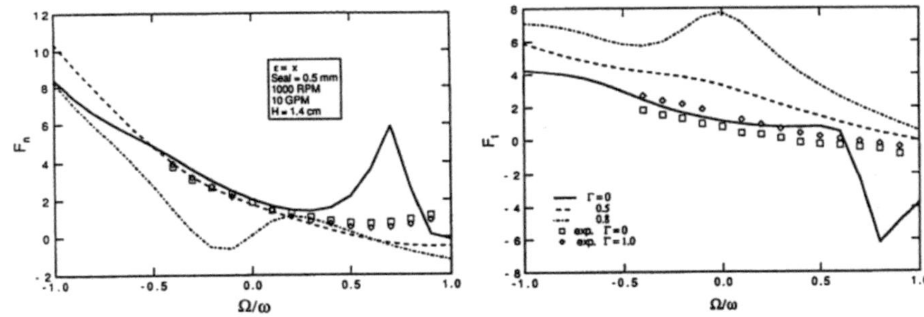

Fig. 16 Comparison with bulk flow calculation, $\varepsilon = 0.0118$ cm, $H = 0.14$ cm, at 1000 rpm

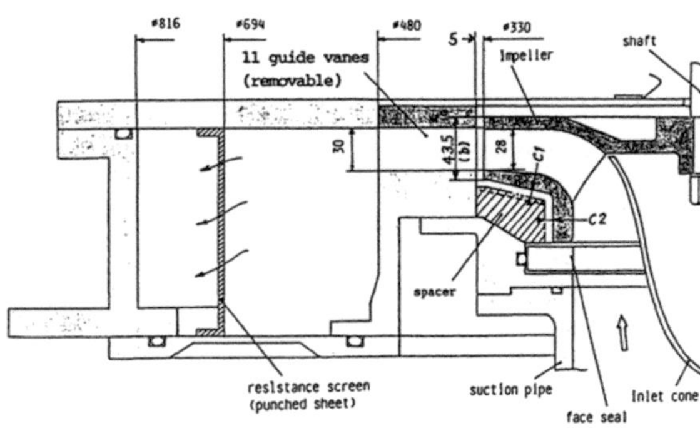

Fig. 17 Casing configuration around a 3-D impeller

$0 < \Omega/\omega < 0.5$ with the spacer. The tangential force and hence the destabilizing region is decreased by using the grooves on the spacer.

This is explained as follows. The tangential velocity of the leakage flow at the clearance approaches one-half of the tangential velocity of the rotor, from the balance of the tangential stresses exerted by the rotor and the casing. Then, the swirling flow will push the rotor in the direction of the whiling motion if the whirl speed ratio Ω/ω is smaller than the tangential velocity ratio 0.5 of the swirling leakage flow. This is the reason why the destabilizing region was found for the smaller clearance. This explanation also applies to the destabilizing forces on journal bearings and seals. If we have grooves on the casing, the tangential velocity ratio of the leakage flow will be decreased. This is the reason why the upper limit of the destabilizing region was decreased for the case with the grooves.

The characters of rotordynamic forces on centrifugal impellers can be summarized as follows:

(1) Rotordynamic forces on a 2-D impeller are basically stabilizing at the design flow coefficient. It can become destabilizing for forward whirl at smaller flow rate caused by rotating stalls in the impeller or the diffuser.

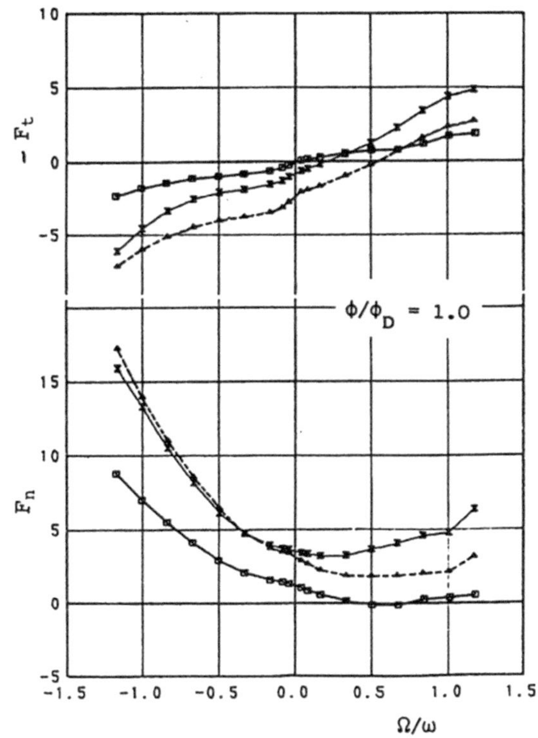

Fig. 18 Influence of clearance configuration on the rotordynamic forces on 3-D impeller shown in Fig. 17

(2) Rotordynamic forces on a 2-D impeller can become destabilizing for forward whirl caused by the interantion with volute or vaned diffuser placed downstream of the impeller.
(3) Rotordynamic forces on a 3-D impeller can become significantly larger than those on 2-D impeller, caused by the forces on the front shroud of the impeller. In addition, the shroud forces can cause forward whirl in the range $\Omega/\omega < 0.5$. The destabilizing region can be decreased by increasing the clearance between the shroud/casing, or by decreasing the tangential flow velocity by grooves on the casing.
(4) Although not shown in the present text, rotordynamic forces promoting forward whirl has been measured also for an open type compressor impeller (Yoshida et al. 1999). The magnitude of the rotordynamic forces is in the same order as the closed type impellers as shown in Fig. 18. It was found that the destabilizing force is caused by the pressure on the hub, caused by the change in the thickness of the flow cannel due to the whirling motion (Hiwata and Tsujimoto, 2002).

4.4 Rotordynamic Problem in a Rocket Turbopump

A rotordynamic problem was found in the fuel turbopump for the main engine LE7A of HIIA rocket and it was suppressed successfully by a small design modification (Motoi et al. 2003). The rotational speed of this pump is 42,200 rpm ($\omega = 703$ Hz), which is between third and fourth critical speeds. The spectra of shaft vibration are shown in Fig. 19 for the LE7 and LE7A turbopumps. For the LE-7 turbopump for HII rocket, no significant frequency peak is found. For the LE-7A turbopump, two peaks are found at 250 Hz and near 500 Hz. These frequencies correspond to the first and the third critical frequencies of the rotor.

The vibration modes of the LE7A turbopump are shown in Fig. 20. For the first and the third modes, for which the frequency peak is found in the data in Fig. 19, a large amplitude is found at the turbine end (to the right). From this result, it was conjectured that the vibration is caused by the forces on the turbine.

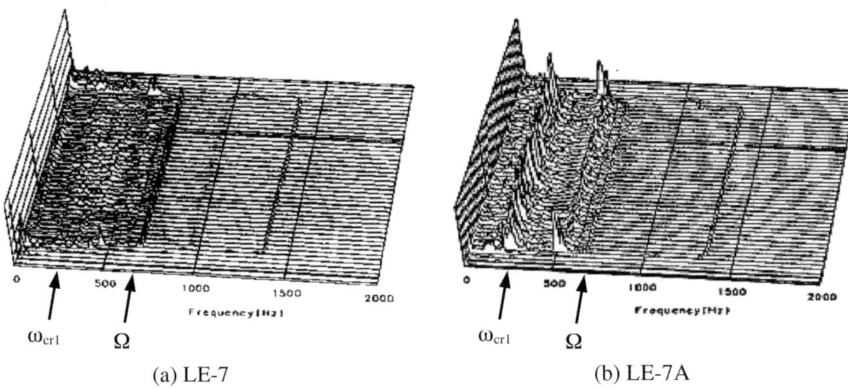

Fig. 19 Spectra of shaft vibration of turbopumps for LE7 and LE7A engines

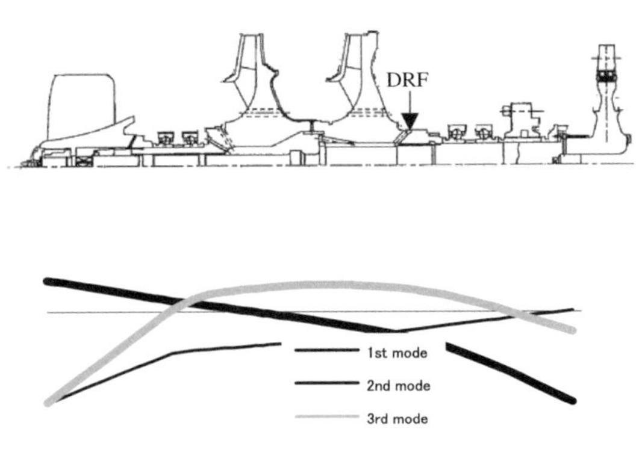

Fig. 20 Vibration modes of the turbopump for LE7A engine

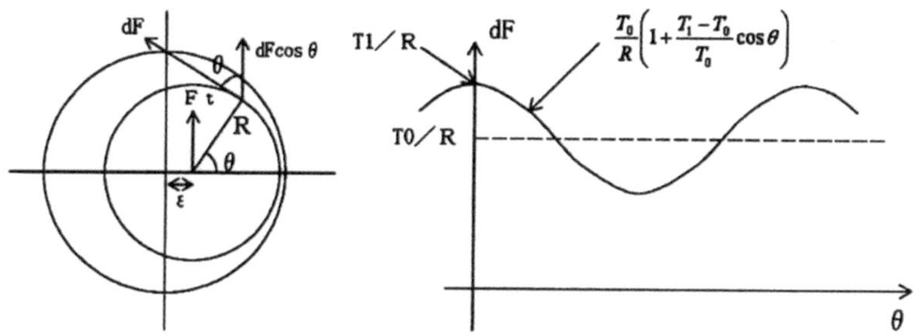

Fig. 21 Evaluation of tangential forces from nonuniform driving force due to eccentricity

For axial flow turbines, it is well known as "steam whirl" that forward whirl is promoted by the difference of the driving forces caused by the difference of the tip clearance due to the eccentricity. Figure 21 explains the tangential rotordynamic forces resulting from the nonuniform driving forces.

It is expected that we have a larger driving torque with smaller tip clearance. The torque with the minimum tip clearance at $\theta = 0$ is represented by T_1 and the average torque with the mean tip clearance by T_0. If we integrate the tangential force component $dF\,cos\theta = (T/R)\,cos\theta(d\theta/2\pi)$ at the mean radius R by assuming a sinusoidal dependence on θ as shown in Fig. 21, the total tangential force can be evaluated to be

$$F_t = [(T_1 - T_0)/T_0](T_0/2R) \tag{19}$$

This shows that $F_t > 0$ and forward whirl is promoted for axial turbines. Thomas (1958) represents the tangential force by

$$F_t = \varepsilon\beta(\mathrm{T}_0/DL), \tag{20}$$

where ε is the eccentricity, D the mean diameter of the rotor, L the length of the turbine blades and β is a nondimensional empirical factor. To determine the values of the turbine torques T_0 and T_1 with nominal and minimum tip clearance, numerical calculations of the flow were made by assuming uniform tip clearance corresponding each case. It was found that the value of the empirical factor for the expression of Eq. (20) is $\beta = 0.56$ for the present case.

To provide sufficient damping to the rotor system, the direction of the injection of the seal liquid (high-pressure liquid hydrogen) to the seal between the turbine and the pump was changed as shown in Fig. 22, in such a way as to decrease the mean tangential flow velocity in the seal. Then the sub-synchronous vibration was successfully suppressed as shown in Fig. 23. This is because the mean tangential velocity of the leakage flow is decreased by the injection of the seal liquid in the direction opposite to the rotation of the rotor and hence the destabilizing force on the seal is decreased.

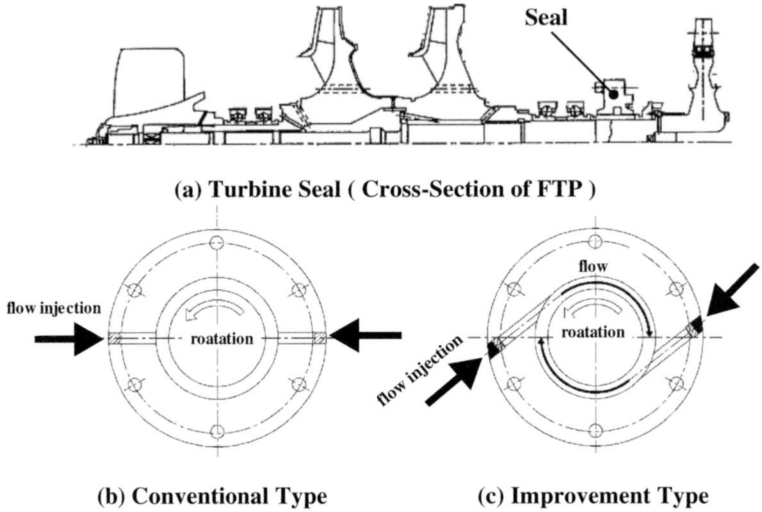

Fig. 22 Modification of the direction of seal fluid injection

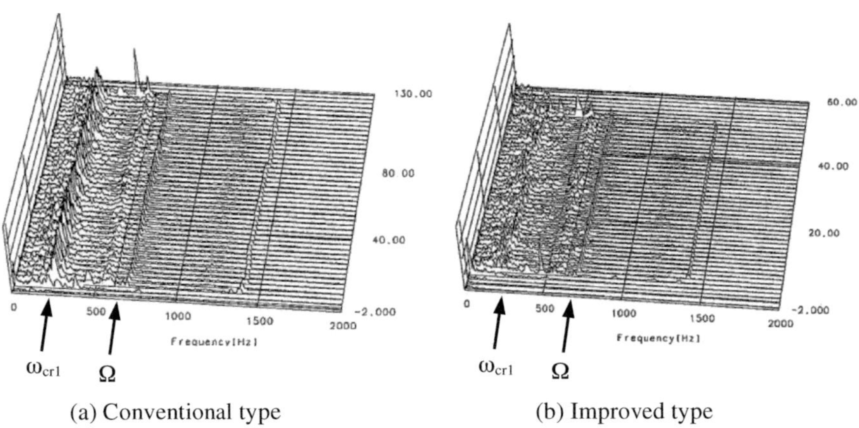

Fig. 23 Shaft vibration in engine firing tests

5 Rotordynamic Problems in Hydraulic Turbines

Severe flexural vibration of the rotor shaft of a Francis turbine generator shown in Fig. 24 was experienced in its test operation and reported by Tomita and Kawamura (1965).

The mechanism of vibration is explained as follows (Ohashi 1991). Figure 25 shows the leakage flow model. As long as the runner remains in the center of the casing, the leakage is uniform around the runner. Once the runner deviates from the

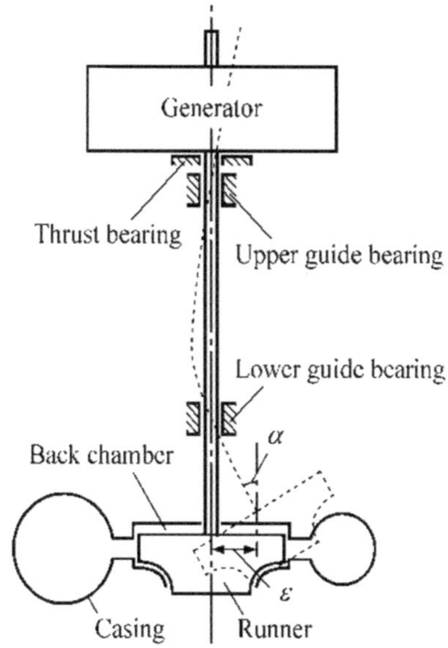

Fig. 24 Flexural vibration of a Francis turbine

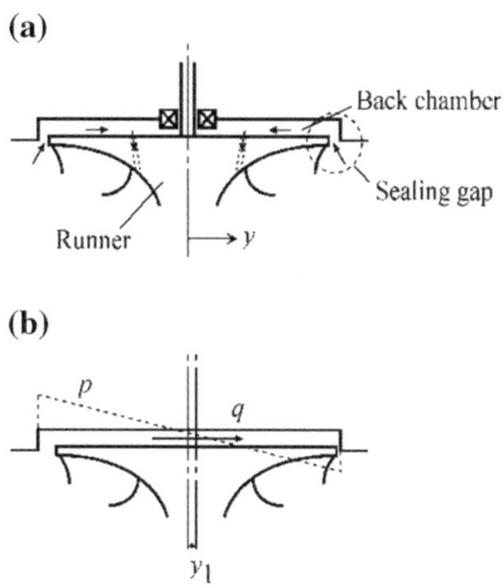

Fig. 25 Leakage flow model

center of the casing by y_1, the uniformity will be broken and the resulting change of the leakage flow, represented by q in Fig. 25 will occur in the back chamber. When the runner vibrates, the variational flow must change its direction accordingly, thus causing the variation of the pressure in the chamber. As the first step of understanding, the rotor is modeled by a plate with an entrance orifice and the leakage flow is modeled by a one-dimensional flow as shown in Fig. 26. We assume that the leakage flow rate q is proportional to the leakage flow velocity U in the entrance orifice and the width B and the height y, i.e., $q = BUy$. The pressure distribution

Fig. 26 One-dimensional flow model

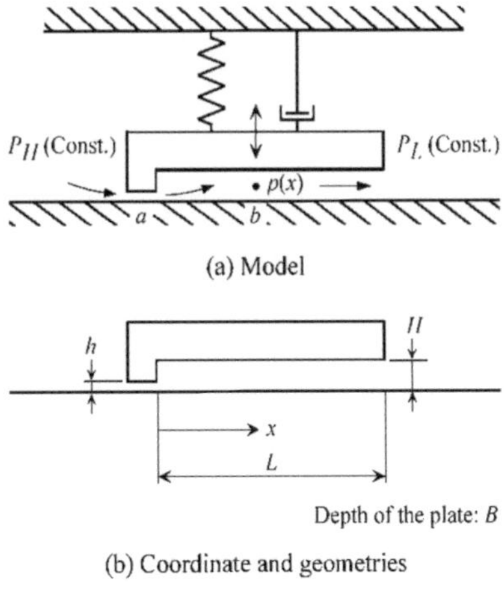

caused by the acceleration of the leakage flow rate is given by $p(x) = \rho(L-x)/(BH) \cdot \dot{q}$. Then the equation of motion of the plate can be expressed by

$$m\ddot{y} + c\dot{y} + ky = \int_0^L p(x)B\,dx = \rho\frac{L^2}{2H}\dot{q} = \rho\frac{L^2}{2H}BU\dot{y} \tag{21}$$

This equation shows clearly that the pressure distribution caused by the acceleration of the leakage flow causes negative damping for the vibration of the plate.

Then, a lateral vibration model of an overhung rotor is proposed as shown in Fig. 27. The upper end is rigidly supported and a bearing is placed at y_2. The lateral force on the rotor is represented by $P_h = A_f\dot{y}_1$ and the moment on the rotor by $M_h = A_m\dot{y}_1$. Here, the influence factors $\xi_{i,j}$ giving the lateral displacement at the node i due to the force at j, and $\eta_{i,j}$ giving the lateral displacement at the node i due to the moment at j are introduced. Then, the equation of motion can be expressed by

$$\begin{aligned} y_1 &= \xi_{11}\left(-M\ddot{y}_1 + A_f\dot{y}_1\right) + \xi_{12}(-c\dot{y}_2) + \eta_{11}(A_m\dot{y}_1) \\ y_2 &= \xi_{21}\left(-M\ddot{y}_1 + A_f\dot{y}_1\right) + \xi_{22}(-c\dot{y}_2) + \eta_{21}(A_m\dot{y}_1) \end{aligned} \tag{22}$$

From these equations, it can be shown that the lateral vibration grows when

$$c\xi_{21}^2 < \left(A_f + \frac{\eta_{11}}{\xi_{11}}A_m\right)\xi_{11}^2 \tag{23}$$

This result shows that fluid moment on the rotor can cause lateral vibration through structural coupling.

Thus, the possible mechanism of the vibration has been shown by the simplified models. However, rotors generally exhibit whirling and precession motion and it is

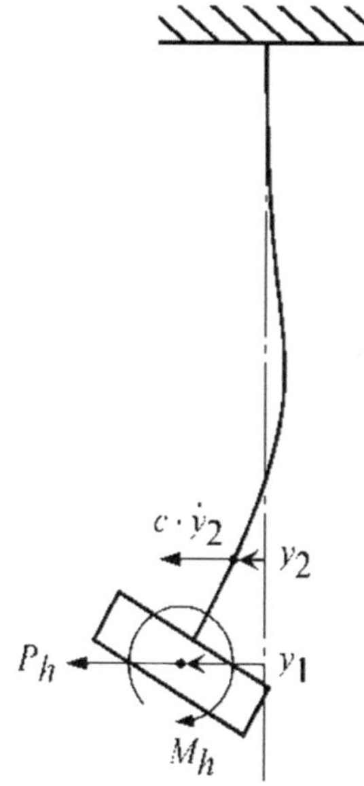

Fig. 27 Lateral vibration model

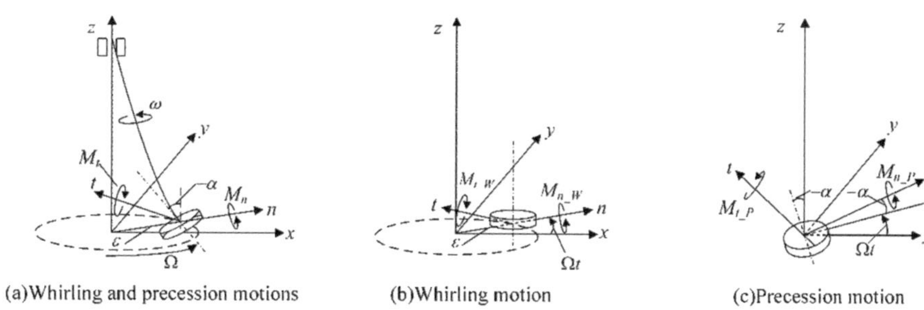

Fig. 28 Coordinate system

not clear under what conditions the self-excited vibration occurs. The rotordynamic analysis presented in Sect. 3 was developed to obtain the criterion of the whirl instability caused by whirl moments.

5.1 Rotordynamic Fluid Force Moment

As shown in Fig. 28, the vibration of an overhung rotor shown in (a) can be decomposed to the whirling motion shown in (b) and the precession motion shown in (c). When the overhung is large compared with the lateral displacement, the displacement due to precession motion can be neglected as compared with the displacement due to whirling motion. This assumption was made also in the

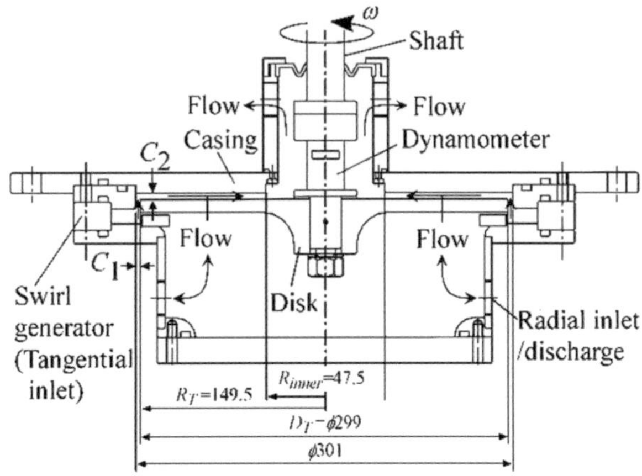

Fig. 29 Schematic of experimental facility

analysis in Sect. 3. The fluid force moment due to the leakage flow was measured on the facility shown in Fig. 29.

The backshroud of the runner is modeled by a disk set close to the casing. The forced whirl motion was given by a special bearing system supporting the shaft and the leakage flow was given by an external pump. Figure 30 shows the whirl moment normalized by the reference moment $M_{0\omega} = \rho|\omega||\varepsilon|/C_2 \cdot \left(v_l \pi R_T^4\right)$ where $C_2 = 4$ mm is the axial clearance, v_l is the mean leakage flow velocity in the radial clearance $C_1 = 1$ mm and $R_T = 149.5$ mm is the radius of the disk (Song et al. 2010a, 2010b). The moment was measured by a force balance attached to the shaft and also evaluated by the unsteady pressure distribution on the casing. The results agree qualitatively. The moment was evaluated also numerically using a bulk flow model (Childs 1989) in which the flow in the clearance is averaged over the clearance and the effect of wall stress is taken into account. Although quantitative difference is found, the tendency can be reproduced by the model. The rotordynamic analysis in Sect. 3 shows that the whirl instability occurs when the normal moment M_n has the same sign as the whirl frequency Ω. The result in Fig. 30 shows that the moment can destabilize both forward and backward whirl at any whirl frequency. This is quite different from the rotordynamic forces which destabilize only forward whirl with the frequency less than a certain percentage of the rotor frequency, as discussed in Sect. 4.

5.2 *Moment Whirl*

To demonstrate the moment whirl, free vibration tests were carried out using a facility similar to that shown in Fig. 2. The length of the main shaft 1000 mm is significantly larger than the seal radial clearance $C_1 = 1$ mm of the disk with the diameter of 299 mm. The maximum axial displacement of the disk 0.15 mm =

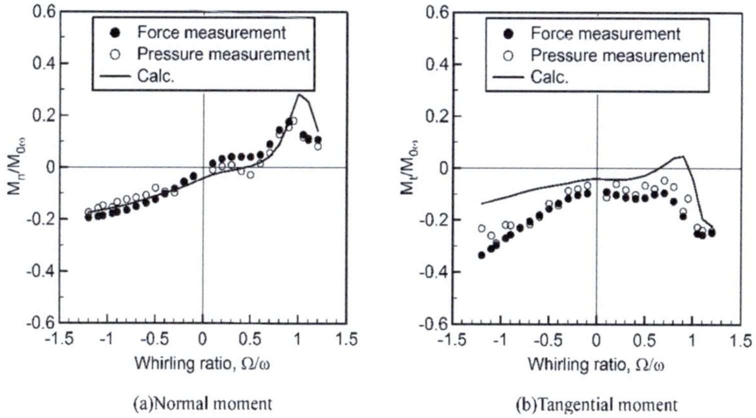

Fig. 30 Fluid force moments obtained by the force sensor, the unnsteady pressure, and the computation

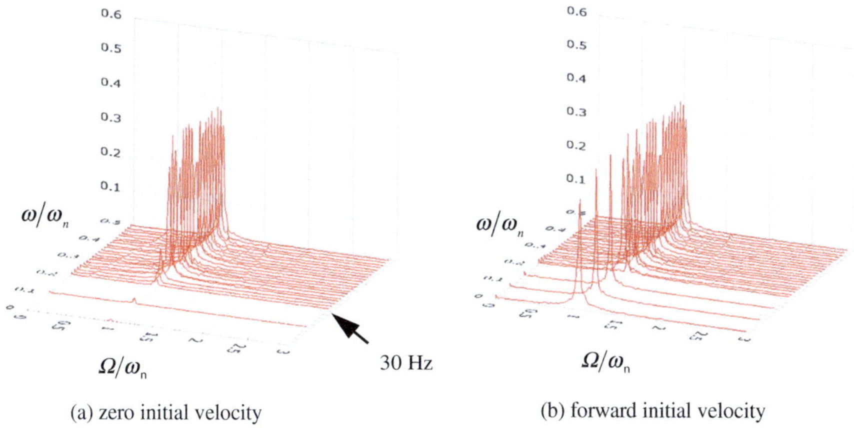

Fig. 31 Spectrum of shaft vibration, Leakage flow rate 9 l/min, natural frequency $\omega_n/2\pi = 2.35$ Hz

150 mm/1000 is significantly smaller than the axial clearance between the disk and the back casing, $C_2 = 4$ mm. This verifies the assumption that the tilting motion of the disk can be neglected as compared with the lateral motion, which was made in the simplified analysis in Sect. 3 as Eq. (5).

The spectrum of shaft vibration with the leakage flow rate of 9 l/min and the natural frequency of the vibration $\omega_n/2\pi = 2.35$ Hz is shown in Fig. 31 for two different initial condition of the shaft vibration.

The results of the test with zero initial velocity without displacement are shown in (a). The rotational speed ω and the frequency Ω of shaft vibration are normalized by the natural frequency ω_n of the shaft. The amplitude of natural frequency component with becomes larger when the rotational speed exceeds 30 rpm or $\omega/\omega_n = 0.21$. The direction of whirl was found to be the same as the disk rotation and the amplitude was the maximum value allowed by the restrictor on the shaft. We can observe the rotational frequency component with $\Omega/\omega_n = \omega/\omega_n$ although the magnitude is much smaller.

The results with the initial velocity in the direction of forward whirl with maximum displacement are shown in (b). Similar results are obtained also with the backward initial velocity. By giving the initial velocity, the vibration with natural frequency with $\Omega/\omega_n = 1$ occurs at all rotational speed ω/ω_n of the disk. The direction of whirl is the same as the initial velocity.

These results show that the moment due to the leakage flow causes both forward and backward whirl, at all rotational speed of the disk. This is consistent with the rotordynamic moment shown in Fig. 30a and is quite different from the rotor whirl due to rotordynamic forces discussed in Sect. 4, in which only forward whirl occurs when the rotational frequency is sufficiently higher than the first critical speed. The whirl direction is the same as that given as the initial condition. When no initial velocity is given, impeller rotation triggers forward whirl if the rotational speed is sufficiently high. This may be caused by the rotordynamic forces due to the interaction of the swirling flow induced by the disk rotation with the uneven clearance due to displacement.

Similar tests were conducted also for outward flow, with inner and outer seals. It was confirmed that the self-excited vibration occurs when the seal is placed at the entrance of the leakage flow, outer seal with inward flow and inner seal with outward flow. However, exit seal can be destabilizing if the seal is designed so that the seal clearance is increased with the deflection with which the clearance of normal design is decreased.

5.3 *Examples of Vibrations in a Large Pump and a Hydraulic Turbine*

Two examples of vibrations at higher frequency than the rotational frequency are introduced here. In the Deriaz pump (Ruud 1976) shown in Fig. 32, violent counter-rotation shaft whirl with the frequency of 3.3–7 Hz occurred during the shutdown transient from the nominal rotating speed of 120 rpm or 2 Hz, after breaking the siphon downstream and the power shutoff. Even when the shaft is stopped, the vibration at 5 Hz was observed although the amplitude is decreased. This is quite different from the rotordynamic problems in high speed centrifugal pumps discussed in Sect. 4 where forward whirl is expected at the rotational speed higher than the first critical speed. Since a rotating pressure pattern was observed in the space between the head cover and the top of the impeller when it is filled with water, air was injected into the space from the ports (a)–(e) in Fig. 32. Injection from ○a and ○e helped to mitigate the vibration. At nominal speed of 120 rpm, it

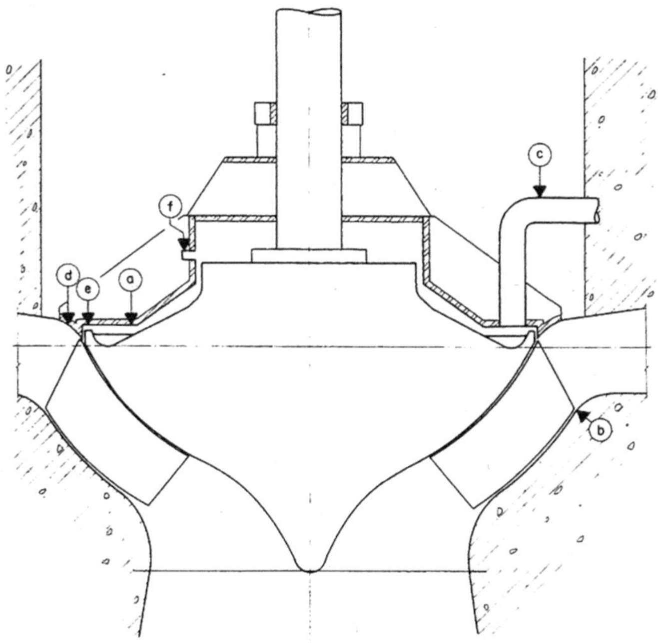

Fig. 32 Deriaz pump

is expected that there is sufficient pumping action to remove water from this space but when the rotational speed decreased during shut down, the pumping action is decreased and the space is flooded resulting in the vibration. Thus, it is quite possible that the vibration was caused by the moments caused by the leakage flow.

Next example is the vibration of a hydraulic turbine runner (Crawford and Ruud 1967) around 11.5–8 Hz, at the rotational speed of 200 rpm or 3.33 Hz. It is stated "The typical whirl was in the direction of rotation, while counter-rotational whirl was relatively rare." The turbine has a fixed plate behind the runner, as shown in Fig. 33. In the space between the fixed plate and the runner, outward flow occurs due to the centrifugal force on the swirling flow in the space driven by the runner. An orifice is placed at the inner radius of the fixed plate. If the runner moves to the right, the clearance is decreased and the pressure in the space would decrease and results in the self-excited vibration as explained in Sect. 5. However, the authors of the paper focus on the disturbance of flow due to the change in the thickness of the gap between the fixed plate and the runner and the gyroscopic effect of the runner. We need to make quantitative analysis to determine which of the mechanisms is more important. Air injection was used also for this case to mitigate the vibration.

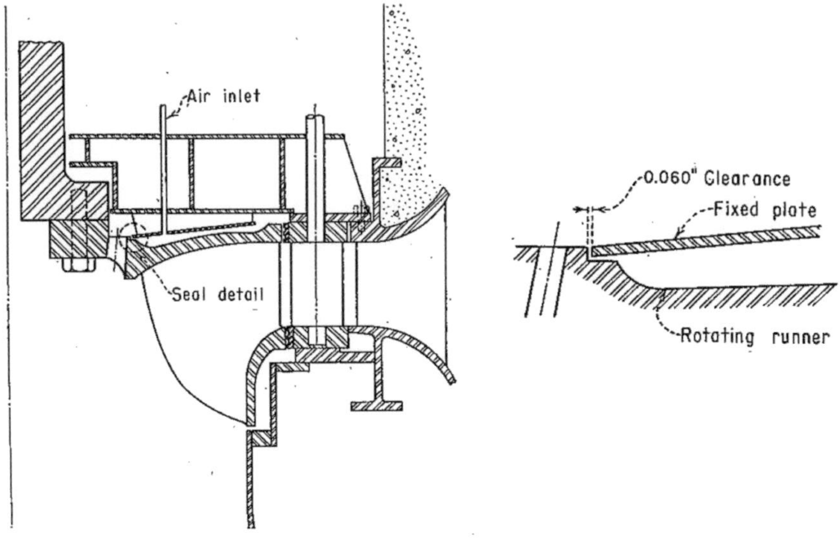

Fig. 33 Deriaz pump

6 Conclusion

Rotordynamic forces on pump impellers and turbine runners are reviewed. In many cases where the leakage flow swirl is important, rotordynamic forces causes forward whirl at supercritical speeds where the rotor is running more than twice the first critical speed. For this case the destabilizing region can be reduced by reducing the swirl velocity. However, when the change of the leakage flow due to displacement is important, the fluid force moment can cause rotor whirl even when the rotor speed is significantly lower than the first critical speed. The instability can be avoided by placing the seal at the exit of the leakage flow or changing the opening/closing convention of the seal associated with the deflection of the rotor.

So far it was difficult to predict the rotordynamic forces by CFD since it is usually small as compared with steady force component. However, it is becoming possible to predict the forces by careful application of CFD. Further progress is expected in this field making use of CFD.

References

Brennen, C. E. (1994). *Hydrodynamics of Pumps*. Concepts ETI, Oxford University Press.

Chamieh, D. S., Acosta, A. J., Brennen, C. E., & Caughy, T. K. (1985). Experimental measurements of hydraulic radial forces and stiffness matrices for a centrifugal pump impeller. *ASME Journal of Fluids Engineering, 107–3,* 307–315.

Childs, D. W. (1989). Fluid structure interaction forces at pump-impeller-shroud surfaces for rotordynamic calculations. *ASME Journal Vibration, Acoustics, Stress and Reliability in Design, 111*, 216–225.

Childs, D. (1993). *Turbomachinery rotordynamics*. New York, USA: John Wiley.

Crawford, C. C., & Ruud, F. O. (1967). Self-excited vibration of a hydraulic turbine. *ASME Journal of Engineering for Power, 1967–10,* 573–576.

Ehrich, F., & Childs, D. (1984). Self-excited vibration in high performance turbomachinery. In *Mechanical Engineering* pp. 66–69.

Ehrich, F. F. (1992). *Handbook of Rotordynamics*. New York, USA: McGRAW-Hill.

Franz, R., & Arndt, N. (1986). Measurements of Hydrodynamic Forces on a Two-Dimensional Impeller and a Modified Centrifugal Pump. In *NASA Report E 249* p. 4.

Guinzburg, A., Brennen, C. E., Acosta, A. J., & Caughy, T. K. (1994). Experimental results for the rotordynamic characteristics of leakage flows in centrifugal pump. *ASME Journal of Fluids Engineering 116*(1) 110–115.

Hiwata, A., & Tsujimoto, Y. (2002). Theoretical analysis of fluid forces on an open-type centrifugal impeller in whirling motion. *ASME Journal of Fluids Engineering, 124*(2), 342–347.

Jery, B., Acosta, A. J., Brennen, C. E., & Caughy, T. K. (1985). Forces on centrifugal pump impellers. In *Proceedings of the Second International Pump Symposium* (pp. 21–32) *Houston, TX, USA*.

Motoi, H., Kitamura, A., Sakazume, N., Uchiumi, M., Uchida, M., Saiki, K., et al. (2003). Sub-synchronous whirl in the le-7a rocket engine fuel turbopump. In *ISCORMA-2, Gdansk, Poland*.

Ohashi, H., & Shoji, H. (1987). Lateral fluid forces on whirling centrifugal impeller (2nd report: experiment in vaneless diffuser). *ASME Journal of Fluids Engineering, 109*(2), 100–106.

Ohashi, H., Imai, H., & Tsuchihashi, T. (1991). Fluid force and moment on centrifugal impellers in precessing motion. In *ASME FE*D (Vol. *119*, pp. 112–115).

Ohashi, H. (1991). *Vibration and oscillation of hydraulic machinery (international hydraulic machinery series)*. Aldershot: Avebury.

Ruud, F. O. (1976). Vibration of deriaz pumps at dos amigos pumping plant. In *ASME Journal of Fluids Engineering*, 674–679.

Shoji, H., & Ohashi, H. (1981). Hydraulic forces on whirling centrifugal impeller (in Japanese). *Trans JSME, B, 47–719,* 1187–1196.

Song, B., Horiguchi, H., Ma, Z., & Tsujimoto, Y. (2010a). Rotordynamic moment on the backshroud of a francis turbine runner under whirling motion. In *ASME Journal of Fluids Engineering*, *132*, 071102_1–071102_9.

Song, B., Horiguchi, H, Nishiyama, Y., Hata, S., Ma, Z., & Tsujimoto, Y. (2010b). Fluid force moment on the backshroud of a francis turbine runner in precession motion. *ASME Journal of Fluids Engineering*, *132*, 051108_1–051108_8.

Song, B., Horiguchi, H., Ma, Z., & Tsujimoto, Y. (2010c). Rotordynamic instabilities caused by the fluid force moments on the backshroud of a francis turbine runner. *International Journal of Fluid Machinery and Systems, 3*(1), 67–79.

Thomas, H. J. (1958). Instabile eigenschwingungen von turbinenfaufern, angefacht durch die spaltstromungen stopfbuschen un beschaufelungen. In *AEG-Sonderdruck*.

Tomita, H., & Kawamura, M. (1965). Self-excited vibration in francis turbines. *Toshiba Review, 20*(8), 787–791. (in Japanese).

Tsujimoto, Y., & Acosta, A. J. (1987). Theoretical study of impeller and/or vaneless diffuser attributed rotating stalls and their effects on whirling instability of a centrigugal impeller. In *IAHR Work Group on the Behavior of Hydraulic Machinery Under Steady Oscillating Conditions*, *Lille, France*.

Tsujimoto, Y., Acosta, A. J., & Brennen, C. E. (1988a). Theoretical study of fluid forces on a centrifugal impeller rotating and whirling in a volute. *ASME Journal of Vibration, Acoustics, Stress and Reliability in Design*, *110*(3), 263–269.

Tsujimoto, Y., Acosta, A. J., & Yoshida, Y. (1988b). A theoretical study of fluid forces on a centrifugal impeller rotating and whirling in a vaned diffuser. *NASA CP, 3026,* 307–322.
Vance, J. M. (1988). *Rotordynamics of turbomachinery*. New York: Wiley.
Yoshida, Y., Tsujimoto, Y., Murakami, Y., & Ito, T. (1992). Self excited and forced vibrations of centrifugal impeller operating at supercritical speed. In *Proceedings of 4th ISROMAC* (pp. 11–23).
Yoshida, Y., Tsujimoto, Y., Ishii, N., Ohashi, H., & Kano, F. (1999). The rotordynamic forces on an open-type centrifugal compressor in whirling motion. *ASME Journal of Fluids Engineering, 121–2,* 259–265.

On the Preliminary Design and Performance Prediction of Centrifugal Turbopumps—Part 1

Luca d'Agostino, Dario Valentini, Angelo Pasini, Lucio Torre, Giovanni Pace and Angelo Cervone

Abstract A reduced-order model for the preliminary design and performance prediction of radial turbopumps is illustrated. The model expresses the 3D, incompressible, inviscid, irrotational flow through helical blades with slow axial variations of their pitch and backsweep angles by superposing a 2D axial vorticity correction to a fully guided forced vortex flow with axisymmetric stagnation velocity in the meridional plane. Application of the relevant governing equations allows for the closed-form definition of the impeller geometry and flowfield in terms of a reduced number of controlling parameters. Mass and momentum conservations are used for coupling the flow leaving the impeller with the 2D reduced-order models of the flow in the diffuser and/or the volute, as well as for the evaluation of the mixing losses in the transfer between successive components of the machine. This information completes the geometric definition of the turbopump and determines its ideal noncavitating performance in accordance with the resulting flowfield. As a consequence of the neglect of viscous effects, the slip factor predicted by the present model exceeds those obtained from theoretical/semi-empirical formulas reported in literature for centrifugal pumps, but correctly captures their trend.

The present work has been supported by the European Space Agency under Contract No. 40001025856/10/NL/SFe. The authors would like to express their gratitude to Dr. Giorgio Saccoccia of ESA-ESTEC for his constant and friendly encouragement.

L. d'Agostino (✉) · D. Valentini · A. Pasini · G. Pace
Università di Pisa, Pisa, Italy
e-mail: luca.dagostino@ing.unipi.it

L. Torre
Alta S.P.a, Ospedaletto (Pisa), Pisa, Italy

A. Cervone
T.U. Delft, Delft, The Netherlands

L. d'Agostino and M.V. Salvetti (eds.), *Cavitation Instabilities and Rotordynamic Effects in Turbopumps and Hydroturbines*, CISM International Centre for Mechanical Sciences 575, DOI 10.1007/978-3-319-49719-8_6

1 Introduction

The range of applications of turbomachines is so wide that even relatively minor gains in their efficiency and performance translate into major economic impacts worldwide (Laskshminarayana 1985). More specifically, in space transportation systems turbopumps represent one of the most crucial components of all primary propulsion concepts powered by liquid-propellant rocket engines, where stringent limitations are associated with the design of high power density, dynamically stable machines capable of meeting the extremely demanding pumping, suction and reliability requirements of the propellant feed systems (Stripling and Acosta 1962). In these applications turbopumps often employ an inducer upstream of the centrifugal stage in order to pressurize the flow sufficiently for the main pump, usually one or more centrifugal stages, to avoid unacceptable cavitation, improve its suction performance, and reduce the pressure and weight of the propellant storage system.

Significant analogies exist between the impeller geometry of centrifugal turbopumps and compressors, as they rely on similar physical phenomena for raising the pressure of the working fluid. In both cases the structural resistance of the blades under the loads imposed by centrifugal and fluid dynamic forces represents the main limiting factor affecting the structural design of these components. However, in centrifugal turbopumps and compressors the relative importance of these forces is reversed because of the widely different densities of their working fluids. Centrifugal forces prevail in radial compressors, allowing for the adoption of a larger number of slender blades. On the other hand, fewer and thicker blades are used in radial turbopumps in order to sustain the higher bending loads generated by liquids (Brennen 1994).

Because of their relative simplicity, inviscid methods have been the first to be developed for describing the flow in turbopumps. They can be broadly classified in streamline curvature, potential, and Euler methods.

Streamline curvature methods essentially derive from the original idea of Wu (1952), of projecting the equations for the steady, ideal flow relative to the impeller on two pseudo-orthogonal meridional and circumferential surfaces. The 3D flow problem is thus split into two coupled two-dimensional flow problems in the hub-to-shroud and blade-to-blade planes, which are then solved by a number of methods, including finite differences, finite elements, and (again) streamline curvature methods (Senoo and Nakase 1971; Bosman and El-Shaarawi 1976; Adler and Krimerman 1978; Hirsch and Warzee 1979).

Potential methods generate the solution of the irrotational ideal fluid equations for the velocity potential, in most cases by means of finite differences, finite elements, or finite volume algorithms in two or three dimensions. They are relatively fast, intrinsically accurate, and can treat unsteady flow problems precisely both in two and three dimensions. However, the hypothesis of irrotationality severely their applicability to turbomachinery since the relative flow in impellers is necessarily rotational, and even for stationary elements inlet flow prerotation is a common occurrence.

Finally, Euler methods use similar numerical approaches to the solution of the ideal fluid equations, without necessarily requiring the flow to be irrotational.

Clearly, all inviscid methods are inherently uncapable to account for real fluid effects and dissipative phenomena such as turbulence, boundary layers, separation, flow reversal, and secondary flows. In order to address these aspects of turbomachinery flows, viscous methods must be used, which can be generally divided into distributed loss, boundary layer, and Navier–Stokes methods.

Distributed loss methods are based on the idea of correcting the inviscid flow models by accounting for dissipative effects in complex turbomachinery flows on an averaged basis, without detailed consideration of the specific mechanisms and locations where energy dissipation actually occurs (Ainley and Mathieson 1951; Dunham and Came 1970; Horlock and Marsh 1971; Bosman and Marsh 1974; Kacker and Okapuu 1982). The appropriate intensity of dissipative effects strongly depends on the operational conditions and must be either deduced from experimental observations or, in more sophisticated models, estimated from the analysis of the various sources of dissipation. Clearly, a sizable amount of information is missed in this intrinsically phenomenological approach. Therefore, distributed loss methods are of little or no use when accurate results are required as, for instance, in direct optimization of turbopump design where the sensitivity of the loss model to minor changes of the controlling parameters is indispensible. However, when supported by adequate experimental data, these models represent an economic and rapid way to include in first approximation the global effects of energy dissipation in turbomachinery analyses.

Boundary layer methods are based on Prandtl's original intuition that viscous effects in unseparated flows at high Reynolds numbers are confined to relatively thin layers of the fluid adjacent to the solid surfaces, while the rest of the flow is virtually inviscid. Therefore, the flow in the two regions can be studied with separate approaches and the solutions matched together in order to obtain an approximate description of the entire flow field. Boundary layers have been studied extensively and a number of reliable boundary integral methods exist for their efficient evaluation in the steady 2D case (White 2006; White and Christoph 1972). The effects of surface roughness, free stream turbulence and, with some uncertainty, turbulent transition can also be included. The problem of steady viscous/inviscid coupling between the two flow regions can either be neglected in the case of weak interaction, or otherwise treated iteratively in order to satisfactorily match the inner and outer flow solutions (Carter 1979; Le Balleur 1981; Whitfield et al. 1981). Boundary layer methods in two dimensions are relatively simple and, in the absence of flow separation, accurate and computationally efficient. On the other hand, extensions to three-dimensional and/or unsteady cases are more difficult and uncertain because of the lack of reliable experimental correlations for solving the integral boundary layer equations and because of the complexity of the computations.

Navier–Stokes methods solve the viscous flow equations, and therefore represent the most general and comprehensive approach to the analysis of turbopump flows. Current methods typically employ either the Reynolds-Averaged Navier–Stokes

(RANS) equations or the large-eddy simulation (LES) equations with a suitable turbulence model, in order to reduce the computational requirements to within affordable levels. A very wide variety of algorithms have been proposed in order to better circumvent computational stability problems of Navier–Stokes solvers and improve their overall efficiency. Nowadays they are extensively used in the simulation of turbopump flows, being the most realistic and promising alternative to direct experimentation in the analysis of complex viscous flow phenomena.

The above methods make use of the machine geometry as an input to evaluate its performance (direct methods), and therefore they do not explicitly provide any guidance for its most efficient design. Their use to this purpose in direct optimization loops is very severely limited by the prohibitive cost of exhaustive nonlinear searches over the large number of free parameters that should conceivably be used to define the geometry of the machine in a sufficiently general way. This is especially the case when the complexity and time requirements of each computation are increased for improving the accuracy of the results. Inverse (or indirect) methods, on the other hand, yield the optimum geometry of the machine under given requirements but, in spite of their theoretical appeal, the difficulties associated with their application to the generation of realistic turbopump geometries still make them rather impractical.

The use of the above methods for the fluid dynamic design of the blading and flow path in centrifugal turbopumps satisfying assigned requirements and specifications typically starts by sizing the main components and evaluating their performance by means of simplified 2D or quasi-1D flow models, possibly with empirical corrections for major sources of flow losses (Laskshminarayana 1985). A first approximation of the machine geometry is then generated and a relatively elaborate optimization process is carried out, usually with the objective of attaining maximum efficiency within the assigned specifications, requirements, and operational constraints. At increasing levels of complexity, this process may involve the use of blade-to-blade and/or hub-to-tip flow models, boundary layer calculations, and three-dimensional inviscid/viscous, steady/unsteady numerical simulations. The real fluid performance of the machine is evaluated from the flow velocity field by accounting for energy losses either explicitly, using empirical correlations of specific databases, or indirectly from boundary layer computations or viscous flow simulations. Finally, the geometries of the blade channels and of the other components are iteratively modified in order to attain the desired velocity/pressure distribution along the flow path, reduce residual spurious and/or detrimental effects, and optimize the overall performance of the machine.

Generally desirable features of practical methods for preliminary design of mixed-flow centrifugal turbopumps are

- the capability of defining the geometry of the machine consistently with the principles of its operation and of predicting its performance in terms of a relatively small number of significant parameters;
- rapid execution times and adequate accuracy and sensitivity to geometrical changes, for more effective iterative optimization of the machine.

The observation that none of the above methods satisfactorily matches these requirements justifies the search for more efficient alternatives. In this context, the development of accurate 3D, closed-form, reduced-order models capable of jointly predicting the geometry and performance of radial impellers and turbopumps is of particular interest to rocket engineers in their search for effective, rapid, and possibly accurate tools for the preliminary design of these machines. However, no such model has been proposed so far, mainly due to the difficulty of adequately describing the 3D flowfield through the impeller. Hence, even today turbopump designers still refer to simple "rules of thumb" or the general indications of specific manuals (Douglass 1973) for the preliminary definition of their machines.

In rocket engine turbopumps the attainment of high power/weight ratios is invariably obtained by running the impeller at the maximum allowable speed and lowest shaft torque. Operation under limited cavitation conditions with lighter—but also more flexible—shafts is therefore tolerated, exposing rocket propellant feed turbopumps to the onset of dangerous fluid dynamic and rotordynamic instabilities (d'Agostino 2013a, b). However, the development of cavitation inside the blade channels should possibly be avoided, because the highly compressible and reverberating nature of the flow would greatly promote the onset of potentially lethal self-sustained flow instabilities at frequencies susceptible to become resonant with the flutter oscillations of the impeller blades. The occurrence of this phenomenon in the inducer of the LE-7 engine has actually been identified as the likely cause of the catastrophic failure of the Japanese H-II launcher in November, 1999 (NASDA 2000a, b). It is therefore advisable to design the impellers, both axial and radial, for cavitation to concentrate on the initial part of the blades at all operational conditions. This can be realized if, for any positive value of the inlet flow incidence, the fully wetted flow pressure on the suction sides of the blades is minimum at the leading edges, and monotonically increasing further downstream. Hence, in particular, the local value of the blade lift vanishes identically at zero incidence, when the stagnation point is located at the leading edge.

These considerations provided the fundamental basis for the selection and justification of the flowfield assumptions adopted by d'Agostino and his collaborators in the development of their closed-form methods for jointly defining the geometry and performance of high-head radial inducers (d'Agostino et al. 2008a, b) and mixed-flow impellers (d'Agostino et al. 2011) optimized for operation under limited cavitation conditions. Specifically, they first showed that in helical inducers the condition of minimum flow pressure at the leading edges, when associated with the preservation of radially uniform axial flow velocity through the rotor (as commonly required in axial turbomachines), results in a functional relation between the local values of the blade angle and hub radius. Later, with a similar approach d'Agostino et al. (2011) demonstrated that, when applied to mixed-flow centrifugal impellers, the same condition links together the axial variations of the blade pitch and backsweep angle.

The present formulation of the preliminary design of mixed-flow turbopumps represents therefore the natural extension of the earlier model developed by some of the authors for the simultaneous geometry definition and noncavitating performance prediction of high-head axial inducers for liquid-propellant rocket engine feed systems. This model has been used for designing several tapered-hub, variable-pitch inducers with different hydrodynamic features, and successfully validated in a series of dedicated characterization experiments (d'Agostino et al. 2008a; Torre et al. 2009, 2011; Cervone et al. 2012; Pace et al. 2013).

More specifically, the present approach is suitable for application to centrifugal pumps with uniform axial inlet flow, variable impeller tip/hub radii, and helical blades with slow axial changes of the pitch and backsweep angles. Following the same approach used in the case of inducers, the 3D incompressible, inviscid, irrotational flow field inside the blade channels is expressed by superposing a 2D cross-sectional axial vorticity correction to a fully guided flow with axisymmetric stagnation velocity in the meridional plane. This choice allows for a radially uniform axial velocity distribution on each plane orthogonal to the axis of the machine, with the additional advantage of providing hub and tip profiles potentially less prone to develop flow separation. Moreover, the assumed flowfield through the blade channels intrinsically accounts for the influence of slip-flow effects, which are known to be one of the major factors adversely affecting the pumping performance of centrifugal turbopumps.

Due to space limitations, the inclusion of flow losses as first proposed d'Agostino et al. (2012) and the results of dedicated validation experiments on a turbopump designed in accordance with the present model are illustrated in a companion paper of the present volume (d'Agostino et al. 2017). Here, for preliminary assessment of the proposed approach, the slip factors predicted by the present model are compared with those obtained from some of the most popular theoretical/semi-empirical formulas reported in the literature for centrifugal pumps (Stodola 1927; Busemann 1928; Wiesner 1967).

2 Turbopump Flow and Geometry

The present model jointly describes the flow and geometry of the three main components of a radial machine: the impeller, the diffuser, and the volute, as schematically indicated in Fig. 1.

The radial impeller transfers energy to the working fluid as a consequence of its rotational speed Ω and the hydrodynamic forces developing on its blades. The geometry of its flow channels is defined by the intersection of N blades, with variable helical and backsweep angles γ_h and χ, and the rotational surfaces generated by the axial profiles of the hub and tip radii, r_H and r_T. The exit flow of the impeller is then collected by a vaneless diffuser with constant axial width b_D, and finally guided into the discharge line by a single-spiral volute with elliptical cross sections, continuously varying in the azimuthal direction from a straight segment at the tongue to a circular cross section with radius R_4 at the exit.

As mentioned in the introduction, the flow through the machine is considered as incompressible, inviscid, and irrotational. For relatively large values of the blade

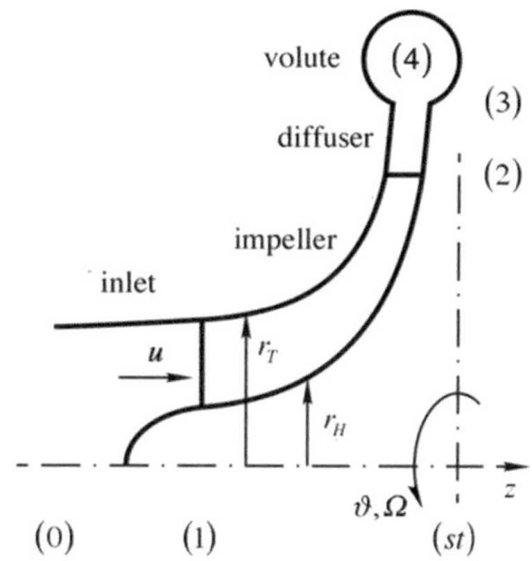

Fig. 1 Radial turbopump schematic and nomenclature

solidity the 3D velocity field inside the blade channels is approximated as the superposition of a fully guided axisymmetric flow and a 2D axial vorticity correction on each impeller cross-sectional orthogonal to the centerline. The meridional component of the fully guided axisymmetric flow is chosen as the velocity field of an incompressible axisymmetric stagnation flow with stagnation plane located at the station (*st*), as illustrated in Fig. 1. Perfect mixing is assumed to take place at the exit of the impeller, so that the flow in the diffuser can be considered steady, axisymmetric, and axially uniform. Finally, steady axisymmetric flow is also assumed in the volute at design flow rate. Consistently with the ideal nature of the present model, mixing losses at the entrance of the diffuser and the volute are neglected, even if they could be evaluated without explicit consideration of real flow effects.

2.1 *Impeller Flow and Geometry*

In the stated assumptions, the ideal flow through the impeller is held by the continuity and irrotationality equations for the velocity $\boldsymbol{u} = \hat{\boldsymbol{u}} + \tilde{\boldsymbol{u}}$, sum of the (rotational) fully guided axisymmetric flow field $\hat{\boldsymbol{u}}$, which generates the volumetric flux through the machine and most of its total head rise, and a 2D cross-sectional slip velocity correction $\tilde{\boldsymbol{u}}$ (Fig. 2), which does not contribute to the flowrate but is necessary to satisfy the irrotationality condition and decreases the fully guided head rise.

Fully Guided Flow and Geometry of the Impeller. With reference to Fig. 3, the azimuthal velocity component of the fully guided flow is expressed by

$$\hat{v}(r,z) = \Omega r - \hat{w}(z)\tan\gamma(r,z) - \hat{u}(r)\tan\chi(z)$$

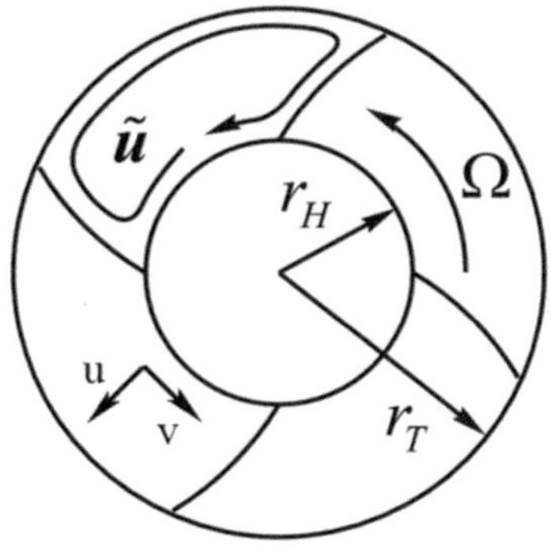

Fig. 2 Schematic of the 2D cross-sectional slip flow in the impeller blade channels

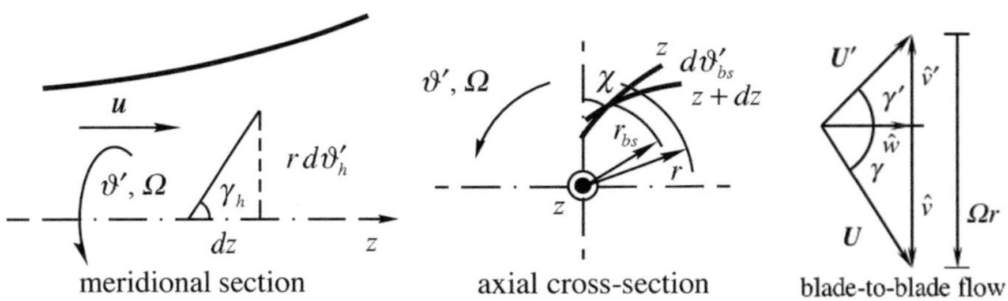

Fig. 3 Helical angle in the meridional plane (*left*), backsweep angle in the axial cross-sectional plane (*center*), and velocity triangle in the blade-to-blade plane (*right*)

in terms of the radial and axial velocities $\hat{u}$ and $\hat{w}$, where the angle γ, measured from the axial direction, includes the effects of the axial variations of the helical pitch P_h and of the logarithmic backsweep angle χ of the blades:

$$\tan\gamma = \frac{2\pi r}{P_h} + r\frac{d}{dz}\left[\ln\left(\frac{r}{r_{bs}}\right)\tan\chi\right] = \frac{2\pi r}{P}$$

In this expression $r_{bs}(z)$ is the radius where the azimuthal coordinate ϑ'_{bs} (associated to the axial variation of χ) vanishes (Fig. 3) and P is an equivalent helical pitch of the blades accounting for the combined axial changes of P_h and χ.

The relatively narrow range of variation of the radial coordinate $r_H \le r \le r_{\max} = \min\{r_T, r_2\}$ in the blade channels justifies the approximation $\ln r \simeq \ln \bar{r}_\chi = \ln\sqrt{r_H r_{\max}}$, so that P is only function of the axial coordinate z. Hence, in particular, the equivalent pitch of the blades reduces to $P \simeq P_h(z)$ when choosing $r_{bs} \simeq \bar{r}_\chi$.

The 2D slip velocity components on each axial cross section of the impeller are most synthetically expressed and solved in terms of a scalar axisymmetric stream function $\tilde{\psi}(r', \vartheta', z')$ in the rotating cylindrical coordinates $r' = r$, $\vartheta' = \vartheta - \Omega t$, $z' = z$.

Moreover, in the proposed approximation the axial variation of the slip flow is neglected, so that $\partial\tilde{\psi}/\partial z \simeq 0$. Imposition to the assumed velocity field $\boldsymbol{u} = \hat{\boldsymbol{u}} + \tilde{\boldsymbol{u}}$ of the irrotationality condition along the radial, azimuthal, and axial coordinates yields

$$\frac{\partial}{\partial z}\left(4\pi\frac{\hat{w}}{P} - \frac{d\hat{w}}{dz}\tan\chi\right) = 0$$

$$\frac{d^2\hat{w}}{dz^2} = 0$$

$$\frac{1}{r}\frac{\partial}{\partial r}\left(r\frac{\partial\tilde{\psi}}{\partial r}\right) + \frac{1}{r^2}\frac{\partial^2\tilde{\psi}}{\partial\vartheta'^2} = 2\Omega - 4\pi\frac{\hat{w}}{P} + \frac{d\hat{w}}{dz}\tan\chi$$

Successive integration of the azimuthal vorticity and continuity equations, with the conditions $\hat{w}(z_1) = \hat{w}_1$ at the inlet Sect. (1), $\hat{w}(z_{st}) = 0$ at the stagnation plane (st) and $\hat{u} = 0$ on the axis, determines the meridional components of the fully guided flow through the impeller:

$$\hat{w}(z) = \hat{w}_1\frac{z - z_{st}}{z_1 - z_{st}}$$

and

$$\hat{u}(r) = -\frac{1}{2}r\frac{d\hat{w}}{dz} = -\frac{1}{2}r\frac{\hat{w}_1}{z_1 - z_{st}}$$

which, in turn, define the profiles of the hub and tip radii $r_H(z)$ and $r_T(z)$ as the corresponding streamlines through r_{H1}, z_1 and r_{T1}, z_1.

The blade surfaces are defined by the azimuthal coordinate ϑ_B' resulting from the combined effects of the helical pitch and backsweep:

$$\vartheta_B' = \vartheta_h' + \vartheta_{bs}' = \vartheta_{B1}' - \int_{z_1}^{z}\frac{2\pi}{P_h}dz - \ln\left(\frac{r}{r_{bs}}\right)\tan\chi$$

where ϑ_{B1}' (possibly function of r) is the azimuthal position of the blade at the inlet section $z = z_1$. The adoption of this blade shape yields a forced vortex flow design of the impeller with

$$\hat{v}' = \hat{v} - \Omega r = -\frac{2\pi r}{P}\hat{w} + \frac{1}{2}r\frac{d\hat{w}}{dz}\tan\chi \sim r$$

It is worth noting that the radial component of the vorticity equation implies that the azimuthal components of the fully guided flow velocities do not depend on the axial coordinate:

$$\frac{\partial}{\partial z}\left(4\pi\frac{\hat{w}}{P}-\frac{d\hat{w}}{dz}\tan\chi\right)=0 \quad\Rightarrow\quad \frac{\partial\hat{v}}{\partial z}=\frac{\partial\hat{v}'}{\partial z}=0$$

Furthermore, integration of this equation in z with the pertinent initial condition at the inlet station (1) determines the backsweep angle $\chi(z)$:

$$\tan\chi=\tan\chi_1+\frac{4\pi}{d\hat{w}/dz}\left(\frac{\hat{w}}{P}-\frac{\hat{w}_1}{P_1}\right)$$

as a function of the axial schedule of the blade pitch $P(z)$ (or viceversa). For simplicity, in the present analysis a cubic variation of $1/P$ has been assumed, with

- finite inlet pitch (P_1) and vanishing first and second axial derivatives of $\tan\chi$ at the blade leading edge ($z=z_1$);
- infinite pitch ($1/P_{st}=0$) at the stagnation plane $z=z_{st}$.

The first condition assures a second-order smooth transition of the blading into a helical surface of assigned pitch at the impeller eye, while the second has been imposed to control the angle of the blade root w.r.t. the hub surface at the rotor exit. Although these conditions seemed suitable for first validation of the proposed model, they have not been optimized and further refinements can possibly lead to better results in the generation of more efficient and realistic impeller bladings.

Impeller Slip Flow. The Poisson's boundary value problem for the stream function $\tilde{\psi}(r,\vartheta')$ of the slip flow on each axial cross section of the blade channels (see Fig. 4), together with the impermeability condition $\tilde{\psi}=0$ on the boundaries, is transformed in a rectangular domain, where it can be solved in closed form by standard spectral methods (Hildebrand 1976).

In particular, using as comparison functions the (orthogonal) eigenfunctions of the corresponding homogeneous problem for the Laplace's equation, the solution writes

$$\tilde{\psi}=\sum_{m=1}^{+\infty}\sum_{n=1}^{+\infty}C_{m,n}\sin\left[n\pi\frac{\ln(r/r_H)}{\ln(r_{max}/r_H)}\right]\sin\frac{(2n-1)N\left(\vartheta'-\vartheta'_{Bk}\right)}{2}$$

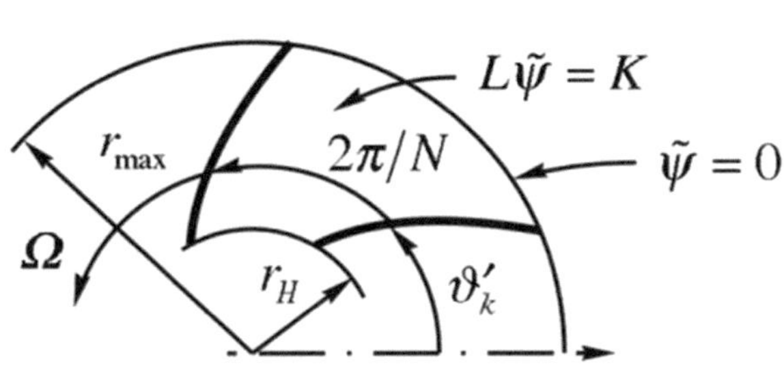

Fig. 4 Blade channel cross section

with

$$C_{m,n} = -\frac{\left[2\Omega - 4\pi\hat{w}_1/P_1 + (d\hat{w}/dz)\tan\chi_1\right]r_H^2 m\left[1-(-1)^m r_{\max}^2/r_H^2\right]}{(n-\frac{1}{2})\left[1+m^2\pi^2/\ln^2(r_{\max}^2/r_H^2)\right]\left[m^2\pi^2(n-\frac{1}{2})^2N^2\ln^2(r_{\max}/r_H)/\cos^2\chi\right]}$$

From this solution for $\tilde{\psi}(r', \vartheta')$ the radial and tangential slip velocity components are readily computed, thus completing the definition of the impeller flow.

2.2 *Diffuser Flow and Geometry*

In the stated assumptions, the steady, axisymmetric, axially uniform flow in the diffuser is held by the continuity and momentum equations. In order to satisfy the condition of axially uniform flow in the diffuser, perfect mixing of the impeller discharge flow is assumed to occur at the diffuser inlet, with flow velocity equal to the mass-averaged velocity at rotor discharge, station (2). The radial and azimuthal components of the velocity inside the diffuser can then be computed by radially integrating the continuity and azimuthal momentum equations in cylindrical coordinates. The resulting flow inside the diffuser is characterized by constant axial velocity and log spiral streamlines with flow angle (φ_D) with respect to the radial direction in a cross-sectional plane. Finally, the flow streamlines through the hub and tip surfaces of the impeller at its discharge station (2) define the lateral surfaces of the diffuser in the meridional plane.

2.3 *Volute Flow and Geometry*

With reference to Fig. 5, the geometry of the spiral volute is designed for alignment of the tongue with the flow leaving the diffuser at nominal operational conditions and smooth transition to azimuthal flow discharge at the exit section, station (4).

The meridional cross sections of the volute are assumed to be segments of ellipses of width $b_3 = b_D$ at the diffuser exit radius r_3, transitioning from a (degenerate) straight segment at the tongue ($\vartheta = 0$) to a segment of a circle of radius R_4 at the exit of the volute (station 4, $\vartheta = 2\pi$), as schematically illustrated in Fig. 6.

In the assumption of axisymmetric flow in the volute at design flow conditions, integration of the azimuthal momentum equation results in a free vortex distribution of the azimuthal velocity component. The volumetric flux can then be computed on each meridional cross section of the volute and equated to the radial inlet flow rate from the diffuser. If, in addition, the external radius r_V of the volute is assigned, the above condition fully defines the geometry of the volute as a function of the azimuthal angle ϑ. Here, in particular, the following expression for

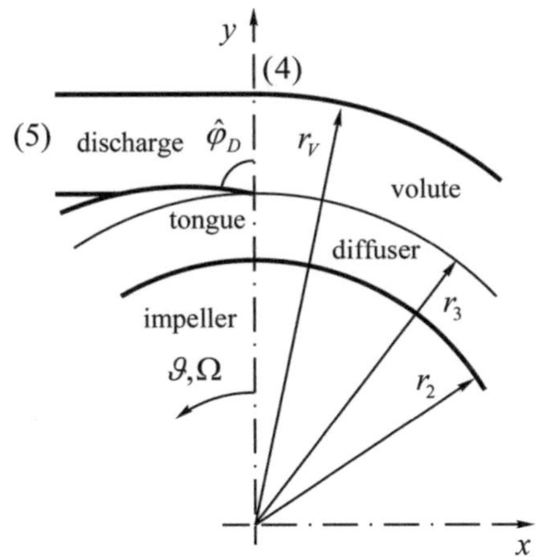

Fig. 5 Volute schematic and nomenclature

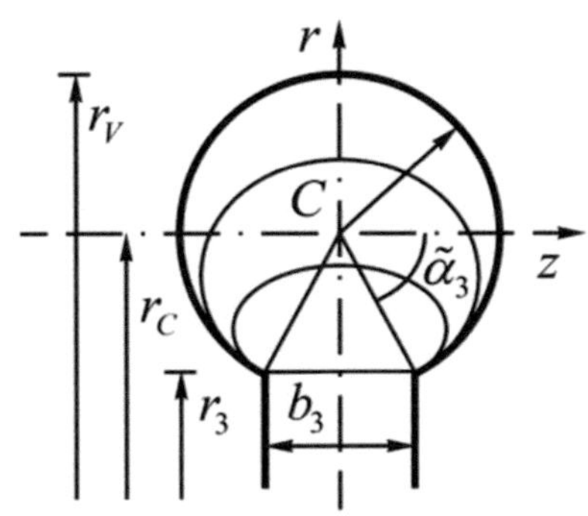

Fig. 6 Elliptical meridional cross sections of the volute

$$r_V = r_3 \exp\left\{\frac{1}{\tan\varphi_D}\left[\vartheta - \frac{\vartheta^{\kappa+1}}{(2\pi)^\kappa(1+\kappa)}\right]\right\}$$

with

$$\kappa = \left[1 - \frac{2\pi}{\tan\varphi_D \ln(r_{V4}/r_3)}\right]^{-1}$$

has been used, corresponding to a smooth variation of the spiral angle β_V of the volute from φ_D at the tongue ($\vartheta = 0$) to zero at the exhaust cross section ($\vartheta = 2\pi$).

The average flow velocities in the volute have been approximated as

$$\tan\bar{v} = u_3 \frac{A_3}{A_4}\frac{r_{V4}}{r_V} \quad \text{and} \quad \bar{u} = \bar{v}\tan\beta_V$$

Finally, in the assumption of a perfect mixing in the outlet duct, the flow at the discharge section (station 5) of the machine is taken uniform with velocity $u_5 = \dot{V}/(\pi R_4^2)$.

3 Turbopump Performance

In the present work the pumping performance of the machine has been evaluated neglecting all sources of energy dissipation, consistently with the ideal flow assumption originally introduced to solve for the machine geometry and flowfield. Clearly, the major contributions to fluid dynamic losses (typically arising from viscous and secondary flow effects, turbulent mixing, and flow incidence at the leading edges of wetted surfaces) play a crucial role for realistic performance prediction and optimization of turbopumps. Their inclusion is illustrated in detail in a companion paper of the present volume (d'Agostino et al. 2017), where it is shown to lead to excellent predictions of the measured performance of the machine over a wide range of operation above and below design conditions.

Under the stated assumptions, the impeller is the only element affecting the total pressure of the flow through the machine. Hence, in the reference frame r, ϑ', z rotating with the impeller the (steady) pressure of the (absolutely) irrotational flow in the blade channels is obtained from the Bernoulli's equation:

$$p + \frac{1}{2}\rho \boldsymbol{u}' \cdot \boldsymbol{u}' - \frac{1}{2}\rho \boldsymbol{r} \cdot \boldsymbol{r} = p_{t0}$$

where $\boldsymbol{u}' = \boldsymbol{u} - \boldsymbol{\Omega} \times \boldsymbol{r}$ is the relative velocity and p_{t0} is the total pressure on the centerline ($\boldsymbol{r} = 0$) at the upstream station (0).

In the absence of losses the total pressure of the flow downstream of the impeller is uniform. Therefore, using the Euler's equation, the pressure in the diffuser and the volute is computed from the relevant flow velocity $\boldsymbol{u}$ by means of

$$p + \frac{1}{2}\rho \boldsymbol{u} \cdot \boldsymbol{u} = p_{t0} + \rho \Omega r_2 \bar{v}_2 = p_{t5}$$

where $\bar{v}_2$ is the mass-averaged azimuthal velocity at the impeller discharge.

Finally, with the above results the total head coefficient is expressed by

$$\Psi = \frac{p_{t5} - p_{t0}}{\rho \Omega^2 r_2^2}$$

In summary, under the stated assumptions and approximations the requirement for axisymmetric stagnation flow in the meridional plane at design conditions determines the profiles of the impeller tip and hub radii, while the irrotationality of the flow completes the geometric definition of the blades by specifying the dependence between the axial schedules of their backsweep and helical pitch angles. The relevant conservation equations and assumptions also parametrically determine the shape of the remaining components (diffuser and volute) and the total head rise of the machine. Both the geometry and performance of the turbopump have therefore been defined in terms of a reduced number of controlling parameters.

4 Model Discussion

In order to illustrate the versatility of the proposed design approach, three examples of pump geometries, indicated as A, B, and C, and representative of mixed-flow machines with different shapes and number of blades have been generated for suitable inputs of the free design parameters of the model. The 3D renderings of the three pump samples A, B, and C and the meridional cross sections of their impellers are shown in Figs. 7, 8, and 9, while Table 1 summarizes their main geometrical characteristics.

Fig. 7 3D rendering of the impeller of the first sample of centrifugal pump geometry (A, *left*) and meridional cross section of the impeller (*right*)

Fig. 8 3D rendering of the second sample of centrifugal pump geometry (B, *left*) and meridional cross section of the impeller (*right*)

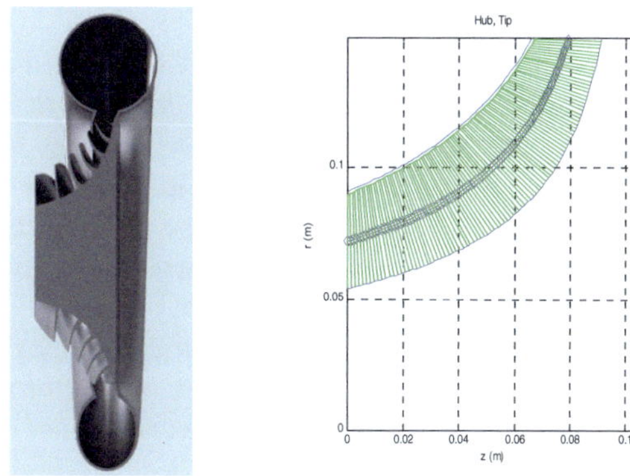

Fig. 9 Cut-out rendering of the third sample of centrifugal pump geometry (C, *left*) and meridional cross section of the impeller (*right*)

Table 1 Main characteristics of the three sample geometries of radial turbopumps obtained by means of the present model

Pump geometry	A	B	C
Number of blades	6	8	10
Inlet tip radius (mm)	71.7	60	90
Inlet hub radius (mm)	46	24	54
Stagnation plane (mm)	91.7	60	105
Inlet tip blade angle (deg)	53.1	54	63
Impeller discharge radius (mm)	150	150	150
Diffuser width (mm)	12.3	8.1	24.2
Diffuser discharge radius (mm)	180	195	165

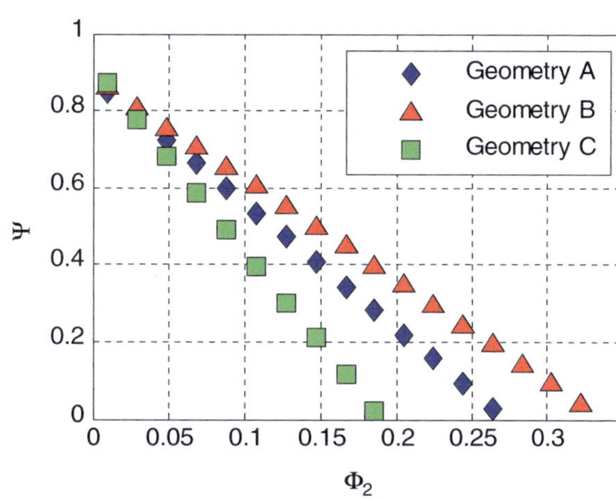

Fig. 10 Comparison between the predicted noncavitating performance of the sample turbopump geometries *A*, *B*, and *C*

Figure 10 illustrates the ideal pumping performance of the sample machines. Since all sources of losses have been neglected, the predicted head characteristics exhibit the linear trend typical of the ideal performance of turbopumps as function of the flow rate.

In the present inviscid flow approximation, the deviation of the flow leaving the impeller w.r.t. the exit angle of the blades is the major source of head degradation in centrifugal turbopumps. In particular, flow deviation at the impeller outlet is mainly due to the slip flow effect generated by the irrotationality condition and the imperfect guidance of the flow in the blade channels, especially at higher loads (Peterson and Hill 1992). Of these two effects, the former is actually predicted by the present model and can be assessed here, while the latter is rather small in high solidity bladings like those typical of centrifugal impellers with large backsweep angles, and in first approximation can be neglected.

In order to preliminarily assess the performance of the ideal flow model before inclusion of fluid dynamic losses and final validation against experimental data from relevant turbopump geometries, the predicted slip factors have been compared to the results of the theoretical/semi-empirical methods reported in literature for centrifugal pumps (Stodola 1927; Busemann 1928; Wiesner 1967; Dixon 1978; Ferguson 1963; Wislicenus 1947).

In particular, as shown in Figs. 11 and 12, the comparison has been focused on the influence of the main parameters affecting the slip factor, such as the number of blades and the exit backsweep angle. The results indicate that the model usually slightly overestimates the slip factor at low number of blades and small exit backsweep angle. However, it is worth noticing that the geometry obtained by the model is more complex than the infinitely thin, logarithmic spiral blades used by Busemann to compute his theoretical slip factors. Moreover, the approximate formula proposed by Stodola provides a reasonable first approximation of the more exact results of Busemann, but underestimates the slip factor when applied to radial impellers with a small number of blades. Both of these methods do not account, even indirectly, for viscous effects:

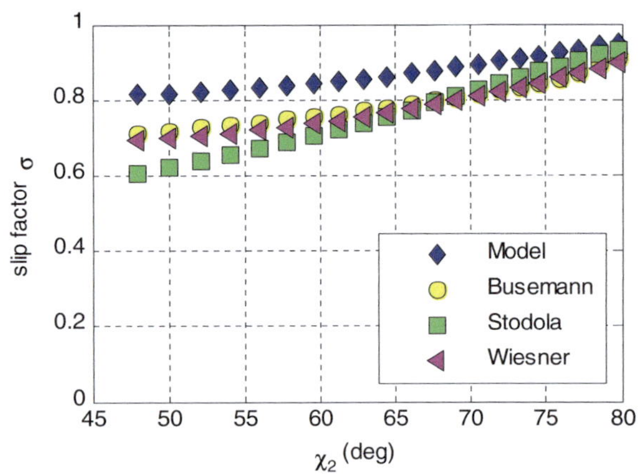

Fig. 11 Comparison between the slip factor computed from the model and the available semi-empirical formulas for radial impellers based on the hub and tip profiles of sample geometry C with variable χ_2

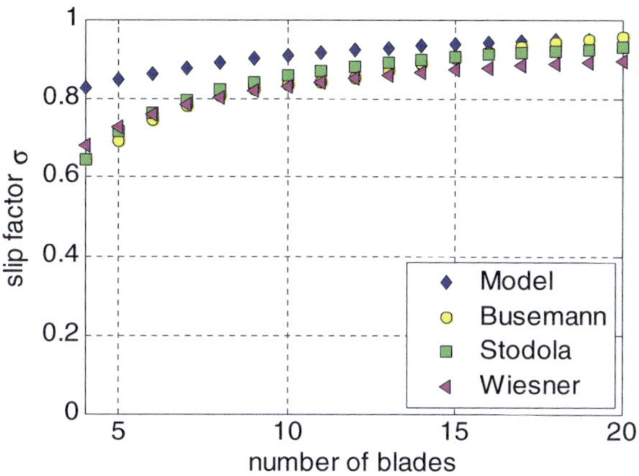

Fig. 12 Comparison between the slip factor computed from the model and the available semi-empirical formulas for radial impeller geometries based on geometry C with variable number of blades

$$\sigma_B = \frac{A_B + B_B\phi_2 \tan\chi_2}{1-\phi_2\tan\chi_2} \quad \text{Busemann}$$

$$\sigma_S = 1 - \frac{0.63\pi/N}{1-\phi_2\tan\chi_2} \quad \text{Stodola}$$

$$\sigma_W = 1 - \frac{\sqrt{\cos\chi_2}/N^{0.7}}{1-\phi_2\tan\chi_2} \quad \text{Wiesner}$$

On the other hand, the approach proposed by Wiesner is based on empirical slip factor data. It provides better agreement with the experimental measurements and implicitly accounts for the additional contributions of viscous effects. Figures 11 and 12 show that the predictions of the present model well reproduce the trend of Wiesner's results but, not surprisingly, are systematically higher because of the neglect of viscous effects. The introduction of the fluid viscosity decreases the value of the slip factor as a consequence of the acceleration and the additional deviation of the relative flow at the rotor exit induced by blockage and asymmetric boundary layer displacement effects on the two sides of the impeller blades. Both of these effects produce in fact the systematic reduction of the azimuthal component of the absolute flow velocity at the exit of the impeller, and therefore of the head developed by the machine, justifying the observed discrepancy between Wiesner's semi-empirical data and the theoretical results of the present inviscid flow model.

5 Conclusions

The present theoretical model proved to be capable of rapidly and efficiently providing quantitative indications for the geometry definition, the 3D flowfield description, and the prediction of the ideal noncavitating pumping characteristics of radial turbopumps with complex and realistic geometries in terms of a relatively small number of controlling parameters. As illustrated in detail in the companion paper of the present volume (d'Agostino et al. 2017), because of these features the present inviscid flow model is especially suited for easy inclusion of the major forms of flow losses in centrifugal turbopumps and integration in a parametric optimization procedure, with the purpose of generating an effective tool for rapid identification of the machine geometry and performance best satisfying the given set of requirements, specifications, and constraints.

The limitations of the model are mostly related to the ideal flow assumption and simplifying approximations introduced in order to attain a practical closed-form solution. The experimental validation presented in the companion paper of the present volume (d'Agostino et al. 2017) clearly demonstrates that these limitations can be effectively removed and that major improvements can be gained by introducing the main sources of flow losses in centrifugal turbopumps. Hence, the capability of the present ideal flow model of rapidly and rationally defining the shape of the machine confirms its effectiveness as a design tool of radial turbopumps.

References

Adler, D., & Krimerman, Y. (1978). The complete 3-dimensional calculation of the compressible flow field in turbo impellers. *Journal of Mechanical Engineering Science*, *20*, 149.

Ainley, D. G., & Mathieson, G. C. R. (1951). A method of performance estimation for axial flow turbines. *British ARC, R&M, 2974,* 923–952.

Bosman, C., & Marsh, H. (1974). An improved method for calculating the flow in turbomachines, including a consistent loss model. *Journal of Mechanical Engineering Science*, 16, 23–31.

Bosman, C., & El-Shaarawi, M. A. I. (1976). *Quasi-three-dimensional numerical solution of flow in turbomachines*. ASME Paper 76-FE-23.

Brennen, C. E. (1994). *Hydrodynamics of pumps*. Oxford University Press.

Busemann, A. (1928). *Das Förderhöhenverhältnis Radialer Kreiselpumpen mit Logarithmisch-Spiraligen Schaufeln*. Weinheim: WILEY-VCH Verlag GmbH & Co. KGaA.

Carter, J. E. (1979). *A new boundary layer inviscid interaction technique for separated flow*. AIAA Paper 79–1450.

Chung-Hua, W. (1952). *A general theory of three-dimensional flow in subsonic and supersonic turbomachines of axial*. NACA, TN: Radial and Mixed Flow Types. 2604.

Cervone, A., Pace, G., Torre, L., Pasini, A., Bartolini, S., Agnesi, L., & d'Agostino, L. (2012). Effects of the leading edge shape on the performance of an axial three bladed inducer. In *14th international symposium on transport phenomena and dynamics of rotating machinery*, ISROMAC-14, Honolulu, HI, USA.

d'Agostino, L., Pasini, A., Valentini, D., Pace, G., Torre, L., & Cervone, A. (2012). A reduced order model for optimal centrifugal pump design. In *14th international symposium on transport phenomena and dynamics of rotating machinery*, ISROMAC-14, February 27th–March 2, Honolulu, HI, USA.

d'Agostino, L. (2013a). Turbomachinery developments and cavitation. In *STO-AVT-LS-206, paper NBR 12–1, VKI Lecture series on fluid dynamics associated to launcher developments, von Karman institute of fluid dynamics*, Rhode-Saint-Genèse, Belgium, April 15–17.

d'Agostino, L. (2013b). On the hydrodynamics of rocket propellant engine inducers and turbopumps. In *6th international conference on pumps and fans with compressors and wind turbines* (IPCF 2013), Sep. 19–22, Beijing, China.
d'Agostino, L., Torre, L., Pasini, A., & Cervone, A. (2008a). On the preliminary design and noncavitating performance of tapered axial inducers. *ASME Journal of Fluids Engineering*, *130*(11), 111303-1/111303-8.
d'Agostino, L., Torre, L., Pasini, A., Baccarella, D., Cervone, A., & Milani A. (2008b). A reduced order model for preliminary design and performance prediction of tapered inducers: comparison with numerical simulations. In *AIAA Paper 5119, 44th AIAA/ASME/SAE/ASEE joint propulsion conference and exhibit*, Hartford, CT, USA, July 21–23.
d'Agostino, L., Pasini, A., & Valentini, D. (2011). A reduced order model for preliminary design and performance prediction of radial turbopumps. In *47th AIAA/ASME/SAE/ASEE joint propulsion conference and exhibit*, July 31–August 3, San Diego, California, USA.
d'Agostino, L., Valentini, D., Pasini, A., Torre, L., & Pace, G. (2017). On the preliminary design and performance prediction of centrifugal turbopumps—Part 2. In this volume, CISM Courses and Lectures No. 1408. In L. d'Agostino & M. V. Salvetti (Eds.), *International centre for mechanical sciences*. Vien and New York: Springer.
Dixon, S. (1978). *Fluid mechanics*. Thermodynamics of Turbomachinery: Pergamon Press.
Douglass, H. W. (1973). *Liquid rocket engine centrifugal flow turbopumps*. NASA SP-8109.
Dunham, J., & Came, P. M. (1970). Improvements of the ainley/mathieson method of turbine performance prediction. *ASME Journal of Engineering for Power*, 252–270.
Ferguson, T. B. (1963). *The centrifugal compressor stage*. London: Butterworths.
Hildebrand, F. B. (1976). *Advanced calculus for applications*. Prentice Hall.
Hirsch, C., & Warzee, G. (1979). An integrated quasi-3d calculation program for turbomachinery flows. *ASME Journal of Engineering for Power, 101,* 141.
Horlock, J. H., and Marsh, H. (1971). Flow models for turbomachines. *Journal of Mechanical Engineering and Science*, *13*, 358–368.
Kacker, S. C., & Okapuu, U. (1982). A mean line prediction method for axial flow turbine efficiency. *ASME Journal of Engineering for Power, 104,* 111–119.
Laskshminarayana, B. (1985). *Fluid dynamics and heat transfer of turbomachinery*. New York, USA: Wiley.
Le Balleur, J. C. (1981). Strong matching method for computing transonic viscous flows including wakes and separations; lifting airfoils. In *La Recherche Aerospatiale*, No. 1981-3, English ed. (pp. 21–45).
NASDA. (2000a). *Report No. 94*, May 2000.
NASDA (2000b). *Report No. 96*, June 2000.
Pace, G., Torre, L., Pasini, A., Valentini, D., & d'Agostino, L. (2013). Experimental characterization of the dynamic transfer matrix of cavitating inducers. In *49th AIAA/ASME/SAE/ASEE joint propulsion conference*, San Jose, California, USA.
Peterson, C., & Hill, P. (1992). *Mechanics and thermodynamics of propulsion*. Reading, (MA) USA: Addison—Wesley Publishing Company.
Senoo, Y., & Nakase, Y. (1971). *A blade theory of an impeller with an arbitrary surface of revolution*. ASME Paper 71-GT-17.
Stodola, A. (1927). *Steam and gas turbines—volumes I and II*. New York: McGraw-Hill.
Stripling, L., & Acosta, A. (1962). Cavitation in turbopumps—Part 1. *ASME Journal of Basic Engineering*, *84*, 326–338.
Torre, L., Pasini, A., Cervone, A., & d'Agostino, L. (2009). Experimental performance of a tapered axial inducer: comparison with analytical predictions. In *AIAA paper 2009-4955, 45th AIAA/ASME/SAE/ASEE joint propulsion conference and Exhibit*, Denver, Colorado, USA, Aug. 2–5.
Torre, L., Pasini, A., Cervone, A., & d'Agostino, L. (2011). Experimental characterization of the rotordynamic forces on space rocket axial inducers. *ASME Journal of Fluids Engineering*, *133*(10).
White, F. M. (2006). *Viscous fluid flow* (3rd ed.). New York: USA, McGraw-Hill.

White, F. M., & Christoph, G. H. (1972). A simple theory for the two-dimensional compressible boundary layer. *ASME Journal of Basic Engineering, 94,* 636–642.
Whitfield, D. L., et al. (1981). Calculation of turbulent boundary layers with separation and viscous-inviscid interaction. *AIAA Journal, 19,* 1315–1322.
Wiesner, F. J. (1967). A review of slip factors for centrifugal impellers. *ASME Journal of Engineering for Power, 89,* 558–576.
Wislicenus, G. F. (1947). *Fluid mechanichs of turbomachinery*. New York: McGraw Hill.

On the Preliminary Design and Performance Prediction of Centrifugal Turbopumps—Part 2

Luca d'Agostino, Dario Valentini, Angelo Pasini, Lucio Torre, Giovanni Pace and Angelo Cervone

Abstract The ideal flow model for the preliminary design and performance prediction of radial turbopumps presented in the companion paper of the present volume (d'Agostino et al. 2017) is here interfaced with the calculation of the boundary layers inside the blade channels and other major forms of flow losses, with the aim of developing an effective tool for rapid parametric optimization of the machine performance and geometry under appropriate design constraints, such as assigned values of the specific speed, flow coefficient, and blade solidity. A mixed-flow turbopump, with a six-bladed impeller, a vaneless diffuser, a single-spiral volute and nondimensional performance characteristics similar to those typically used in liquid propellant rocket engine feed systems, has been designed, parametrically optimized and manufactured in accordance with the indications of the present model. The pumping and suction performance of the machine has been determined in a series of tests in the cavitating pump rotordynamic test facility (CPRTF). Under fully wetted flow conditions the measured pumping characteristics of the machine (hydraulic head and efficiency as functions of the flow coefficient) proved to be in excellent agreement with the model predictions, thus successfully confirming the validity of the proposed model as an effective tool for rapid and efficient design of high-performance centrifugal turbopumps.

The present work has been supported by the European Space Agency under Contract No. 40001025856/10/NL/SFe. The authors would like to express their gratitude to Dr. Giorgio Saccoccia of ESA-ESTEC for his constant and friendly encouragement.

L. d'Agostino (✉) · D. Valentini · A. Pasini · G. Pace
Dipartimento di Ingegneria Civile ed Industriale, Università di Pisa, Pisa, Italy
e-mail: luca.dagostino@ing.unipi.it

L. Torre
Alta S.p.A, Ospedaletto, Pisa, Italy
e-mail: luxtorre@gmail.com

A. Cervone
T.U. Delft, Delft, The Netherlands
e-mail: A.Cervone@tudelft.nl

L. d'Agostino and M.V. Salvetti (eds.), *Cavitation Instabilities and Rotordynamic Effects in Turbopumps and Hydroturbines*, CISM International Centre for Mechanical Sciences 575, DOI 10.1007/978-3-319-49719-8_7

1 Introduction

Performance, simplicity, reliability, operating range and cost-effectiveness are some of the assets that make centrifugal pumps a very popular component of an almost endless variety of liquid management systems. More specifically, high power-density turbopumps currently represents the most weight-effective solution for liquid propellant feed systems of rocket engines for primary space propulsion. These applications are in fact characterized by the joint requirements for high total and specific impulses, where large amounts of usually cryogenic propellants are to be injected at high pressures in the engine's combustion chamber. Under these conditions, turbopump-fed systems are preferable, because the mass reduction allowed by the use of low-pressure tanks far outbalances the additional weight of the machine.

Typically, in modern liquid propellant rocket engine feed systems an axial inducer is also present upstream of the main pump in order to increase its suction performance, while the desired delivery pressure is mostly generated by a suitable number of mixed-flow centrifugal stages. Severe limitations are associated with the design of the high power-density, dynamically stable turbomachines capable of meeting the extremely demanding pumping, suction, and reliability requirements of modern space transportation systems (Stripling and Acosta 1962). In these applications, operation under limited cavitation conditions is usually tolerated in order to minimize the weight of the propellant storage system, exposing rocket engine feed turbopumps to the onset of dangerous fluid dynamic and rotordynamic instabilities (d'Agostino 2013a, b).

The design of centrifugal turbopumps determines about 80% of their final cost (Busby et al. 2008), and therefore its importance cannot be underestimated. Usually, it starts with the development a 2D flow model, which is progressively refined in a sequence of iterative trades between architectural, fluid dynamic, mechanical, structural, rotordynamic, manufacturing, and assembling aspects of the pump, until the relevant requirements and specifications are met. The definition of the inducer/impeller bladings, as well as of the flow paths in the diffuser (if any) and the volute, is typically finalized at this stage of the design process. Next, steady-state 3D fluid dynamic and structural analyses are carried out and additional details are included, as required by the specific application. Off-design, transient, and unsteady flow analyses finally complete the design process.

Optimization strategies based on the radial basis neural network (RBNN) have recently been applied by many researches to the design of turbopumps (Huppertz et al. 2007; Kim et al. 2010; Choi et al. 2010; Rai and Madyavan 2000; Kim et al. 2011). Other systematic optimization approaches commonly used in the refinement of the design of turbopumps are based on the mean streamline analyses (Oh and Kim 2011), three-dimensional inverse methods coupled with optimization algorithms (Ashihara and Goto 2011), and steady/unsteady numerical flow simulations (Visser et al. 2000). Significant efforts are currently being developed for accurately predicting and improving the performance of turbopumps with the help of complex

3D simulations and numerical optimization methods, which are extremely demanding in terms of computational time and power (Laskshminarayana 1985).

However, the expensive iterative nature of this approach intrinsically imposes significant limitations to its effective use for preliminary design of turbopumps. On the other hand, the availability of rational and efficient tools, capable of rapidly providing a realistic and accurate definition of both the machine geometry and performance since the early stages of its design, translates into very significant advantages for turbopump engineers and manufacturers. Up to date, however, efficient methods with these characteristics are lacking, and their development has therefore been selected as the main focus of the work illustrated in this publication.

The present activity starts from the observation that the ideal flow model illustrated in a companion paper (d'Agostino et al. 2017) proved to be quite efficient in making use of the relevant conservation equations for rationally deducing realistic geometries of mixed-flow centrifugal turbopumps specifically designed for locating the occurrence of cavitation on the initial portion of the blades, where its destabilizing effects on the flow field are less pronounced. On the other hand, the ideal flow approximation clearly prevented this model from being equally effective in the accurate prediction of the pumping performance of the machine.

In order to overcome this limitation, in the present approach the above flow model has been interfaced with the calculation of the boundary layers inside the blade channels and of other major forms of fluid dynamic losses (d'Agostino et al. 2012), with the aim of developing an effective tool for rapid parametric optimization of the machine geometry and performance under appropriate design constraints, such as, for instance, assigned values of the specific speed, flow coefficient, and blade solidity. The combined model has then been used, in conjunction with a standard parametric maximization algorithm, to design and optimize a mixed-flow turbopump (named VAMPIRE) with a six-bladed impeller, a vaneless diffuser, a single-spiral volute and nondimensional performances similar to those typical of space rocket applications (Pasini et al. 2011). The pumping and suction characteristics of the VAMPIRE pump have been determined in a series of steady operation tests in the CPRTF. Finally, the measured pumping characteristics (hydraulic head and efficiency) of the machine have been compared with the theoretical predictions, with the aim of assessing and confirming the validity of the proposed model.

2 Hydraulic Losses and Efficiency

The main fluid dynamic losses to be considered for more realistic performance prediction and optimization of mixed-flow centrifugal turbopumps typically arise from flow incidence at the leading edges of the impeller blades and the volute tongue, from viscous and secondary flow effects in the impeller, and from turbulent mixing in the diffuser and the volute, as schematically indicated in Fig. 1.

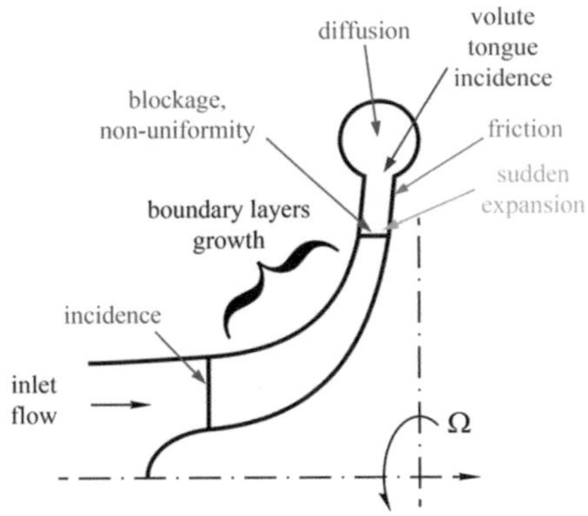

Fig. 1 The main sources of losses included in the model and their locations in the machine

Their contributions have been included in the present model and are illustrated in detail in this Section.

Notice that, in order to provide effective indications for parametric optimization of the machine, the evaluation of fluid dynamic losses must be sensitive to specific and relatively small changes of the geometry of the various components. Standard empirical loss correlations, such as those commonly used for preliminary assessment of viscous, slip flow, and turbulent mixing effects in centrifugal impellers, are therefore inadequate to this purpose, and have therefore been replaced here by more detailed and realistic loss models.

2.1 Impeller

Incidence losses due to the sudden change of the flow direction at the leading edge of the inducer blades are expressed in terms of the nondimensional equivalent length $L_{eq} = D_{ch}\Delta\alpha/3$ (function of the turning angle $\Delta\alpha$, in degrees, evaluated on the mean streamline) by means of:

$$\Delta p_{tI1} = f\left(\frac{L_{eq}}{D_{ch}}\right)\frac{1}{2}\rho V_1^{'2},$$

where ρ is the density of the liquid, $V_1^{'}$ is the relative flow velocity at the mean inlet radius and the friction factor f is obtained from Moody's chart using the Reynolds number based on the hydraulic diameter D_{ch} and the relative roughness of the blade channel.

Viscous effects developing on the surfaces of the blade channels generate frictional losses, flow blockage and, as a consequence of the asymmetric growth of the boundary layers on the pressure and suction sides of the blades, deviation of the flow at the impeller discharge. Moreover, the effective flow cross-section is also reduced by the blade thickness.

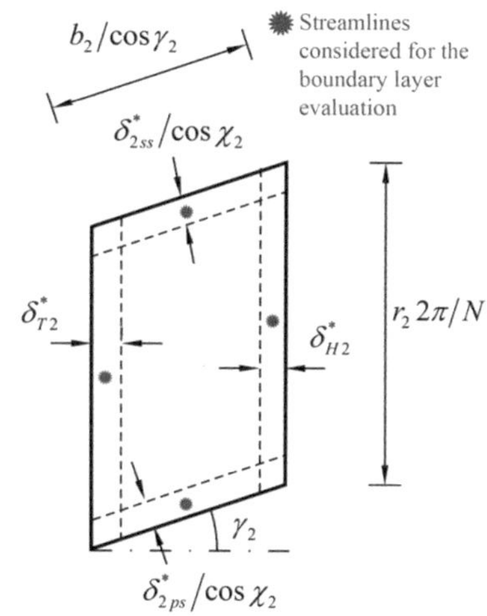

Fig. 2 Schematic for the approximate evaluation of BL blockage effects at the exit section of the impeller blade channels

In order to account for the influence of these aspects on the pumping performance of the machine, the evolution of the boundary layers has been evaluated along the four streamlines passing through the mid-points of the blade channel sides at the impeller exit. The results so obtained have been considered as representative of the average viscous effects on the corresponding surfaces, as schematically illustrated in Fig. 2.

Consequently, the flow blockage at the impeller exhaust has been approximated by the following expression:

$$B_{BL2} \simeq \frac{1}{A_{ch}}\left[\delta^*_{H2}\left(\frac{2\pi r_2}{N} - \frac{\delta^*_{2ss}}{\cos\chi_2}\right) + \delta^*_{T2}\left(\frac{2\pi r_2}{N} - \frac{\delta^*_{2ps}}{\cos\chi_2}\right) + \frac{\delta^*_{2ss}\left(b_2 - \delta^*_{T2}\right)}{\cos\chi_2} + \frac{\delta^*_{2ps}\left(b_2 - \delta^*_{H2}\right)}{\cos\chi_2}\right],$$

where δ^*_{2ss}, δ^*_{2ps} and δ^*_{H2}, δ^*_{T2} are respectively, the displacement BL thicknesses on the suction and pressure sides of the blades and on the impeller hub and tip, while b_2 is the exhaust axial blade height, N is the blade number and A_{ch} is the nominal channel area. Then the flow velocity at the impeller exit, corrected for blockage effects, is:

$$\boldsymbol{u}'_{2b} = \frac{\boldsymbol{u}'_2}{(1 - B_2)}$$

where $B_2 = B_{B2} + B_{BL2}$ is the total blockage at the impeller exit $(r = r_2)$, sum of the contributions generated by the blade thickness and the boundary layers.

The flow deviation generated by the boundary layer growth at the impeller exhaust mainly affects the azimuthal velocity and is expressed by:

$$\delta_2^{\circ} = \frac{1}{2}\left(\frac{d\delta_{ss}^*}{dx}\bigg|_2 \sin \lambda_{ss2} - \frac{d\delta_{ps}^*}{dx}\bigg|_2 \sin \lambda_{ps2} \right),$$

where x is the curvilinear coordinate along the streamline and λ is the angle between the rotational axis and the streamline in the meridional plane. With these corrections, the relative azimuthal velocity at the impeller discharge becomes:

$$v'_{2b\delta^{\circ}} = -u'_{2b} \tan\left[-\tan^{-1}\left(v'_{2b}/u'_{2b} \right) + \delta_2^{\circ} \right]$$

Similarly, with the same approach used for the evaluation of viscous blockage effects (Fig. 2), the total pressure loss at the impeller exhaust due to boundary layer friction is expressed by:

$$\Delta p_{tBL2} \simeq \frac{\theta_{H2}}{A_{ch}}\left(\frac{2\pi r_2}{N} - \frac{\theta_{2ss}}{\cos\chi_2} \right)\frac{\rho U_{H2}'^2}{2} + \frac{\theta_{T2}}{A_{ch}}\left(\frac{2\pi r_2}{N} - \frac{\theta_{2ps}}{\cos\chi_2} \right)\frac{\rho U_{T2}'^2}{2}$$
$$+ \frac{\theta_{2ss}(b_{I2} - \theta_{T2})}{A_{ch}\cos\chi_2}\frac{\rho U_{ss2}'^2}{2} + \frac{\theta_{2ps}(b_{I2} - \theta_{H2})}{A_{ch}\cos\chi_2}\frac{\rho U_{ps2}'^2}{2},$$

where θ_{2ss}, θ_{2ps} and θ_{H2}, θ_{T2} are the momentum boundary layer thicknesses on the suction and pressure sides of the blades and on the impeller hub and tip and U' is the streamwise velocity of the free stream in the relative frame.

The streamlines for the evaluation of the boundary layers along the impeller blade channels have been determined by numerical integration of the relevant ordinary differential equations:

$$\frac{dr}{dx} = \frac{u'}{\sqrt{u'^2 + v'^2 + w'^2}}$$
$$\frac{d\vartheta'}{dx} = \frac{v'}{r\sqrt{u'^2 + v'^2 + w'^2}}$$
$$\frac{dz}{dx} = \frac{w'}{\sqrt{u'^2 + v'^2 + w'^2}}$$

using the relative velocity field obtained from the ideal flow model and the appropriate (final) conditions at the impeller discharge (Fig. 3).

The governing equations for the laminar incompressible boundary layers on the impeller blades have then been obtained by writing the relative continuity and momentum equations:

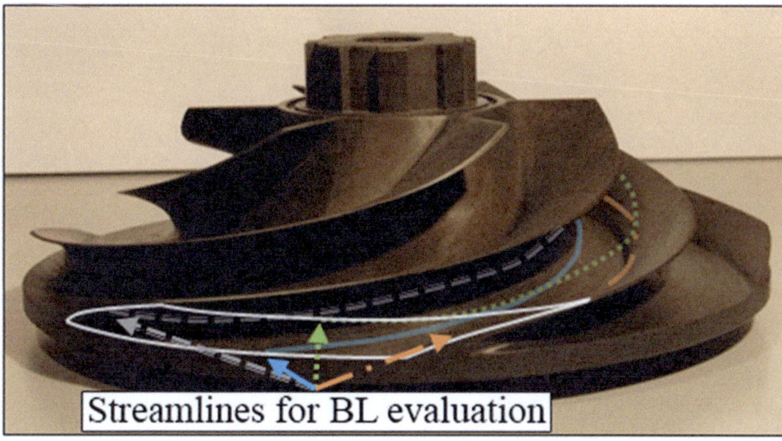

Fig. 3 Relative streamlines of the inviscid flow on the impeller blade channels for turbulent BL computations

$$\nabla \cdot \boldsymbol{u}' = 0$$

$$\boldsymbol{u}' \cdot \nabla \boldsymbol{u}' = -\frac{1}{\rho}\nabla p + \nu \nabla \cdot \left[\nabla \boldsymbol{u}' + \left(\nabla \boldsymbol{u}'\right)^T\right] - 2\Omega \times \boldsymbol{u}' + \Omega^2 \boldsymbol{r}$$

in body-fitted coordinates x and y along the streamwise and normal directions. Coriolis forces are intrinsically small in the inner layers where the relative flow velocity is low, and approximately accounted for in the outer layers by the matching with the freestream velocity, which includes the vorticity correction. Hence, with usual quasi-parallel flow approximations and neglecting Coriolis forces, obtain:

$$\frac{\partial (r_0 u_x)}{\partial x} + r_0 \frac{\partial u_y}{\partial y} = 0$$

$$u_y \frac{\partial u_x}{\partial y} + u_x \frac{\partial u_x}{\partial x} = -\frac{1}{\rho}\frac{\partial p}{\partial x} + \nu \frac{\partial^2 u_x}{\partial y^2} + \Omega^2 r \sin \lambda \cos \mu,$$

$$0 \simeq -\frac{1}{\rho}\frac{\partial p}{\partial y} + \Omega^2 r_0 \cos \lambda$$

where r_0 is the radial distance from the axis, λ is the meridional angle of the boundary layer surface from the axial z-direction and μ is the angle of the boundary layer free streamline from the meridional plane.

The pressure can now be eliminated by differentiating the steady Bernoulli's equation:

$$\frac{p}{\rho}+\frac{1}{2}\left(\boldsymbol{U}'\cdot\boldsymbol{U}'-\Omega^2\boldsymbol{r}\cdot\boldsymbol{r}\right)=\text{constant}$$

along a relative streamline:

$$-\frac{1}{\rho}\frac{\partial p}{\partial x}\simeq U\frac{dU}{dx}-\Omega^2 r\frac{dr_0}{dx}\simeq U\frac{dU}{dx}-\Omega^2 r_0\sin\lambda\cos\mu$$

$$-\frac{1}{\rho}\frac{\partial p}{\partial y}=-\Omega^2 r\frac{dr}{dy}\simeq-\Omega^2 r_0\cos\lambda,$$

where U is the relative streamwise velocity of the freestream.

In the stated assumptions, centrifugal effects in the boundary layer and in the freestream balance each other. Hence, the momentum equation in the normal direction y is identically satisfied, while in the streamwise direction x reduces to:

$$u_y\frac{\partial u_x}{\partial y}+u_x\frac{\partial u_x}{\partial x}=U\frac{dU}{dx}+\nu\frac{\partial^2 u_x}{\partial y^2}$$

(independent on Ω).

Finally, the equations for the turbulent boundary layers on the impeller blades can be obtained by standard Reynolds averaging in the form:

$$\frac{\partial(r_0\bar{u}_x)}{\partial x}+\frac{\partial\left(r_0\bar{u}_y\right)}{\partial y}=0$$

$$\bar{u}_x\frac{\partial\bar{u}_x}{\partial x}+\bar{u}_y\frac{\partial\bar{u}_x}{\partial y}=U\frac{dU}{dx}+\frac{1}{\rho}\frac{\partial\tau_t}{\partial y}$$

where bars indicate the mean flow properties, primes the turbulent fluctuations and $\tau_t=\mu\,\partial\bar{u}_x/\partial y-\rho\overline{u_x'u_y'}$ is the turbulent shear stress (equal to τ_w at the wall).

These equations have been solved by extending the method proposed by White (1974) to axisymmetric configurations, as required for application to mixed-flow impellers. The basic idea of White's method consists in eliminating $\bar{u}_y$ with the continuity equation and integrating the boundary layer momentum equation from the blade surface $(y=0)$ to the freestream $(y=\delta)$. The velocity profile is approximated by the logarithmic law of the wall:

$$u^+\simeq\frac{1}{0.4}\ln y^+ +5.5+w(y^+,\xi)$$

expressed in terms of the inner variables:

$$u^+=\frac{\bar{u}_x}{v^*};\quad v^+=\frac{\bar{u}_y}{v^*};\quad y^+=\frac{yv^*}{\nu}$$

with frictional velocity $v^* = \sqrt{\tau_w/\rho}$ and an outer wake $w(x, y^+) = 0.6y^+ \xi(x)$, linear in y^+ and in the pressure-gradient correlation parameter (Mellor 1966):

$$\xi(x) = \frac{\mu U}{\tau_w v^*} \frac{dU}{dx}$$

Upon introduction of the nondimensional streamwise coordinate $x_* = x/2r_2$ and freestream velocity $U_* = U/\Omega r_2$, the following ordinary differential equation for the skin friction parameter $\lambda = U^+$ has been obtained (Valentini 2011, 2015; d'Agostino et al. 2012):

$$\frac{d\lambda}{dx_*} = \frac{U_* \mathrm{Re}_D - \left[\lambda^2 \delta^+ - G(\lambda, \xi)\right] \lambda \frac{d \ln U_*}{dx_*} + \frac{\lambda^4 H(\lambda, \xi)}{\mathrm{Re}_D} \frac{d^2}{dx_*^2}\left(U_*^{-1}\right) - K(\lambda, \xi) \lambda \frac{d \ln r_0}{dx_*}}{G(\lambda, \xi) - \frac{3\lambda^3 H(\lambda, \xi)}{\mathrm{Re}_D} \frac{d}{dx_*}\left(U_*^{-1}\right)}$$

Here $\mathrm{Re}_D = 2\Omega r_2^2/\nu$ is the Reynolds number based on the impeller tip speed Ωr_2 and diameter $2r_2$,

$$G(\lambda, \xi) = \int_0^{\delta^+} u^{+2} dy^+$$

and:

$$H(\lambda, \xi) = \int_0^{\delta^+} \left(u^+ \frac{\partial u^+}{\partial \xi} - \frac{\partial u^+}{\partial y^+} \int_0^{y^+} \frac{\partial u^+}{\partial \xi} dy^+ \right) dy^+$$

have the same expressions as in White (1974), while:

$$K(\lambda, \xi) = \int_0^{\delta^+} \left(\frac{\partial u^+}{\partial y^+} \int_0^{y^+} u^+ dy^+ \right) dy^+$$

is an additional function to be included when $dr_0/dx \neq 0$.

The above ordinary differential equation for λ has been numerically integrated along the four relative streamlines passing through the mid-points of the blade channel sides at the impeller exit with the IC $\delta^+(0) \simeq 10$ imposed by the lower limit of validity of the law of the wall (merging with the laminar sublayer), which in turn determines the initial value of $\lambda(0) = \frac{1}{0.4} \ln 10 + 5.5 + 6\xi$.

Finally, at any location x_* the boundary layer nominal, displacement and momentum thicknesses have been computed as:

$$\delta = 2r_2 \frac{\delta^+ \lambda}{U_* \mathrm{Re}_D}; \quad \delta^* = 2r_2 \frac{\delta^{*+} \lambda}{U_* \mathrm{Re}_D} \quad \text{and} \quad \theta = 2r_2 \frac{\theta^+ \lambda}{U_* \mathrm{Re}_D}$$

Fig. 4 Streamwise evolution of the boundary layer thicknesses along the blade channels

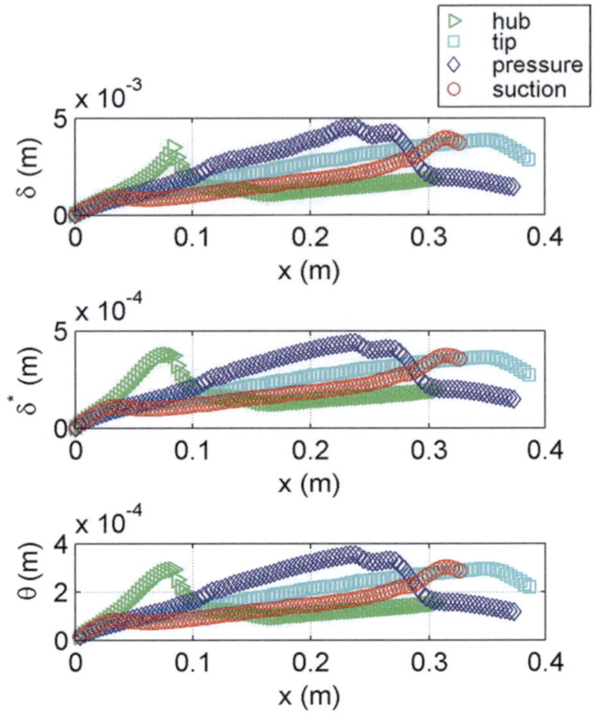

Their evolution along the blade channels of the VAMPIRE pump operating at nominal conditions is illustrated in Fig. 4. The results are similar for δ, δ^* and θ, which in turbulent boundary layers are nearly proportional to each other. Specifically, the development of the boundary layer is predicted to proceed rather regularly on the suction side of the blades and, even if not relevant to the present case of unshrouded impeller as indicated later, also on the casing surface at the blade tip radius. On the other hand, the evolution of δ, δ^* and θ is slightly more irregular on the pressure side of the blades and on the hub surface where, however, flow turning effects tend to oppose the separation of the flow. The regular growth of the boundary layer is indicative of a favorable pressure distribution and is especially significant on the suction side of the blades, where the flow is generally more susceptible to separation. As a whole, the results of Fig. 4 fully confirm the validity of the proposed method for the definition of hydrodynamically efficient impeller geometries.

By supporting the blades on both sides, impeller shrouds can be quite useful to reduce bending stresses and flutter instabilities. But they also add to the manufacturing complexity and weight of impellers, thus reducing their critical speeds. Besides, the narrow recirculating leakage flow between their outer surface and the machine casing increases ventilation losses and generates spurious and uncertain rotordynamic effects. For these reasons, whenever possible unshrouded impellers are typically favored in space applications (Droege et al. 2000). In these impellers the boundary layers developing on the impeller casing are continuously wiped out by passing blades. Therefore, an approximate but practical way to account for the absence of the shroud in the present model consists in simply not considering the boundary layers on the inner surface of the impeller casing.

The above approach allows for the quantitative evaluation of viscous effects (blockage, deviation and losses) taking place in the flow through centrifugal impellers as functions of their geometry and operating conditions. Perfect mixing is assumed to occur at rotor discharge at the inlet cross-section of the diffuser. The resulting total pressure losses have then been be evaluated as the mass averaged integral difference between the dynamic pressures of the mean flowfield at the diffuser inlet and the nonuniform velocity distribution at the impeller discharge station, that is:

$$\Delta p_{tM2}(z) = \frac{N}{4\pi}\rho \int_0^{2\pi/N} \left(\tilde{v}_2^2 - v_S^2\right) d\vartheta'$$

where $\tilde{v}_2$ is the local azimuthal component of the vorticity correction and v_s is the azimuthal slip velocity.

2.2 *Diffuser*

Friction on the side walls and the sudden increase of the flow passage area at the exit of the impeller are the main sources of fluid dynamic losses in vaneless diffusers. Frictional losses can be evaluated along the nominal streamlines (consisting in log-spirals with constant angle φ from the radial direction) by means of the expression:

$$\Delta p_{tF3} = \frac{1}{2}\rho\left(u_3^2 + v_3^2\right) f_D \frac{r_3(r_3 - r_2)}{2b_D r_2 \cos\varphi}$$

where b_D is the width of the diffuser and the friction factor f_D depends on the flow Reynolds number based on the hydraulic diameter $D_{Hdiff} = 2b_D$. Finally, diffusion losses due to the sudden variation of the flow passage area from the impeller discharge to the diffuser can be estimated using the following expression:

$$\Delta p_{tD2} = \frac{1}{2}\rho\left(\bar{u}_{b2}^2 - u_2^2\right) = \frac{1}{2}\rho\left(\bar{u}_{b2}^2 - \bar{u}_2^2 \frac{b_{I2}^2}{b_D^2}\right)$$

where $\bar{u}_{b2}$ and $\bar{u}_2$ are the average values of the radial velocity at the impeller exit with and without corrections for blockage effects.

2.3 Volute

The flow losses occurring in the volute typically include radial diffusion losses at the volute entrance, incidence losses at the tongue, frictional losses on the walls, azimuthal diffusion losses in the volute and mixing losses in the exhaust duct. The volute flowfield is relatively complex, being intrinsically 3D, unsteady, and, in most cases, turbulent. Accurate prediction of the resulting losses would in principle call for the simultaneous simulation of the entire flow through the turbopump, quite an ambitious task, computationally very expensive and time-consuming. As anticipated in the first part of this publication (d'Agostino et al. 2017), such simulations are indeed capable of evaluating the flow properties in the volute with spatial resolution dependent on the size of the computational grid (Van den Braembussche 2006), but do not provide any direct indication for the selection of the machine geometry, which must be assigned as a known input to the computations.

On the other hand, the present model is intended as a rapid and effective tool for simultaneous and accurate definition of the geometry of the machine and prediction of its fluid dynamic performance, as required in the preliminary phase of the design. For most efficient definition of the model it is worth noting that total pressure losses in the volutes of radial turbopumps are usually dominant, especially at off-design conditions, with respect to the contributions from other components. This is because the separated/swirling nature of the flow in the volute promotes large-scale turbulent mixing and because the geometry of the volute can only be optimally adapted to the impeller for one single value of the flow coefficient. Based on these considerations, only incidence losses at the tongue and mixing losses in the volute have been taken into account in the present model. Available evidence demonstrates that this approximation is capable of providing accurate predictions of the magnitude and trends of flow losses occurring in the volute of the test turbopump, thus supporting the conclusion that it can be safely applied to similar machines designed in accordance with the indications of the present model.

The total pressure loss due to the flow incidence at the tongue of the volute has been computed as:

$$\Delta p_{tLVi} \simeq \xi_{tLVi} \frac{1}{2} \rho \left\{ \boldsymbol{u}_3^2 - \left[|\boldsymbol{u}_3| \cos(\varphi_3 - \varphi_T) \right]^2 \right\}$$

corresponding to the dissipation of a constant portion ξ_{tLVi} (here equal to 0.9) of the kinetic energy of the approach velocity component normal to the orientation φ_T of the tongue w.r.t. the azimuthal direction ($\varphi_T = \varphi_{3des}$ at design conditions).

For simplicity, mixing losses in the volute have been included only when the incoming flow velocity is faster than at the exit cross-section and correspond to losing a constant portion ξ_{tLVd} (here equal to 0.9) of the associated dynamic pressure change:

$$\Delta p_{tLVd} = \xi_{tLVd} \frac{1}{2} \rho \left(u_3^2 - u_4^2 \right)$$

Finally, frictional losses along the walls of the volute have been neglected because of their relatively small magnitude.

2.4 *Overall Performance*

With the above results, the pumping performance of the machine becomes:

$$\Psi = \frac{\Delta p_t}{\rho \Omega^2 r_2^2} = \frac{\Delta p_{tE2} - \Delta p_{tLOSS}}{\rho \Omega^2 r_2^2},$$

where the total pressure increase computed from Euler equation includes the effects of flow blockage and deviation generated by the BLs developing along the blade channels:

$$\Delta p_{tE2} = \rho \Omega r_2 \bar{v}_{2b\delta^\circ}$$

and the total pressure loss comprises all of the sources of dissipation considered above:

$$\Delta p_{tLOSS} = \Delta p_{tI1} + \Delta p_{tBL2} + \Delta p_{tM2} + \Delta p_{tD2} + \Delta p_{tF3} + \Delta p_{tLVi} + \Delta p_{tLVd}$$

Finally, the hydraulic efficiency of the machine has been evaluated by means of the expression:

$$\eta_h = \frac{Q \Delta p_t}{\Omega \tau}$$

where Q is the volumetric flow rate, Δp_t is the total pressure increase, Ω is the rotational speed and τ is the torque applied to the pump shaft.

3 Experimental Apparatus

The Test Facility. The experimental activity reported in the present paper has been carried out in the Cavitating Pump Rotordynamic Test Facility (CPRTF, Fig. 5).

This facility is specifically designed for characterizing the performance of cavitating and noncavitating turbopumps in a wide variety of alternative configurations (axial, radial or mixed flow, with or without an inducer (Cervone et al. 2012; Pace et al. 2012; Rapposelli et al. 2002). It operates in water at temperatures from room

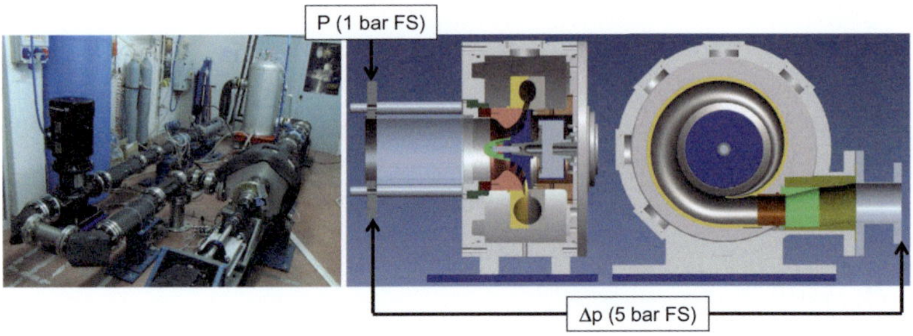

Fig. 5 The cavitating pump rotordynamic test facility (*left*) and cut-out of the test section showing the pump assembly and the locations of the pressure transducer (*right*)

conditions up to 90 °C and is intended as a flexible apparatus readily adaptable to conduct experimental investigations under hydrodynamic, Reynolds, and thermal cavitation similarity conditions on virtually any kind of fluid dynamic phenomena relevant to high-performance turbopumps. The test section can be equipped with a rotating dynamometer, for the measurement of the forces and moments acting on the impeller, and with a mechanism designed for adjusting and rotating the eccentricity of the impeller axis in the range 0 ÷ 2 mm and ±3000 rpm. The inlet section and mounting elements of the turbopump in the housing can be easily replaced in order to allow for testing inducers and machines with different diameters and geometries.

In present experiments, the CPRTF has been equipped with the rotating dynamometer in order to eliminate the uncertainties related to friction in the seals and bearings of the shaft in the measurement of the torque applied to the pump impeller. The dynamometer is realized in one solid piece of AISI 630 H1025 phase hardening steel and consists of two flanges connected by four square pillars acting as flexing elements in a classical squirrel-cage configuration. The deformation of the pillars is measured by 40 semiconductor strain gauges arranged in 10 full Wheatstone bridges, which provide redundant measurements of the instantaneous forces and moments acting on the impeller along the axial and lateral directions.

Each bridge is temperature self-compensated, with separate bipolar excitation and read-out for better reduction of cross-talking. The inlet pressure and the pressure rise, necessary for the characterization of the pump performance, have been measured by means of an absolute pressure transducer (Druck, model PMP 1400, 0 ÷ 1.5 bar, 0.25% precision) installed about two impeller diameter upstream of the inlet cross-section and a differential pressure transducer (Druck, model PMP 5073, range 0–5 bard, 0.1% precision) installed between the inlet and the outlet sections of the test pump (with the low-pressure tap at the same location as the absolute pressure tap and the high pressure tap more than one impeller diameter downstream of the exit cross-section of the volute).

The desired values of the blade tip clearance have been obtained by adjusting the inner diameter of the interchangeable casing of the impeller (see Fig. 5).

This feature is particularly useful when conducting rotordynamic tests in order to tailor the clearance as necessary for accommodating the imposed eccentricity of the rotor and avoiding rotor/stator contact.

Two electromagnetic flowmeters (mod. 8732E by Fisher-Rosemount, range 0–100 l/s, 0.5% precision), mounted on the suction and discharge lines, measure the pump's inlet/outlet flow rates and a thermocouple monitors the temperature of the working fluid with ±0.5 K precision.

A custom-made Silent Throttle Valve, designed for NASA by Innerspace Corp., Covina, California (USA), has been used for adjusting the pump load.

The Test Pump. The pump, named VAMPIRE, used in present experiments has been designed by means of the present model and parametrically optimized as documented in d'Agostino et al. (2012). It has been manufactured in 7075-T6 aluminum alloy, and comprises a six-bladed unshrouded impeller (see Fig. 6), a vaneless diffuser and a single-spiral volute. Its main operational parameters and geometrical characteristics are summarized in Table 1.

Fig. 6 The six-bladed centrifugal impeller of the VAMPIRE turbopump

Table 1 Main operational parameters and geometrical characteristics of the VAMPIRE turbopump

Parameter	Units	Symbol	Value
Design flow coefficient	–	Φ	0.092
Number of blades	–	N	6.00
Outlet radius	mm	r_2	105.0
Inlet tip radius	mm	r_{T1}	57.2
Inlet hub radius	mm	r_{H1}	31.8
Axial length (fully developed blade)	mm	z_{H2}	46.4

(continued)

Table 1 (continued)

Parameter	Units	Symbol	Value
Inlet tip blade angle	deg	γ_{T1}	56.60
Inlet backsweep angle	deg	χ_1	0.00
Diffuser outlet radius	mm	r_3	126.0
Rotational speed	rpm	Ω	1500
Design volumetric flowrate	l/s	V_D	16.8
Tip solidity	–	σ_T	2.26
Incidence tip angle @ design	deg	α	17.40
Outlet tip blade mean angle	deg	γ_{T2}	67.78
Outlet tip backsweep angle	deg	χ_2	66.00
Exit blade height	mm	b_2	10.5
Exit cross-section volute radius	mm	R_4	38.2
Volute maximum radial dimension	mm	r_{V4}	201.5

4 Results and Discussion

A series of tests has been conducted on the VAMPIRE pump with the aim of experimentally validating the proposed model for preliminary design and noncavitating performance prediction of centrifugal pumps.

The model has been generalized in order to account for the results of a parallel test campaign aimed at investigating the influence of thermal cavitation effects on the rotordynamic whirl forces on the VAMPIRE pump with different values of the tip clearance (Pace et al. 2014, 2016; Valentini et al. 2016). The presence of a tip clearance in unshrouded impellers significantly increases the complexity of the flow structure. Since the detailed treatment of this aspect is too demanding in terms of additional complexity and/or computational cost of the model, a simplified approach has been adopted to approximately describes its influence on the turbopump performance.

Tip clearance changes have long been known to affect both the head and flow coefficients of the machine, as recently confirmed by some of the authors in the case of inducers (Torre et al. 2011). Specifically, the presence of blade tip clearance in unshrouded rotors involves three major effects since it:

- affects the secondary flow in the blade channels;
- induces the occurrence of tip leakage flows as a consequence of the pressure difference on the two sides of the blades;
- modifies the intermittent "scraping" caused by the blade passage on the boundary layers developing on the inner surface of the impeller casing (Koshide and Nielson 1973).

With reference to Fig. 7, the operating point of the test pump is defined by the intersection between its head characteristic and the load curve of the facility, which can be considered constant for given setting of the silent throttle valve and

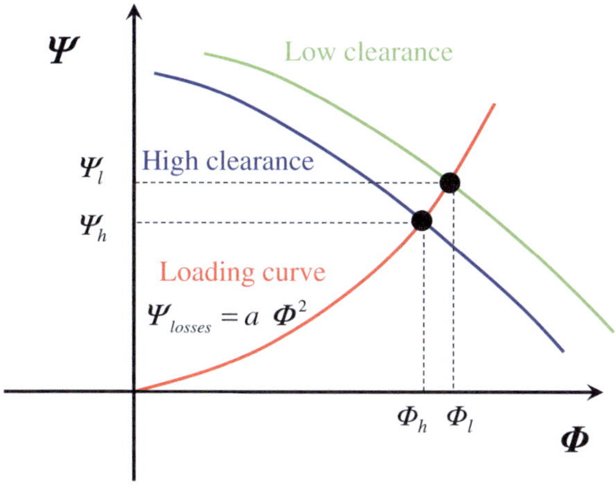

Fig. 7 Noncavitating operating points (*solid circles*) of a turbopump respectively operating with higher (subscript *h*) and lower (subscript *l*) impeller blade tip clearances at constant settings of the test facility

approximately quadratic in the flow rate at Reynolds numbers higher than 10^6 based on the impeller tip diameter and speed. Then, all other operating conditions being the same, the relation between the head and flow coefficients of the machine for two values (subscripts l for lower and h for higher) of the blade tip clearance becomes:

$$\Phi_l = \Phi_h \sqrt{\frac{\Psi_l}{\Psi_h}}$$

This equation allows for scaling the noncavitating performance of the machine for different tip clearances once the associated head change is known. To this purpose it is worth noting that, as reported in the literature (Torre et al. 2011; Koshide and Nielson 1973; Brennen 1994), the noncavitating head coefficient is approximately linear in the relative tip clearance $(c_{\%})$, defined as the ratio between the tip clearance (c) and the average height of the impeller blades (b_a).

With reference to Fig. 8 (where the head coefficient has been nondimensionalized with its nominal design value Φ_0), the interpolating straight line for the VAMPIRE pump has been evaluated from the experimental results for two different tip clearances (squares). For any given value of the relative clearance, the corrected head coefficient is then obtained from this correlation and the corresponding flow coefficient from the above equation. In turn, these coefficients have been used to modify the pumping characteristics predicted by the model in order to account for the actual value of the blade tip clearance.

The relative tip clearances of the reference turbopump configurations used for the definition of the above correction procedure are $c_{\%} = 5.6\%$ and 11.2%, corresponding to 1 and 2 mm blade tip clearances, respectively. Each experimental point has been obtained by averaging the readings of the differential pressure and flow rate acquired over 2 s at 5,000 sps (samples per second). In order to attain fully developed turbulent flow conditions, rotational speed of the pump and the water

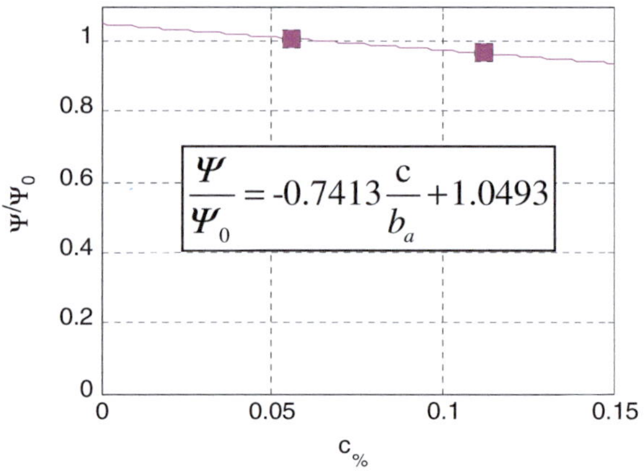

Fig. 8 Normalized nondimensional head of the VAMPIRE turbopump as a function of the relative tip clearance. The linear correlation of the diagram (*solid line*) has been obtained from two experiments at different blade tip clearances (*solid squares*)

temperature have been kept respectively constant at 1500 ± 3 rpm and 19.5 ± 1 °C, corresponding to a Reynolds number $\mathrm{Re} = 2\Omega r_2^2/\nu = 5.01 \times 10^6$.

In order to validate the above procedure for tip clearance correction, two different series of tests have been conducted, where the pumping performance of the VAMPIRE pump has been measured for two different clearances equal to 1 mm ($c_\% = 5.6\%$) and 0.16 mm ($c_\% = 0.9\%$). Figure 9 shows the comparison of the experimental pumping performance (squares and upward-pointing triangles) with the corrected predictions of the analytical model (circles and stars).

Figure 9 also shows the hydraulic efficiency characteristics for the same two configurations of the pump. It is worth recalling that in the experimental evaluation of the hydraulic efficiency η_h the shaft power has been obtained by means of the rotating dynamometer and is therefore not affected by the uncertain influence of frictional torque in the bearings and seals.

Due to numerical limitations of the boundary layer solver under severe flow separation on the suction/pressure sides of the impeller blades (see White 1974), the predictions of the model are restricted to flow coefficients in the range from 65 to 160% of the design value. Anyway, the accessible operational range of the model covers with wide margins the conditions typically encountered by liquid propellant feed turbopumps in space propulsion applications.

The relative deviations of the measured head coefficients from the theoretical results are illustrated in Fig. 10 for two values of the relative blade tip clearance. The shaded area indicates the ±5% bandwidth above and below the observed performance. In the same figure, the dashed and dotted lines show for each clearance setting the mean deviation between the experimental and the theoretical results. In both cases, the calculated pumping performance at design conditions is quite accurate and, in particular, in the low clearance case it is less than 0.5% off the measured value. As a whole, the results of the present investigation clearly indicate that the calculated performance closely matches the experimental measurements and successfully confirm the predictive capability of the proposed model.

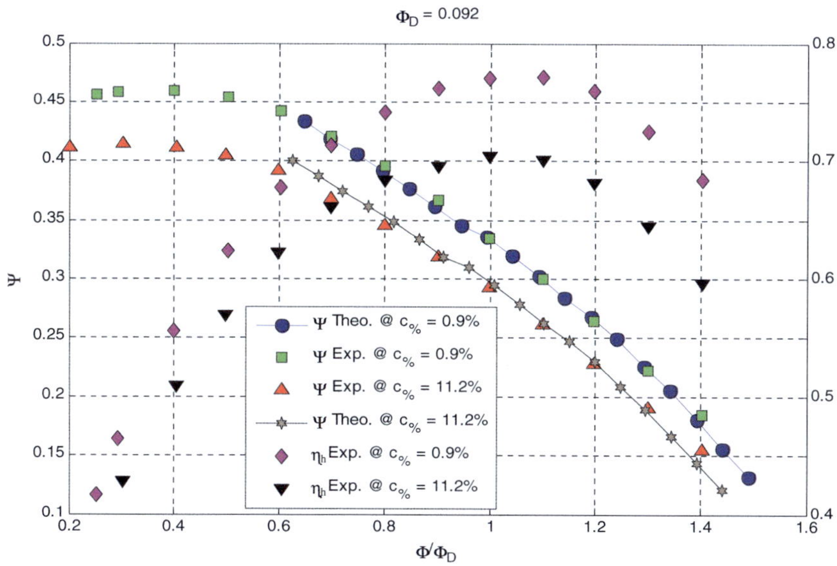

Fig. 9 Normalized comparison between the experimental and predicted (theoretical) noncavitating performance of the VAMPIRE pump operating at Ω = 1500 rpm and Re = 5.01 × 10^6 for two different values of the blade tip clearance (*left scale*). The figure also shows the corresponding hydraulic efficiencies (*right scale*)

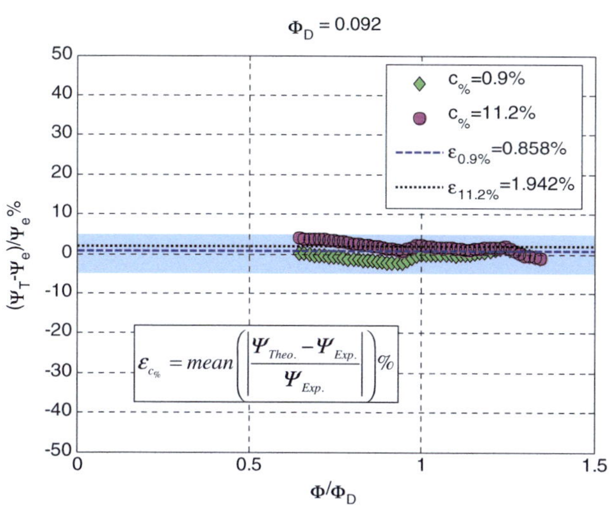

Fig. 10 Relative deviation between the predicted and experimental head performance of the VAMPIRE pump for two values of the relative blade tip clearance

5 Conclusions

The noncavitating tests conducted on the VAMPIRE turbopump, designed, and parametrically optimized by means of the reduced-order model recently developed by some of the authors and extended in the present work with the inclusion of the major sources of fluid dynamic losses, successfully confirmed that the proposed model is indeed capable of efficiently defining the geometry of mixed-flow centrifugal turbopumps and accurately predicting their noncavitating performance.

By expressing the performance of these machines in terms of a relatively small number of controlling parameters, the proposed model is especially suited for rapid and efficient optimization. This feature has been fully and successfully exploited in the design of the VAMPIRE pump to given requirements and specifications.

For lack of space, the illustration of the optimization procedure used for the design of the test pump has been postponed to a later publication, but the quality of its results is well documented here by the very close match between the predicted and measured values of the hydraulic performance (head and efficiency) of the VAMPIRE pump. To this effect, special attention has been paid to obtaining an accurate measurement of the torque actually applied to the impeller by making use the CPRTF rotating dynamometer, thus eliminating the uncertain and spurious influence of seal and bearing torques.

The tests conducted on the VAMPIRE pump are also in excellent agreement with the semi-heuristic approach used to account for the influence on the machine performance of the impeller blade tip clearance, which is not explicitly incorporated in the model.

As a whole, the results of present experiments successfully confirm the validity of the proposed model as a very effective tool for rapid and efficient preliminary definition of the geometry and hydraulic performance of mixed-flow centrifugal turbopumps, as required in the conceptual, feasibility and initial stages of their development and design.

References

Ashihara, K., & Goto, A. (2011). *Turbomachinery blade design using 3-d inverse design method, CFD and optimization algorithm*. ASME Turbo Expo, 2001.

Brennen, C. E., (1994). *Hydrodynamics of pumps*. Oxford University Press.

Busby, J., Batton, W., Furst, R., & Ashvin, H. (2008). High suction specific speed lox pump: design, analysis and testing. In *44th AIAA/ASME/SAE/ASEE joint propulsion conference and exhibit*, Hartford, (CT), USA.

Cervone, A., Pace, G., Torre. L., Pasini, A., & d'Agostino, L. (2012). Effects of the leading edge shape on the performance of an axial three bladed inducer. In *14th international symposium on transport phenomena and dynamics of rotating machinery*, Honolulu, Hawaii, USA.

Choi, K. J., Kim, J. H., & Kim, K. Y. (2010). *Design optimization of circumferential casing grooves for a transonic axial compressor to enhance stall margin*. ASME Turbo Expo 2010.

d'Agostino, L., Pasini, A., Valentini, D., Pace, G., Torre, L., & Cervone, A. (2012). A reduced order model for optimal centrifugal pump design. In *14th international symposium on transport phenomena and dynamics of rotating machinery*, ISROMAC-14, February 27th–March 2, Honolulu, HI, USA.

d'Agostino, L. (2013a). Turbomachinery developments and cavitation. In *STO-AVT-LS-206, Paper NBR 12-1, VKI lecture series on fluid dynamics associated to launcher developments, von karman institute of fluid dynamics, Rhode-Saint-Genèse*, Belgium, April 15–17.

d'Agostino, L. (2013b). On the hydrodynamics of rocket propellant engine inducers and turbopumps. In *6th international conference on pumps and fans with compressors and wind turbines (IPCF 2013)*, Sep. 19–22, Beijing, China.

d'Agostino, L., Valentini, D., Pasini, A., Torre, L., & Pace, G. (2017). On the preliminary design and performance prediction of centrifugal turbopumps—Part 1. In this volume, CISM Courses and Lectures No. 1408. In L. d'Agostino & M. V. Salvetti (Eds.), *International centre for mechanical sciences*. Vien and New York: Springer.

Droege, A. R., Williams, R. W., & Garcia, R. (2000). *Unshrouded impeller technology development status*. NASA Technical Reports Server.

Huppertz, A., Flassig, P. M., Flassig, R. J., & Swoboda, M. (2007). *Knowledge-based 2d blade design using multi-objective aerodynamic optimization and a neural network*. ASME Turbo Expo.

Kim, J. H., Choi, K. J., & Kim, K. Y. (2010). Surrogate modeling for optimization of a centrifugal compressor impeller. *International Journal of Fluid Machinery and Systems, 3*(1), 29–38.

Kim, J.-H., Jin, C.-H., & Kim, K.-Y. (2011). *ASME-JSME-KSME joint fluids engineering conference performance improvement of a mixed-flow pump by optimization techniques*. Japan: Hamamatsu.

Koshide, R. K., & Nielson, C. E. (1973). *Study of blade clearance effects on centrifugal pumps*. NASA CR-120815 R-8806.

Laskshminarayana, B. (1985). *Fluid dynamics and heat transfer of turbomachinery*. New York: Wiley.

Mellor, G. L. (1966). The effects of pressure gradients on turbulent flow near a smooth wall. *Journal of Fluid Mechanics, 24*(2), 255–274.

Oh, H. W., & Kim, K. Y. (2011). Conceptual design optimization of mixed-flow pump impellers using mean streamline analysis. *Proceedings of the Institution of Mechanical Engineers, Part A: Journal of Power and Energy, 215*(1), 133–138.

Pace, G., Pasini, A., Torre, L., Valentini, D., & d'Agostino, L. (2012). The cavitating pump rotordynamic test facility at ALTA S.p.A.: Upgraded capabilities of a unique test rig. In *Space propulsion conference*, Bordeaux, France, 2012.

Pace, G., Valentini, D., Torre, L., Pasini, A., & d'Agostino, L. (2014). Experimental characterization of rotordynamic forces acting on space turbopumps. *Space propulsion 2014, May* (pp. 19–22). Germany: Cologne.

Pace G., Valentini D., Pasini A., Torre L., Hadavandi R., & d'Agostino L. (2016). Inducer and centrifugal pump contributions to the rotordynamic fluid forces acting on a space turbopump. In *ISROMAC 2016, international symposium on transport phenomena and dynamics of rotating machinery*, Honolulu, Hawaii, USA, April 10–15.

Pasini, A., Torre, L., Cervone, A., & d'Agostino, L. (2011). Characterization of the rotordynamic forces on tapered axial inducers by means of a rotating dynamometer and high-speed movies. In *WIMRC 3rd international cavitation forum 2011,* University of Warwick, UK, 4th–6th July 2011.

Rai, M. M., & Madyavan, N. K. (2000). Aerodynamic design using neural networks. *AIAA Journal, 38*(1), 173–182.

Rapposelli, E., Cervone, A., & d'Agostino, L. (2002). A new cavitating pump rotordynamic test facility. In *AIAA joint propulsion conference and exhibit*, Indianapolis, USA.

Stripling, L., & Acosta, A. (1962). Cavitation in turbopumps—part 1. *ASME Journal of Basic Engineering*, *84*, 326–338.

Torre, L., Pasini, A., Cervone, A., Pace, G., Miloro, P., & d'Agostino, L. (2011). Effect of tip clearance of the performance of a three-bladed axial inducer. *AIAA Journal of Propulsion and Power, 27*(4), 890–898.

Valentini, D. (2011). *Sviluppo di un Modello per la Progettazione di Turbopompe Centrifughe*, M. S. Thesis, Università di Pisa, Pisa, Italy, October 10, 2011.

Valentini, D. (2015). *Modelling and testing of chemical propulsion rocket subsystems*. Ph.D. Thesis, Università di Pisa, Pisa, Italy, December 14, 2015.

Valentini, D., Pace, G., Pasini, A., Torre, L., Hadavandi, R., & d'Agostino, L. (2016). Fluid-induced rotordynamic forces on a whirling centrifugal pump. In *ISROMAC 2016, international symposium on transport phenomena and dynamics of rotating machinery*, Honolulu, Hawaii, USA, April 10–15.

Van den Braembussche, R. A. (2006). Flow and loss mechanism in volutes of centrifugal pumps. In *Design and analysis of high speed pumps. RTO-EN-AVT-143, Paper 12* (pp. 12-1–12-26). Neuilly-sur-Seine, France.

Visser, F. C., Dijkers, R. J. H., & op de Woerd, J. G. H. (2000). Numerical flow-field analysis and design optimization of a high-energy first-stage centrifugal pump. *Computing and Visualization in Science, 3*(1–2), 103–108.

White, F. M. (1974). *Viscous fluid flow* (1st ed.). New York: McGraw-Hill Inc.

Numerical Simulation of Cavitating Flows with Different Cavitation and Turbulence Models

Eric Goncalves

Abstract The simulation of cavitating flows is a challenging problem both in terms of modelling the thermodynamic path during the phase transition and the complex interaction with the turbulence. Based on a one-fluid compressible Reynolds-Averaged Navier-Stokes (RANS) solver, a numerical study is proposed to investigate and to compare different turbulence and cavitation models (with and without thermal effects). Steady and unsteady numerical results are given for 1D rarefaction cases and 2D Venturi geometries for which experimental data are available.

1 Introduction

Several physical and numerical models have been developed to investigate cavitating flows within the framework of averaged two-phase model. For the averaged model, there are different approaches according to the assumptions made on the local thermodynamic equilibrium and the slip condition between phases. A hierarchy of models exists, with the numbers of equations ranging from seven to three only. The full non-equilibrium two-fluid models with relaxation procedures have been tested on inviscid high-speed applications (see for example Petitpas et al. 2009; Zein et al. 2010), whereas one-fluid models have been massively used for industrial applications and turbomachineries.

An important class of reduced models is formed by the five-equation models, in which velocity equilibrium and pressure equilibrium are considered. The archetype five-equation model is that of Kapila et al. (2001). It is composed of four conservation laws: two for masses, one for the mixture momentum and one for the mixture energy. It is completed by an equation for a non-conservative quantity describing the flow

E. Goncalves (✉)
LEGI Laboratory, University of Grenoble, Grenoble, France
e-mail: Eric.Goncalves@ensma.fr

E. Goncalves
Institut Pprime, UPR 3346 CNRS, ENSMA, Futuroscope, Poitiers, France

L. d'Agostino and M.V. Salvetti (eds.), *Cavitation Instabilities and Rotordynamic Effects in Turbopumps and Hydroturbines*, CISM International Centre for Mechanical Sciences 575, DOI 10.1007/978-3-319-49719-8_8

topology, usually the void ratio. Such a model has been used for inviscid high-speed cavitating applications and cavitation pocket in fuel injector nozzles (Saurel et al. 2008, 2009).

By assuming the thermal equilibrium between phases, a four-equation model can be expressed. A very popular formulation has been developed to simulate turbulent cavitating flows (Merkle et al. 1998; Kunz et al. 2000; Senocak and Shyy 2002; Singhal et al. 2002; Venkateswaran et al. 2002; Vortmann et al. 2003; Hosangadi and Ahuja 2005; Utturkar et al. 2005; Wu et al. 2005; Zhang et al. 2008; Ji et al. 2013). It is composed by three conservation laws for mixture quantities (mass, momentum, energy) plus a mass equation for the vapour or liquid density including a cavitation source term. The main difficulty is related to the formulation of the source term and the tunable parameters involved for the vaporisation and condensation processes (different sets of parameters are presented in Utturkar et al. 2005). Moreover, the thermodynamic coherence was not attested and that constitutes a major problem (Goncalves and Patella 2011). Recently, another formulation for the mass transfer between phases was proposed using a void ratio transport-equation model. The mass transfer was closed assuming its proportionality to the divergence of the homogeneous velocity field (Goncalves 2013, 2014; Goncalves and Charriere 2014). Another popular model devoted to ebullition problems uses a mass fraction equation with a relaxation term (Homogeneous Relaxation Model). The source term involves a relaxation time, that is the time for the system to regain its thermodynamic equilibrium state. This time is very difficult to determine and is estimated from experimental data (Downar-Zapolski et al. 1996; Bilicki et al. 1996; Barret et al. 2002).

With the assumption of complete thermodynamic equilibrium between phases (local temperature, pressure and free Gibbs enthalpy equality between phases), we obtain the 3-equation models or homogeneous equilibrium models (HEM). Vaporisation or condensation processes are assumed to be instantaneous. An equation of state (EOS) is necessary to close the system. Different closure relations (tabulated EOS or combination of pure phase EOSs) that link the pressure to the thermodynamic variables have been proposed (Delannoy and Kueny 1990; Saurel et al. 1999; Schmidt et al. 1999; Clerc 2000; Edwards and Franklin 2000; Ventikos and Tzabiras 2000; Shin et al. 2003; Liu et al. 2004; Vuyst et al. 2005; Schmidt et al. 2006; Sinibaldi et al. 2006; Ihm and Kim 2008; Goncalves and Patella 2009; Goncalves et al. 2010b; Hu et al. 2011).

In addition, the turbulence modelling plays a determinant role in the capture of unsteady behaviours. Cavitation sheets that appear on solid bodies are characterized by a closure region, which always fluctuates with the existence of a re-entrant jet and the propagation of pressure waves. Moreover, compressibility effects on turbulence are involved. These effects and interactions with two-phase structures are not yet well known and understood. For usual applications, the Reynolds decomposition is often used with an averaged statistical processing resulting in the RANS equations for the mean flow quantities. The limitation of the turbulent viscosity evaluated with transport-equation turbulence models (through the Boussinesq assumption) is a key point to capture realistic cavitation sheets. Different methods have been investigated to limit or to correct standard turbulence models. One of the most popular limiter was proposed by Reboud to reduce the turbulent viscosity (Reboud et al. 1998), and has successfully been used by different authors (Coutier-Delgosha et al. 2002; Chen et al. 2006; Zhou and Wang 2008; Srinivasan et al. 2009; Goncalves 2011; Ahn and Kwon 2013).

The final goal of the present work is to compare various 3- and 4-equation models for the simulation of realistic turbulent cavitating flows involving unsteady sheet cavities. A comparison with incompressible numerical simulation is also proposed. Moreover, the turbulence modelling is investigated, especially the compressibility effects and the eddy viscosity limitation. The considered test cases are the following. First, 1D inviscid rarefaction tubes are computed allowing a comparison of models without the turbulence model influence. Second, two Venturi geometries are considered involving unsteady partial cavities for which the dynamic behaviour is different (an aperiodic quasi-stable sheet and a self-sustained periodic sheet). A good model should be able to simulate both dynamics. Finally, another Venturi case is studied for which the running fluid is freon R-114 in order to compare cavitation models including thermal effects.

This paper is organized as follows. We first review the theoretical formulation including cavitation models, equation of state and mass transfer formulations. Turbulence equations are detailed followed by the description of the 1-fluid solver. The preliminary studies carried out in inviscid test cases are presented. This is followed by sets of results on two turbulent Venturi flows and discussions.

2 The 1-Fluid Inviscid Formulation

The homogeneous mixture approach is used to model two-phase flows. The phases are assumed to be sufficiently well mixed and the disperse particle size are sufficiently small thereby eliminating any significant relative motion. The phases are strongly coupled and moving at the same velocity. In addition, the phases are assumed to be in thermal and mechanical equilibrium: they share the same temperature T and the same pressure P. The evolution of the two-phase flow can be described by the conservation laws that employ the representative flow properties as unknowns just as in a single-phase problem.

To obtain the one-fluid equations, we use the phase average of a phase quantity ϕ_k defined by Ishii and Hibiki (2006):

$$\overline{\overline{\phi_k}} = \frac{1}{T_k} \int \phi_k d\tau \qquad (1)$$

where T_k is the time period of phase k. The mass-weighted average or Favre average is commonly used for compressible flows. In the following, $\overline{\overline{()}}$ and $\tilde{()}$ will denote a phase average and a Favre average, respectively. Fluctuations with respect to the average will be $()'$ and $()''$, respectively. Using this definition and starting from the instantaneous conservation equations, the equations for each phase can be expressed. The equation for mixture quantities are obtained by the summation of the separate equations of the phase quantities.

We introduce α_k the void fraction or the averaged fraction of presence of phase k. The density ρ, the center of mass velocity u and the internal energy e for the mixture are defined by

$$\rho = \sum_k \alpha_k \overline{\overline{\rho_k}} \quad (2)$$

$$\rho u_i = \sum_k \alpha_k \overline{\overline{\rho_k}}\, \widetilde{u}_{k,i} \quad (3)$$

$$\rho e = \sum_k \alpha_k \overline{\overline{\rho_k}}\, \widetilde{e}_k \quad (4)$$

The mass gain per unit volume due to phase change (evaporation or condensation) is noted $\Gamma_k = \rho_k(u_I - u_k).n_k\delta_I$ where δ_I is a Dirac distribution having the different interfaces as a support, u_I the interface velocity and n_k the vector normal to the interface directed outward from phase k.

2.1 The Conservations Laws

The 3-equation system consists in mixture balance laws for density, momentum and total energy. In order to simplify the formulation, we present below the inviscid one-dimensional equations, expressed in conservative variables $w = (\rho, \rho u, \rho E)$:

$$\frac{\partial \rho}{\partial t} + \frac{\partial \rho u}{\partial x} = 0 \quad (5)$$

$$\frac{\partial(\rho u)}{\partial t} + \frac{\partial(\rho u^2 + P)}{\partial x} = 0 \quad (6)$$

$$\frac{\partial(\rho E)}{\partial t} + \frac{\partial(\rho u H)}{\partial x} = 0 \quad (7)$$

where $E = e + u^2/2$ denotes the mixture total energy and $H = h + u^2/2$ the mixture total enthalpy.

2.2 The Pure Phases EOS

In the present study, we used the convex stiffened gas EOS for the pure phases (see Metayer et al. 2004):

$$P(\rho, e) = (\gamma - 1)\rho(e - q) - \gamma P_\infty \tag{8}$$
$$P(\rho, T) = \rho(\gamma - 1)C_v T - P_\infty \tag{9}$$
$$T(\rho, h) = \frac{h - q}{C_p} \tag{10}$$

where $\gamma = C_p/C_v$ is the heat capacity ratio, C_p and C_v are thermal capacities, q the energy of the fluid at a given reference state and P_∞ is a constant reference pressure. The speed of sound c is given by

$$c^2 = \gamma \frac{P + P_\infty}{\rho} = (\gamma - 1)C_p T \tag{11}$$

2.3 Mixture EOS

Stiffened gas EOS On the basis of the stiffened gas EOS for each pure phase, an expression for the pressure can be deduced from the mechanical equilibrium assumption (Saurel et al. 2008). These expressions are available in all possible fluid states, function of the void ratio $\alpha = \alpha_v$ and the vapour mass fraction $Y = Y_v = \alpha\rho_v/\rho$:

$$P(\rho, e, \alpha, Y) = (\gamma(\alpha) - 1)\rho(e - q(Y)) - \gamma(\alpha)P_\infty(\alpha) \tag{12}$$
$$\frac{1}{\gamma(\alpha) - 1} = \frac{\alpha}{\gamma_v - 1} + \frac{1 - \alpha}{\gamma_l - 1} \tag{13}$$
$$q(Y) = Yq_v + (1 - Y)q_l \tag{14}$$
$$P_\infty(\alpha) = \frac{\gamma(\alpha) - 1}{\gamma(\alpha)} \left[\alpha \frac{\gamma_v}{\gamma_v - 1} P_\infty^v + (1 - \alpha) \frac{\gamma_l}{\gamma_l - 1} P_\infty^l \right] \tag{15}$$

The 3-equation system form a system of conservation laws having a hyperbolic nature. Without heat and mass transfer, the eigenvalues of the system are $u - c_{wallis}$, u and $u + c_{wallis}$.

Without mass transfer, the propagation of acoustic waves follows the Wood or Wallis speed of sound (Wallis 1967). This speed c_{wallis} is expressed as a weighted harmonic mean of speeds of sound of each phase c_k:

$$\frac{1}{\rho c_{wallis}^2} = \frac{\alpha}{\rho_v c_v^2} + \frac{1 - \alpha}{\rho_l c_l^2} \tag{16}$$

Sinusoidal EOS A sinusoidal law (Delannoy and Kueny 1990; Goncalves and Patella 2009) is considered for the mixture. This law is characterized by its maximum slope $1/c_{baro}^2$. The quantity c_{baro} is an adjustable parameter of the model, which can be interpreted as the minimum speed of sound in the mixture.

When the pressure is smaller than $P_{vap} + \Delta P$, the following relationship applies:

$$P(\alpha) = P_{vap} + \left(\frac{\rho_l^{sat} - \rho_v^{sat}}{2}\right) c_{baro}^2 \,\text{Arcsin}\,(1 - 2\alpha) \tag{17}$$

where ΔP represents the pressure range of the law and, for a void ratio value of 0.5, the pressure is equal to the saturation pressure P_{vap} at the reference temperature. This law introduces a small non-equilibrium effect on the pressure. The cavitation phenomenon is assumed to be isothermal and thermodynamic effects are neglected.

The void ratio is computed with saturation values of densities as follows:

$$\alpha = \frac{\rho - \rho_l^{sat}}{\rho_v^{sat} - \rho_l^{sat}} \tag{18}$$

The speed of sound in the mixture can be computed easily as follows:

$$c^2 = \left(\frac{\partial P}{\partial \rho}\right)_s = \left(\frac{\partial P}{\partial \rho}\right)_T = \frac{c_{baro}^2}{2\sqrt{\alpha(1-\alpha)}} \tag{19}$$

The system is hyperbolic with eigenvalues $(u - c, u, u + c)$.

Properties of the model (such as convexity conditions of the EOS) and the influence of the parameter c_{baro} have been studied in Goncalves and Patella (2009).

2.4 A Void Ratio Transport-Equation Model

The model consists in three conservation laws for mixture quantities (mass, momentum and total energy) and an additional equation for the void ratio. It is obtained from a reduction of a 5-equation model (Goncalves 2013) assuming the liquid is at its saturation state. The expression for the void ratio equation is

$$\frac{\partial \alpha}{\partial t} + \text{div}\,(\alpha V) = \left(\underbrace{\left(\frac{\rho_l c_l^2 - \rho_v c_v^2}{\frac{\rho_l c_l^2}{1-\alpha} + \frac{\rho_v c_v^2}{\alpha}} \right)}_{=K} + \alpha \right) \text{div}\, V + \underbrace{\left(\frac{\frac{c_v^2}{\alpha} + \frac{c_l^2}{1-\alpha}}{\frac{\rho_l c_l^2}{1-\alpha} + \frac{\rho_v c_v^2}{\alpha}} \right)}_{=1/\rho_I \text{ the interfacial density}} \Gamma_v \tag{20}$$

where Γ_v is the mass transfer between phases and V the velocity vector. The term K involves the speed of sound of pure phases c_k and it reflects the effects of changes in volume of each phase. Without mass transfer the system is hyperbolic, eigenvalues are (u-c_{wallis}, u, u, u+c_{wallis}).

By assuming that the mass transfer is proportional to the divergence of the velocity, it is possible to build a family of models in which the mass transfer is expressed as Goncalves (2013)

$$\Gamma_v = \frac{\rho_l \rho_v}{\rho_l - \rho_v}\left(1 - \frac{c^2}{c^2_{wallis}}\right) \operatorname{div} V \tag{21}$$

When heat and mass transfer effects are involved in the flow, the sound speed c decreases to the thermodynamic equilibrium one (Petitpas et al. 2009). This limit speed is evaluated with the assumption of local thermodynamic equilibrium: equalities of pressure, temperature and free enthalpy.

With this generic form for the mass transfer, we remark that all models in which the mixture speed of sound is the Wallis one cannot produce void ratio during the phase transition (that is the case of most of models proposed in the literature). The void ratio is only modified through the term $K \operatorname{div} V$.

Using different mixture EOS it is possible to built different models. Two models were built using the stiffened gas mixture equation and the sinusoidal EOS.

2.5 *Other Formulations for the Void Ratio Equation*

A class of cavitation models introduces a mass transfer between phases involving a separate contribution for vaporisation and condensation processes. Two tunable parameters are associated for each process. This empirical source term can be calibrated with experimental data base. Different formulations and sets of parameters are presented in Utturkar et al. (2005).

This kind of model reproduces propagation of acoustic disturbance at the Wallis speed of sound that is not thermodynamically coherent. Moreover, when pure phases are assumed to be constant at their saturation state, the void ratio is therefore proportional to the mixture density:

$$\alpha = \frac{\rho - \rho_l^{sat}}{\rho_v^{sat} - \rho_l^{sat}} \tag{22}$$

The mass and void ratio equations are not two independent equations. The compatibility of the system is not clear.

First model A formulation derived from the model proposed by Hosangadi and Ahuja (2005) can be expressed as:

$$\Gamma_v = C_{prod}\frac{\rho_v}{\rho_l}(1-\alpha)\frac{\text{Min}\,(0, P - P_{vap})}{0.5\rho_{ref}U^2_{ref}} + C_{des}\frac{\rho_v}{\rho_l}\alpha\frac{\text{Max}\,(0, P - P_{vap})}{0.5\rho_{ref}U^2_{ref}} \tag{23}$$

where C_{prod}, C_{des} are constants to calibrate.

With this formulation, the void ratio value can be higher than one. A limiter is added in the solver to clip the void ratio into its physical domain of evolution.

Second model The second model, implemented in the OpenFOAM solver (Weller et al. 1998), is derived from the Kunz model (Kunz et al. 2000). It is based on the incompressible RANS equations for an homogeneous mixture. The phase change is modelled using a transport-equation for the liquid volume fraction α_l that reads

$$\frac{\partial \alpha_l}{\partial t} + u_j \frac{\partial \alpha_l}{\partial x_j} = \Gamma_v \tag{24}$$

Following Kunz development, the source term is expressed as the sum of a vapourisation term m_v and a condensation term m_c:

$$\Gamma_v = m_v + m_c \tag{25}$$

with

$$m_c = \frac{\rho}{\rho_l \rho_v} C_c \frac{\rho_v}{t_\infty} \alpha_{Llim}^2 \frac{\max\left(P - P_{vap}\,;\,0\right)}{\max\left(P - P_{vap}\,;\,0.01\,P_{vap}\right)} \tag{26}$$

$$m_v = \frac{\rho}{\rho_l \rho_v} C_v \frac{\rho_v}{\frac{1}{2}\rho_l U_\infty^2 t_\infty} \min\left(P - P_{vap}\,;\,P_0\right) \tag{27}$$

C_c, C_v, U_∞ and t_∞ are constant set by the user, whereas P_0 and α_{Llim} are included to avoid non- physical values. Usually U_∞ is set to the freestream value, and t_∞ represents a relaxation time not well defined in the literature. For the present computations, the following values are specified:

$$C_c = 10\,;\; C_v = 8000\,;\; t_\infty = 0.005$$

3 The Thermal Effects Modelling

We detailed the formulation in the case of thermo-sensitive fluid and non isothermal thermodynamic path.

3.1 A Mixture of Stiffened Gas EOS

By assuming the thermal equilibrium between phases, the mixture temperature is expressed as

$$T(\rho, h, Y) = \frac{h_l - q_l}{C_{p_l}} = \frac{h_v - q_v}{C_{p_v}} = \frac{h - q(Y)}{C_p(Y)} \quad (28)$$

$$C_p(Y) = YC_{p_v} + (1 - Y)C_{p_l} \quad (29)$$

We assume that the vaporisation pressure varies linearly with the temperature:

$$P_{vap}(T) = P_{vap}(T_{ref}) + \frac{dP}{dT}(T - T_{ref}) \quad (30)$$

The constant quantity dP/dT is evaluated with a thermodynamic table. The speed of sound in the mixture can be expressed as Goncalves (2014):

$$C_1 = \frac{1}{\gamma - 1} + \frac{\rho_v h_v - \rho_l h_l}{(\rho_l - \rho_v)\frac{dP}{dT}} \left(\alpha \frac{d\rho_v}{dT} + (1 - \alpha)\frac{d\rho_l}{dT} \right) \quad (31)$$

$$\rho c^2 = \frac{1}{C_1} \left[\frac{\rho_v \rho_l}{(\rho_l - \rho_v)}(h_v - h_l) \right] \quad (32)$$

Enthalpies of pure phase h_l and h_v are computed with the mixture temperature T. Due to numerical problems, the derivatives of densities with the temperature are not taken into account in this study.

3.2 A Modified Sinusoidal EOS

A modified sinusoidal relation (Goncalves and Patella 2010) can be considered for the mixture. When the pressure is smaller than $P_{vap}(T) + \Delta P$, the following relationship applies:

$$P(\alpha, T) = P_{vap}(T) + \left(\frac{\rho_l^{sat} - \rho_v^{sat}}{2} \right) c_{baro}^2 \,\mathrm{Arcsin}\,(1 - 2\alpha) \quad (33)$$

where ΔP represents the pressure range of the law and, for a void ratio value of 0.5, the pressure is equal to the saturation pressure $P_{vap}(T)$ at the local temperature T. This temperature is evaluated using the relation (29). ρ_l^{sat} and ρ_v^{sat} are pure phases densities at saturation evaluated at the reference temperature T_{ref}. The quantity c_{baro} is a parameter of the model, it is no more the minimum sound speed in the mixture.

As previously we assume that the vaporisation pressure varies linearly with the temperature and follows the relation (30). The speed of sound in the mixture can be evaluated (Goncalves and Patella 2010):

$$c^2 = \frac{\frac{\rho_v \rho_l}{\rho(\rho_l - \rho_v)}(h_v - h_l)\frac{dP}{dT} + \rho C_p c_T^2}{\rho C_p - \frac{dP}{dT}} \quad (34)$$

$$c_T^2 = \left(\frac{\partial P}{\partial \rho}\right)_s = \left(\frac{\partial P}{\partial \rho}\right)_T = \frac{c_{baro}^2}{2\sqrt{\alpha(1-\alpha)}} \tag{35}$$

where c_T is the isothermal speed of sound (i.e. when $dP/dT = 0$).

3.3 A Void Ratio Transport-Equation Model

We use the generic formulation previously presented:

$$\Gamma_v = \frac{\rho_l \rho_v}{\rho_l - \rho_v}\left(1 - \frac{c^2}{c_{wallis}^2}\right) \operatorname{div} V \tag{36}$$

The liquid density ρ_l is assumed to be in its equilibrium state at the reference temperature: $\rho_l = \rho_l^{sat}(T_{ref})$. We did not test to introduce a thermal variation for the liquid density. The vapour density ρ_v varies with the temperature through the pure phase EOS.

Two models can be built using the stiffened gas and sinus EOSs. As the temperature and pressure relations are coupled, an iterative procedure on the temperature is introduced. The temperature is initialized by the liquid temperature computed with the pure phase EOS. Five iterations are done and it has been checked that the numerical solution did not change with a higher number of iteration. More details are given in Goncalves (2014).

4 The 1-Fluid RANS Formulation

We detailed the RANS formulation associated with the Boussinesq approximation and transport-equation turbulence models.

4.1 The Mixture Conservation Laws

The classical form for the homogeneous mixture balance equations are Ishii and Hibiki (2006):

$$\frac{\partial \rho}{\partial t} + \frac{\partial (\rho u_j)}{\partial x_j} = 0 \tag{37}$$

$$\frac{\partial}{\partial t}\left(\rho u_i\right) + \frac{\partial\left(\rho u_i u_j + P\delta_{ij}\right)}{\partial x_j} = \frac{\partial\left(\tau^v_{ij} + \tau^t_{ij}\right)}{\partial x_j} \tag{38}$$

$$\frac{\partial}{\partial t}(\rho E) + \frac{\partial\left[(\rho E + P)u_i\right]}{\partial x_j} = \frac{\partial\left[(\tau^v_{ij} + \tau^t_{ij})u_j - Q^v_i - Q^t_i\right]}{\partial x_j} \tag{39}$$

where E is the mixture total energy. The mixture total stress tensor τ_{ij} (with viscous and turbulent contributions) is evaluated using the Stokes hypothesis, Newton's law and the Boussinesq assumption. The mixture total heat flux vector Q_i is obtained from the Fourier law involving a turbulent thermal conductivity λ_t with the constant Prandtl number hypothesis.

$$\tau_{ij} = \tau^v_{ij} + \tau^t_{ij} = (\mu + \mu_t)\left[\left(\frac{\partial u_i}{\partial x_j} + \frac{\partial u_j}{\partial x_i}\right) - \frac{2}{3}\frac{\partial u_l}{\partial x_l}\delta_{ij}\right] + \frac{2}{3}\rho k\delta_{ij}$$

$$Q_i = Q^v_i + Q^t_i = -\left(\lambda + \lambda_t\right)\frac{\partial T}{\partial x_i} \quad \text{with} \quad \lambda_t = \frac{\mu_t C_p}{P_{rt}} \tag{40}$$

where μ_t is the mixture eddy viscosity, k the mixture turbulent kinetic energy (TKE), C_p the mixture thermal capacities and P_{rt} the turbulent Prandtl number set to 1.

In pure liquid, the viscosity is determined by an exponential law and, in pure vapour, the viscosity follows the Sutherland law. The mixture viscosity is defined as the arithmetic mean of the liquid and vapour viscosities (fluctuations of viscosity are neglected) (Ishii and Hibiki 2006):

$$\mu_l(T) = \mu_{0_l}\exp(B/T) \tag{41}$$

$$\mu_v(T) = \mu_{0_v}\sqrt{\frac{T}{293}}\frac{1 + T_S/293}{1 + T_S/T} \tag{42}$$

$$\mu(\alpha, T) = \sum_k \alpha_k\mu_k \tag{43}$$

where μ_{0_l}, μ_{0_v}, B and T_S are constants.

The mixture thermal conductivity λ is also defined as the arithmetic mean of the liquid and vapour values:

$$\lambda = \sum_k \alpha_k\frac{\mu_k C_{p_k}}{P_{r_k}} \tag{44}$$

4.2 Mixture Turbulence Equations for Two-Phase Flows

The mixture TKE equation Different authors have studied the turbulent kinetic energy equation for a gas-liquid flow. The mixture equation can be obtained by the summation of the separate equations of the phase turbulent quantities with homogeneous mixture assumptions (Besnard and Harlow 1988; Elghobashi and Abou-Arab 1983; Kataoka and Serizawa 1989; Lance et al. 1984). The expression of the mixture TKE equation is

$$\frac{\partial \rho k}{\partial t} + \frac{\partial \rho k u_l}{\partial x_l} = \rho P_k + \rho \Pi - \rho \epsilon + \rho M + \rho D + \Gamma \tag{45}$$

where $k = \sum_k \alpha_k \frac{\widetilde{u''_{k,i} u''_{k,i}}}{2}$, P_k is the mixture production term, Π is the mixture pressure-dilatation term, ε is the mixture TKE dissipation rate which can be split into a solenoidal part ε_s and a dilatational part ε_d, M is the mixture turbulent mass flux, D is the mixture turbulent and viscous diffusions. The last term Γ is specific to multiphase flows. It involves the mass transfer due to phase change and is usually set to zero.

In multiphase flow, even if pure phases are assumed to be incompressible, the divergence of the fluctuating phase velocity is not zero, caused by the presence of interfaces. Indeed, with this assumption, the divergence of the fluctuating phase velocity is

$$\frac{\partial u''_{k,i}}{\partial x_i} = -\frac{\partial \widetilde{u_{k,i}}}{\partial x_i} = \frac{1}{\alpha_k} u''_k . n_k \delta_I \tag{46}$$

Therefore, the dilatational dissipation rate $\varepsilon_d = \frac{4}{3} \sum_k \alpha_k \overline{\overline{\mu_k}}\, \overline{\overline{s'_{k,jj} s'_{k,ll}}}$ is not zero. And the pressure-dilation term $\Pi = \sum_k \alpha_k \overline{\overline{P'_k \frac{\partial u''_{k,i}}{\partial x_i}}}$ is instantaneously present but null in mean as $\overline{\overline{P'_k}} = 0$.

The modelling of these additional terms is very complex to realize in turbulent cavitating flows due to the lack of experimental data for turbulent quantities and DNS of cavitation flows. Only closure relations developed for aerodynamic high-speed flows have been tested.

TKE dissipation rate equation for the mixture Kataoka and Serizawa (1989) derived a transport equation for the mixture turbulent dissipation rate assuming that both phases were considered as incompressible. This equation put in evidence that the turbulent dissipation is affected by interfacial transport terms. Nevertheless, no theoretical or experimental results regarding these terms are available in case of two-phase flows with phase change. Usually, it is used the same equation for ε than in single-phase flows.

4.3 Turbulence Models

Various popular turbulence models are considered: the Smith $k-\ell$ model (KL) (Smith 1994), the one-equation Spalart–Allmaras model (SA) (Spalart and Allmaras 1994), the Wilcox $k-\omega$ model (Wilcox 1988) and the Jones-Launder $k-\varepsilon$ model (KE) (Jones and Launder 1972). For comparisons with the OpenFOAM solver, the Menter $k-\omega$ SST model (Menter 1994) is used, assuming the validity of the Bradshaw assumption in a two-phase turbulent boundary layer.

Compressible terms have been included in both the $k-\varepsilon$ and $k-\omega$ models.

A $k-\varepsilon$ model in compressible form We include the compressible terms Π, M and ε_d into the Jones- Launder $k-\varepsilon$ model:

$$\frac{\partial \rho k}{\partial t}+\frac{\partial}{\partial x_j}\left[\rho k u_j-\left(\mu+\frac{\mu_t}{\sigma_k}\right)\frac{\partial k}{\partial x_j}\right]=P_k-\rho\varepsilon_s+\rho\varepsilon_d+\rho\Pi+\rho M$$
$$\frac{\partial \rho \varepsilon_s}{\partial t}+\frac{\partial}{\partial x_j}\left[\rho \varepsilon_s u_j-\left(\mu+\frac{\mu_t}{\sigma_\varepsilon}\right)\frac{\partial \varepsilon_s}{\partial x_j}\right]=c_{\varepsilon 1}\frac{\varepsilon_s}{k}P_k-\rho c_{\varepsilon 2}f_2\frac{\varepsilon_s{}^2}{k}+E$$

The low Reynolds number term E, the damping functions f_μ and f_2 and all constants follow the Jones–Launder formulation (Jones and Launder 1972).

The additional terms Π and ε_d are modelled according to Sarkar propositions (Sarkar 1992):

$$\rho\Pi=-\alpha_2\rho P_k M_t+\alpha_3\rho\varepsilon_s M_t^2$$
$$\varepsilon_d=\alpha_1\varepsilon_s M_t^2 \tag{47}$$

where $M_t=\frac{\sqrt{k}}{2c}$ is the turbulent Mach number defined with the mixture speed of sound c and α_i are constants to calibrate.

The turbulent mass flux M is closed with the Jones formulation (Jones 1979):

$$\rho M=-\frac{\mu_t}{\rho^2\sigma_p}\frac{\partial\rho}{\partial x_i}\frac{\partial P}{\partial x_i} \tag{48}$$

where σ_p is a turbulent Schmidt number close to one.

A $k-\omega$ model in compressible form Wilcox (1998) proposed a compressible version of his $k-\omega$ model:

$$\frac{\partial \rho k}{\partial t} + \frac{\partial}{\partial x_j}\left[\rho k u_j - \left(\mu + \sigma^* \mu_t\right)\frac{\partial k}{\partial x_j}\right] = P_k - \beta_c^* \rho k \omega \tag{49}$$

$$\frac{\partial \rho \omega}{\partial t} + \frac{\partial}{\partial x_j}\left[\rho \omega u_j - \left(\mu + \sigma \mu_t\right)\frac{\partial \omega}{\partial x_j}\right] = \alpha \frac{\omega}{k} P_k - \beta_c \rho \omega^2 \tag{50}$$

with $\mu_t = \rho \frac{k}{\omega}$. The compressibility effects on turbulence is accounted through the coefficients β_c^* and β_c functions of the turbulent Mach number:

$$\beta_c^* = \beta^* \left(1 + \xi^* F(M_t)\right) \tag{51}$$

$$\beta_c = \beta - \beta^* \xi^* F(M_t) \tag{52}$$

$$F(M_t) = \left(M_t^2 - M_{t0}^2\right) H\left(M_t - M_{t0}\right) \tag{53}$$

with $M_{t0} = 0.25$ and $\xi^* = 1.5$.

4.4 Eddy Viscosity Limitation

Turbulence models always leads to the generation of stable cavities, because very strong turbulent eddy viscosity μ_t inside the cavity avoids the re-entrant jet development and unsteadiness formation. The link to compressibility effects on turbulence is not clear. In cavitating flows, supersonic regime is reached in the mixture area, because of the drastic diminution of the speed of sound. The detailed mechanisms of the interaction between turbulent flows and cavitation have not yet been clearly revealed especially for the phenomena occurring at small scales.

As a remedy to limit the turbulent viscosity, one can use an eddy viscosity limiter in the mixture area. The most famous limiter is the one proposed by Reboud et al. (1998), which has shown its efficiency in the simulation of sheet cavities (Coutier-Delgosha et al. 2002; Chen et al. 2006; Zhou and Wang 2008; Srinivasan et al. 2009; Goncalves 2011; Ahn and Kwon 2013). Other corrections can be used, for example the Shear Stress Transport (SST) correction proposed by Menter (1994) to reduce the eddy viscosity in case of positive pressure gradient and a variant of the latter based on realizability constraints (Durbin 2009). Applications to cavitating simulation can be found in Decaix and Goncalves (2012).

Reboud correction Reboud proposed an arbitrary limiter by introducing a function $f(\rho)$ in the computation of the turbulent viscosity for the $k - \varepsilon$ model:

$$\mu_t = f(\rho) C_\mu \frac{k^2}{\varepsilon} \quad \text{with } f(\rho) = \rho_v + (1-\alpha)^n (\rho_l - \rho_v) \tag{54}$$

where n is a parameter fixed to 10.

This correction can be extended to other turbulence models with the same function $f(\rho)$.

Recent experimental works in a cavitating mixing layer (Aeschlimann et al. 2011) showed that the turbulent kinetic energy decreased when the cavitation developed but the turbulent viscosity (evaluated through the Boussinesq approximation) remained quasi constant. As a consequence, the Reboud limiter is not valid in all cavitating flows.

Menter SST correction The Menter correction is based on the empirical Bradshaw's assumption, which binds the shear stress to the turbulent kinetic energy for two-dimensional boundary layer. The stress ratio predicted by two-equation models scales with the ratio of production P_k to dissipation ε as

$$\frac{-\overline{u'v'}}{k} = \sqrt{\frac{P_k}{\varepsilon}}\sqrt{C_\mu} \qquad \text{with } C_\mu = 0.09 \tag{55}$$

Single-phase experiments show that the quantity $-\overline{u'v'}/k \leq 0.3$. Menter devised his SST limiter from this inequality. The empirically based constraint is expressed as

$$\nu_t = \min\left[\frac{k}{w}, \frac{0.3k}{\sqrt{2} \mid \Omega \mid F_2(y)}\right] \tag{56}$$

where F_2 is a blending function that tends to zero outside the boundary layer, and Ω is the vorticity.

The evolution of the ratio $-\overline{u'v'}/k$ was recently measured by Aeschlimann and Barre (2009) in the case of a cavitating mixing layer. The authors showed that this ratio decreased significantly and continuously when cavitation developed. They observed a factor 2 between the non-cavitating regime and the most severe cavitating case. Yet, the evolution of this ratio for cavitating flows involving sheet cavities is unknown.

Realizability constrains By replacing the vorticity Ω with the stress tensor S in the SST formula, a correction is derived based on the realizability principle (Durbin 2009). The correction involves a parameter c varying between 0 and 1 and set to 0.3.

4.5 Scale-Adaptive Simulation

The Scale-Adaptive Simulation first appeared in the formulation of the KE1E one-equation model for the eddy viscosity derived from the standard $k - \varepsilon$ model (Menter 1997). This model provided a dynamical behaviour similar to a Detached Eddy Simulation (DES) model but without an explicit grid dependence, allowing the development of a turbulent spectrum in the detached regions. Later, Menter and Egorov (2004) showed that the exact transport equation for the turbulent length scale l as derived by Rotta actually introduces the second derivative of the velocity field. They proposed a model for this term which introduces the von Karman length-scale L_{vk}.

This term, referred to as SAS, was inserted in the existing $k - \omega$ SST model and they calibrated it for the decay of a homogeneous isotropic turbulence. Various test cases (Menter 2010a, b; Davidson 2006) have showed the ability of the SST-SAS model to resolve turbulent structures down to the limit grid and have confirmed that the SAS term reduces the turbulent eddy viscosity and predicts resolved fluctuations much larger than standard models.

It is possible to include scale-adaptive terms into the $k - \ell$ and Spalart–Allmaras turbulence models following the works of Menter and Egorov. Details of the formulation are given in Decaix and Goncalves (2012, 2013b). The SAS term is only activated in the mixture area when $\alpha > 0$.

4.6 Wall Functions

For the modelling of flow close to the wall, a two-layer wall law approach is used as follows:

$$\begin{aligned} &u^+ = y^+ && \text{if} \quad y^+ < 11.13 \\ &u^+ = \frac{1}{\kappa} \ln y^+ + 5.25 && \text{if} \quad y^+ > 11.13 \\ &u^+ = \frac{u}{U_\tau} \quad ; \quad y^+ = \frac{yU_\tau}{\nu_w} \quad ; \quad U_\tau = \sqrt{\frac{\tau_w}{\rho_w}} && \end{aligned} \tag{57}$$

where $\kappa = 0.41$ is the von Karman constant and the subscript 'w' is used for a wall value.

With regard to the turbulent transport-equation models, the production of k is computed according to the formulation proposed by Viegas and Rubesin (1983). The value of ℓ in the first cell is computed with the linear relation $l = \kappa y$.

For the one-equation Spalart–Allmaras model, the transported quantity is calculated using the model's closure relations, the velocity profile and a mixing-length formulation for the eddy viscosity. More details concerning the wall law approach are given in Goncalves and Decaix (2012).

We assume that wall functions are similar in a two-phase flow and in a single-phase flow. For unsteady flows, the existence of a wall law is assumed to be valid at each instant. In order to investigate the validity of these assumptions, comparisons were proposed with a thin boundary layer approach (TBLE). This approach is based on the integration of a simplified averaged set of partial equations derived from the Navier–Stokes equations:

$$\frac{\partial u_i}{\partial t} + \frac{\partial u_i u_j}{\partial x_j} + \frac{1}{\rho}\frac{dP}{dx_i} = \frac{\partial}{\partial y}\left[(\mu + \mu_t)\frac{\partial u_i}{\partial y}\right] \tag{58}$$

A fine one-dimensional grid is embedded between the first grid point and the wall (see Fig. 1), and the turbulent boundary-layer equations are discretized and solved in

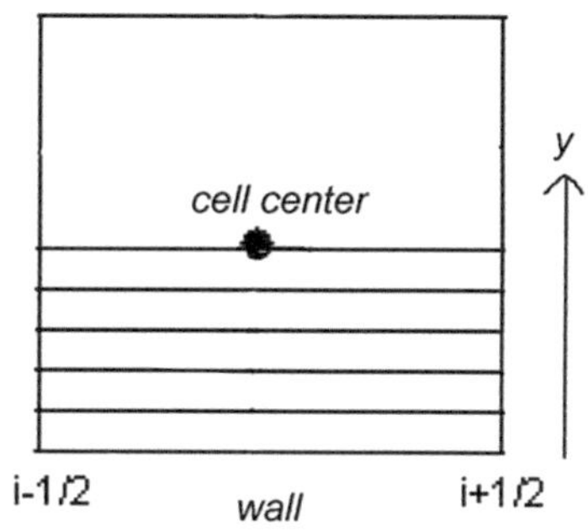

Fig. 1 Embedded grid in the adjacent cells to walls

the embedded mesh. The integration is performed using a Newton algorithm for the wall shear stress τ_w. In the present computations, the number of nodes in the embedded grid is 30. More details can be found in Goncalves and Decaix (2012).

5 The 1-Fluid Filtered Formulation

The major difficulty in modelling multiphase turbulence is the wide range of length and time scales on which turbulent mixing occurs. The Direct Numerical Simulation (DNS) approach, with no modelling, resolves all the scales present in the flow. However, it is not feasible for practical engineering problems involving high Reynolds numbers. The large-eddy simulation (LES) is an intermediate way between DNS and RANS in terms of the fraction of the resolved scales. In LES, the large-scale turbulent motions are resolved while the effects of the smaller ones are modelled. This approach allows a significant decrease in the computational cost over direct simulation and captures more dynamics than a simple RANS model.

A first point is to identify the scales at which the governing equations are to be applied: micro-scales (scales which are small enough to describe individual bubble shapes), meso-scales (which are comparable to bubble sizes) and macro-scales (which entail enough bubbles for statistical representation). When LES is applied at a micro-scale, filtering of turbulent fluctuations needs to be combined with interface tracking methods (for example Liovic and Lakehal (2012)). When LES is applied at a macro-scale, the interface resolution is not considered.

Different authors have studied the 1-fluid filtered equations for incompressible gas–liquid flows (Labourasse et al. 2007; Vincent et al. 2008; Toutant et al. 2009; Larocque et al. 2010; Bois et al. 2010). The scale separation is mathematically obtained by applying a convolution product using a large-scale-pass filter function G. For a quantity ϕ, the filtered variable is defined as $\overline{\phi} = G \circ \phi$ and the Favre filtered variable: $\tilde{\phi} = \overline{\rho\phi}/\overline{\rho}$. Let X_k be the phase indicator function ($X_k = 1$ in phase k and 0 elsewhere). The filtered phase indicator function $\overline{X_k} = G \circ X_k = \alpha_k$ can be interpreted as a filtered volume fraction of phase k.

It is assumed that the filtering operator commutes with time and spatial derivatives. The mass transfer is assuming to be proportional to the velocity divergence through a constant C. Applying the filter G to the governing system leads to the filtered 1-fluid equations system:

$$\frac{\partial \overline{\rho}}{\partial t} + \frac{\partial \left(\overline{\rho}\tilde{u}_i\right)}{\partial x_j} = 0 \tag{59}$$

$$\frac{\partial \overline{\rho}\tilde{u}_i}{\partial t} + \frac{\partial \left(\overline{\rho}\tilde{u}_i\tilde{u}_j + \overline{P}\delta_{ij}\right)}{\partial x_j} = \operatorname{div}\left(2\overline{\mu}\tilde{S} + 2\tau_{\mu S} - 2\tau_{\rho S} - \tau_{\rho uu}\right) \tag{60}$$

$$\frac{\partial \overline{\rho}\tilde{E}}{\partial t} + \frac{\partial \left(\overline{\rho}\tilde{H}\tilde{u}_i + \overline{Q}_i^{v}\right)}{\partial x_j} = \operatorname{div}\left(2\tau_{\mu S u} - 2\overline{\mu}\,\tilde{V}\tau_{\rho S} - 2\overline{\mu}\tilde{S}\tau_{\rho u} + 4\overline{\mu}\tau_{\rho u}\tau_{\rho S} + \overline{\rho}\tilde{H}\tau_{\rho u} - \tau_{\rho H u}\right)$$

$$\frac{\partial \alpha}{\partial t} + \tilde{V}.\nabla\alpha - C\frac{\partial \tilde{u}_i}{\partial x_j} = -\tau_{u\alpha} - C\operatorname{div}\left(\tau_{\rho u}\right) \tag{61}$$

with the appearance of subgrid terms:

$$\begin{array}{lll}
\tau_{u\alpha} = \overline{V.\nabla X_v} - \tilde{V}.\nabla\alpha & ; & \tau_{\rho u} = \tilde{V} - \overline{V} \\
\tau_{\mu S} = \overline{\mu S} - \overline{\mu}\overline{S} & ; & \tau_{\rho S} = \tilde{S} - \overline{S} \\
\tau_{\rho H u} = \overline{\rho H V} - \overline{\rho}\tilde{H}\tilde{V} & ; & \tau_{\mu S u} = \overline{\mu S V} - \overline{\mu}\overline{S V} \\
\tau_{\rho uu} = \overline{\rho}\left(\widetilde{V \otimes V} - \tilde{V} \otimes \tilde{V}\right) & &
\end{array}$$

The subgrid terms must be formally modelled. A part of theses terms is present in standard single-phase compressible LES equations (the subgrid stress tensor), while the subgrid interfacial transport $\tau_{u\alpha}$ is specific to two-phase flows. The magnitude of the different subgrid terms was a priori evaluated in the case of phase separation flows and turbulence bubble interaction (Labourasse et al. 2007; Vincent et al. 2008). Following these studies, the influence of $\tau_{u\alpha}$ is highly dependent on the flow configurations and/or on the chosen two-phase description. Yet, in cavitating flows, the influence and the hierarchy of all these terms have never been investigated.

Following the works of Bois et al. (2010) about the concept of continuous LES in two-phase flows, a supplementary term appears in the viscous stress tensor formulation due to the mass transfer and the jump relations:

$$\tau_{ij} = 2\mu S_{ij} - \frac{2}{3}\mu S_{kk}\delta_{ij} + \mu\Gamma_v\left(\frac{1}{\rho_v} - \frac{1}{\rho_l}\right)\left(2 - \frac{2}{3}\delta_{ij}\right)n_i\delta_I \tag{62}$$

This term has never been studied in cavitation.

6 The Cavitation Software CaviFlow

The numerical simulations are carried out using an implicit compressible solver for multi-domain structured meshes. This solver is based on a cell-centred finite-volume discretization.

6.1 Spatial Discretization

For the mean flow, the convective flux density vector on a cell face is computed with the Jameson scheme (Jameson et al. 1981). The viscous terms are discretized by a second-order space-centred scheme. For the turbulence transport equations, the upwind Roe scheme (Roe 1981) is used to obtain a more robust method. The second-order accuracy is obtained by introducing a flux-limited dissipation.

6.2 The Low Mach Number Preconditioner

For low Mach number applications, a well-known problem concerns the stiffness on the solution convergence. In this situation, the dominance of convection terms renders the system stiff and compressible solvers converge slowly. To overcome this difficulty, a preconditioning method is necessary. The physical acoustic waves are replaced by pseudo-acoustic modes that are much closer to the advective velocity, reducing the stiffness and enhancing the convergence. The method is based on the modification of the derivative term by a premultiplication with a suitable preconditioning matrix. In order to simplify the formulation, we present below the one-dimensional formulation. With the primitive variables $W = (P, u, e, \alpha)$ the preconditioned Euler equations can be expressed as

$$P_e^{-1}\frac{\partial W}{\partial t} + A_e\frac{\partial W}{\partial x} = 0 \tag{63}$$

We use the preconditioning matrix based on the Turkel approach (Guillard and Viozat 1999; Turkel 1987):

$$P_e = \begin{bmatrix} \beta^2 & 0 & 0 & 0 \\ 0 & 1 & 0 & 0 \\ 0 & 0 & 1 & 0 \\ 0 & 0 & 0 & 1 \end{bmatrix} \quad ; \quad A_e = \begin{bmatrix} u & \rho c^2 & 0 & 0 \\ 1/\rho & u & 0 & 0 \\ 0 & P/\rho & u & 0 \\ 0 & -K & 0 & u \end{bmatrix}$$

β is a parameter on the order of the Mach number expressed by Choi and Merkle (1993): $\beta^2 = \min\left[\max\left(M^2, \theta M_\infty^2\right), 1\right]$

This form implies that there is no preconditioning used in transonic and supersonic flow regions (in the mixture). When $\beta^2 = 1$, the preconditioning matrix becomes the identity matrix and the system returns to its classical non-preconditioned form. Moreover, for a very small flow velocity, β^2 is not allowed to be less than a given percentage of the freestream velocity, determined by the coefficient θ.

The eigenvalues of the preconditioned system are

$$u; u \quad ; \quad \lambda_{\pm} = \frac{1}{2}\left[u(1+\beta^2) \pm \sqrt{(\beta^2-1)^2u^2 + 4\beta^2c^2}\right] \tag{64}$$

For the variables $w = [\rho, \rho u, \rho E, \alpha]$, the method involves the preconditioning matrix $P_c^{-1} = \frac{\partial w}{\partial W} P_e^{-1} \frac{\partial W}{\partial w}$ and the Jacobian matrix of the convective fluxes A_c. Expressions of matrices are given in Goncalves and Charriere (2014).

6.3 Temporal Discretization

Time integration is achieved using the dual time stepping approach and a low-cost implicit method consisting in solving, at each time step, a system of equations arising from the linearization of a fully implicit scheme. The derivative with respect to the physical time is discretized by a second-order formula. The main advantage of this method is that the storage of the Jacobian matrix is completely eliminated, which leads to a low-storage algorithm (Goncalves and Patella 2009).

With the preconditioned method, the dissipation matrices are modified and the preconditioned matrix P_c^{-1} remains in the formulation. By judiciously exploiting the idempotence propriety of the matrix M, it is possible to preserve a low-cost system where matrix operations and matrix-vector products can be easily computed.

For the turbulence transport equations, the diffusive flux Jacobian matrix is replaced by its spectral radius. The source term needs special treatment (Merci et al. 2000). Only the negative part of the source term Jacobian matrix is considered and replaced by its spectral radius. The system obtained is solved with a line-alternated Jacobi relaxation algorithm.

6.4 Inlet and Outlet Boundary Conditions

The numerical treatment of the boundary conditions is based on the use of the preconditioned characteristic relations of Euler equations. The number of variables to impose at boundaries is given by the number of characteristics directed into the domain of interest. The characteristic relations obtained for the preconditioned system, in two-dimensional flows, are

$$-c^2(\rho^c - \rho^s) + (P^c - P^s) = 0 \tag{65}$$

$$V_t^c - V_t^s = 0 \tag{66}$$

$$\rho(\alpha^c - \alpha^s) - K(\rho^c - \rho^s) = 0 \tag{67}$$

$$(\lambda_+ - V_n)(P^c - P^s) + \rho\beta^2 c^2(V_n^c - V_n^s) = 0 \tag{68}$$

$$(\lambda_- - V_n)(P^c - P^s) + \rho\beta^2 c^2(V_n^c - V_n^s) = 0 \tag{69}$$

The variables with superscript c denote the variables to be computed at the boundary. Variables with superscript s denote the variables obtained by the current numerical scheme. V_t and V_n are the tangential and the normal component of the mean velocity, respectively.

At inflow, we impose the stagnation pressure P_i, the stagnation temperature T_i, the direction of the velocity and the initial values of the void ratio. The pressure is evaluated with the relation (69) and all variables can be evaluated at the boundary.

At outflow, the static pressure is imposed. The conservative variables are computed with four characteristic relations (65)–(68).

7 Test Cases Used for the Numerical Studies

To investigate and to compare models various test cases are considered as follows:

- 1D inviscid tube cases
  These cases involve a large rarefaction area in a tube which leads to the apparition of cavitation. The ability of cavitation models to reproduce the physical phenomenon without viscosity and turbulence problems can be evaluated.
- Cavitation sheet in Venturi geometries in turbulent flow
  Two Venturi configurations are studied for which the divergent angle is different. Partial sheet cavities appear along the solid wall downstream the Venturi throat. Two dynamics behaviour were experimentally observed: a transitional aperiodic phenomenon and a self-sustained periodic one.
- For the study of thermal effects, a Venturi configuration in which the running fluid is freon R-114.
  For this case, a steady cavitation sheet develops downstream the throat of the Venturi.

7.1 Water–Gas Mixture Expansion Tube at Different Velocities

A double rarefaction tube problem is considered with an initial velocity discontinuity located at the middle of the tube. This test, proposed by Saurel et al. (2008), consists in a one meter long tube filled with liquid water at atmospheric pressure and with

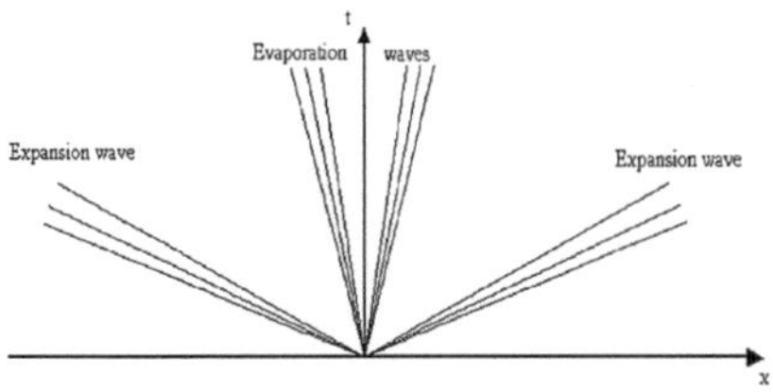

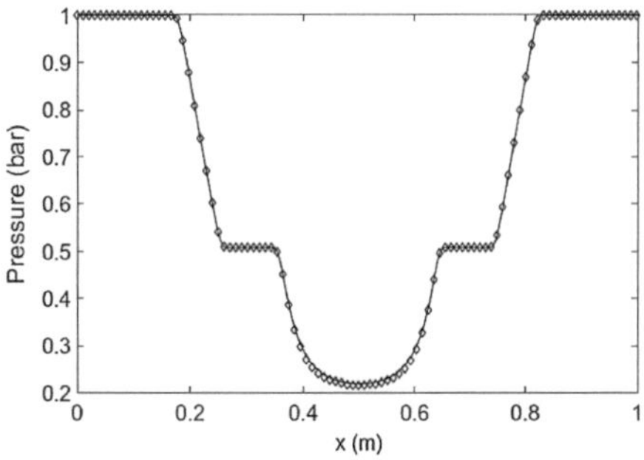

Fig. 2 Expansion waves and pressure evolution in the tube, extracted from Zein et al. (2010)

density ρ_l =1150 kg/m^3. The temperature of water is T_{ref} = 355 K. A weak volume fraction of vapor $\alpha = 0.01$ is initially added to the liquid. The initial discontinuity is set at 0.5 m, the left velocity and the right velocity have the same value but an opposite sign. Two velocities are tested: $u = \pm 2$ m/s and $u = \pm 100$ m/s. The vapour pressure $P_{vap}(T_{ref})$ is set to 51000 Pa at the initial temperature.

The solution involves two expansion waves. As gas is present, the pressure cannot become negative. To maintain positive pressure, the gas volume fraction increases due to the gas mechanical expansion and creates a pocket. Liquid water is expanded until the saturation pressure is reached then evaporation appears and quite small amount of vapor is created. The solution with phase transition is composed of four expansion waves. The extra two expansion waves correspond to the evaporation fronts (see Fig. 2).

These cases were computed in Zein et al. (2010) using a two-fluid model in which instantaneous relaxation processes towards equilibrium are included for the temperature and the Gibbs free energy. The mesh contains 5000 cells. The time step is set to 10^{-7} s. The value of c_{baro} is set to 1.31 m/s for the 4-equation sinus model as presented in Goncalves (2013).

7.2 Venturi Data in Cold Water

Experiments have been performed in a closed loop at CREMHyg (Centre d'Essais de Machines Hydrauliques de Grenoble). Two flow configurations have been tested corresponding to flat upper and side walls, whereas the height of the lower wall varied in order to form a convergent-divergent nozzle. In the present considered experiments, case 1 was constituted by an angle of the divergent part of 4°. Case 2 corresponds to a convergent part of about 18° and the divergent part was 8°. The pressure P_{inlet} was lowered until the desired cavitation number allowing two specific behaviours of the sheet cavity. The cavitation number in the inlet section is defined

Table 1 Flow configuration, Venturi geometries

Experimental parameters	Case 1	Case 2
Angle of the divergent	4°	8°
Inlet velocity U_{inlet}	10.8 m/s	7.04 m/s
Inlet pressure P_{inlet}	0.35 bar	0.52 bar
Cavitation parameter σ_{inlet}	0.547 ± 0.05	2.15 ± 0.06
Reynolds number $Re_{L_{ref}}$	$2.7\ 10^6$	$1.57\ 10^6$
Sheet cavity	Smoothly fluctuating	Self-oscillating
Clouds shedding frequency	None	45 Hz
Time-averaged cavity length	~80 mm	–

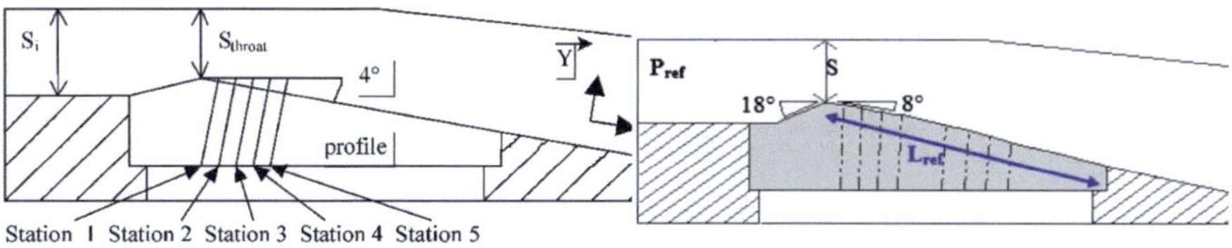

Fig. 3 Venturi design and probe locations, 4° divergence angle (*left*) and 8° angle (*right*)

as: $\sigma_{inlet} = \dfrac{P_{inlet} - P_{vap}}{0.5\rho U_{inlet}^2}$, where P_{vap} is the vapour pressure at 20 °C and P_{inlet}, U_{inlet} are the pressure and velocity respectively at the reference section upstream of the Venturi. The flow conditions are given in Table 1.

With these previous parameters and according to experimental observations (Barre et al. 2009; Aeschlimann et al. 2013), cavitation sheets developed from the Venturi throat (Fig. 3). The obtained cavity length is ranging from 70 to 85 mm for case 1 and having a relatively stable aspect (see Fig. 4). The attached cavity length corresponding to the end of the re-entrant jet is around 30 mm. For this case, no periodic cycles with large shedding were observed.

For case 2, a typical self-oscillation behaviour was observed with quasi-periodic vapour clouds shedding and the maximum cavity length (before the break-off of the cavity) was 45 ± 5 mm. Figure 4 shows an instantaneous photograph of the cavity with a large structure shedding. The cloud shedding frequency was about 45 Hz leading to the classical Strouhal number based on the cavity length of about 0.3. The divergent part was equipped with eight probe holes to take various measurements such as the instantaneous pressure, local void ratio and velocity in the sheet cavity. Specific probe locations are presented in Fig. 3 for the two tested cases.

Numerical data Both grids used for simulations are a H-type topology. For the non-cavitating regime, computations are started from an uniform flow-field using a local time step. For the unsteady cavitating regime, computations are performed with the dual time stepping method and are started from the non-cavitating numerical solution. All numerical values are obtained by a time-averaged treatment. For case 2, a

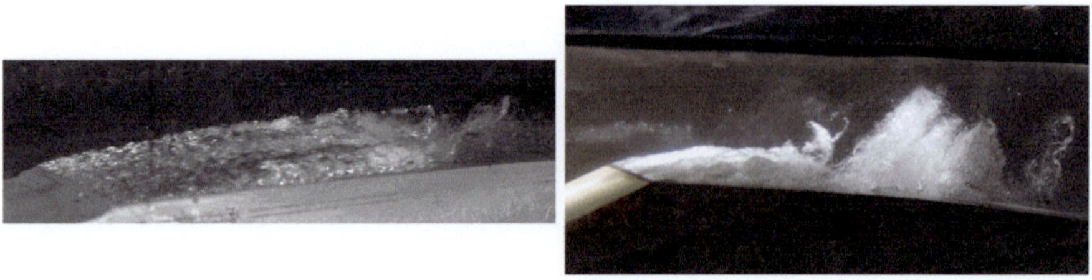

Fig. 4 Photograph of cavities, 4° Venturi (*left*) and 8° Venturi (*right*)

direct Fourier transformation (DFT) of the vapour volume signal was performed to evaluate the frequency of the phenomenon.

7.3 Venturi Data in Freon R-114

The freon R-114 experimental facility of the CREMHyG is a closed loop operating with a reference pressure, obtained by pressurizing a tank with nitrogen gas. The cavitation tunnel was designed to simulate cavitating flows developing on the blades of space turbopump inducers (Stutz 1996; Fruman et al. 1999). The loop is fitted with a test section having the shape of a two-dimensional Venturi, characterized by a divergence angle of 4° (similar to the previous case 1). This geometry is equipped with three probing holes to take various measurements. Optical probes and micro-thermocouples are used to evaluate the local void ratio and the wall temperature, respectively. Flow conditions and experimental parameters are given in Table 2. With these parameters, a cavity length around 80 mm was obtained, with a relatively stable aspect.

The freon R-114 allows to study the thermodynamic effect in ambient conditions. It provides the same temperature depression $\Delta T^* = 1.22$ K in comparison with the liquid hydrogen at $T_{ref} = 22$ K.

All cavitating simulations are steady computations, which are started from the non-cavitating numerical solution. The H-type grid contains 251 nodes in the flow direction and 77 in the orthogonal direction.

Table 2 Flow configuration for the Venturi case with freon R-114

Inlet velocity V_{inlet}	14.4 m/s
Inlet pressure P_{inlet}	265300 Pa
Reference temperature T_{ref}	$\simeq$293 K
Vapour pressure at T_{ref}	181100
Cavitation parameter σ_{inlet}	$\simeq$0.55
Reynolds number $Re_{L_{ref}}$	18.4×10^6

8 Comparison of Turbulence Models

We present here only results obtained on the first Venturi (4° divergence angle). All unsteady simulations were performed with the same cavitation model: the 3-equation sinus model.

8.1 Eddy Viscosity Limitation

A qualitative view of the cavity sheet is plotted in Fig. 5 with the contours of the density gradient modulus (Schlieren-like visualization). On the left, the simulated cavity using the Spalart–Allmaras model with the Reboud correction is in good agreement with the experimental visualization. The numerical sheet presents an attached cavity up to the abscissa $x = 0.03$ m and downstream a recirculating area with small cavitation clouds. The re-entrant jet is well captured. On the contrary, the solution obtained without the Reboud limiter presents an extended attached cavity (up to the abscissa $x = 0.06$ m) followed by a very small recirculation. The re-entrant is not developed.

The effect of the eddy viscosity limiter is shown in Fig. 6 where are plotted the profiles of the ratio μ_t/μ at station 3 (the re-entrant jet reaches this station). On the left, we can clearly observed the drastic reduction of μ_t using the Reboud limiter in comparison with the computation without correction. In the sheet cavity, the ratio is close to zero. On the left, a similar observation is done using the KE model and

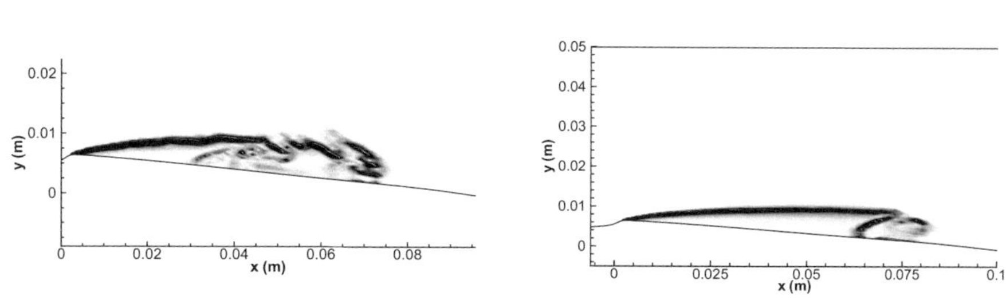

Fig. 5 Contour of the density gradient: with (*left*) and without (*right*) the Reboud limiter

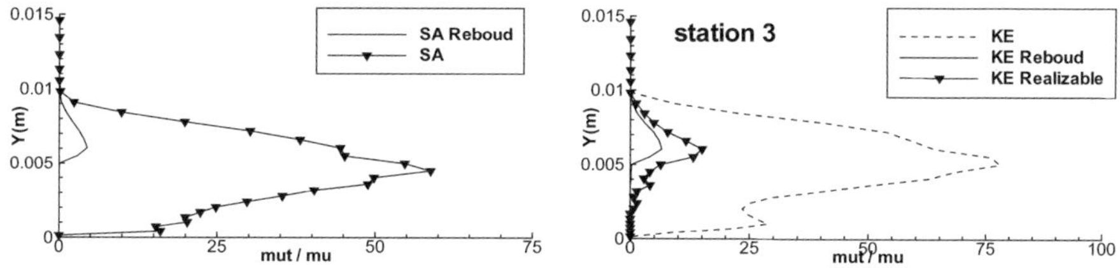

Fig. 6 Time-averaged ratio μ_t/μ at station 3: SA model (*left*) and KE model (*right*)

the Reboud limiter. Moreover, the use of realizability constrains allows similarly to strongly reduce the eddy viscosity and to make possible the re-entrant jet development.

The Reboud limiter can be associated to different transport-equation models with the same value of the parameter $n = 10$. It clearly improves the behaviour of the model and the prediction of the re-entrant jet. Figure 7 shows the time-averaged longitudinal velocity profiles (on the right) and void ratio profiles (on the left) for the experiments and computations using different turbulence models. Four models are tested: Spalart–Allmaras, $k - \ell$, $k - \varepsilon$ and $k - \omega$ SST and three limiters are considered: Reboud, realizability and SST correction. For all simulations, the re-entrant jet is well developed and the intensity of the recirculation is correctly simulated except with the KE Reboud model. For the void ratio profiles, discrepancies appear between models especially for the near-wall value. All models over-estimate the void ratio value at wall. The thickness of the cavity is well predicted by all simulations. More comparisons can be found in Goncalves (2011).

Using the Wilcox $k - \omega$ turbulence model, we observed a strong influence of the value of the parameter n. Contours of the density gradients are illustrated in Fig. 8 for two values of the parameter n. On the left, with $n = 4$, we obtained a sheet cavity in correct agreement with the experiments. Using the reference value $n = 10$, a large shedding of two-phase structures is simulated and the re-entrant jet is too extended. Moreover, the calibration of n depends on the configuration. For the second Venturi

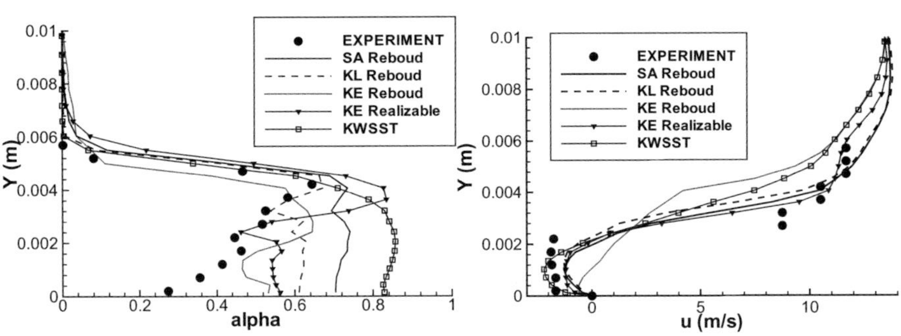

Fig. 7 Void fraction (*left*) and velocity (*right*) profiles at station 3

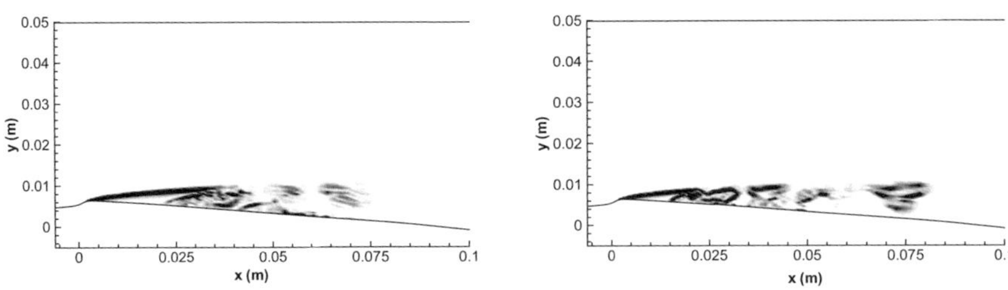

Fig. 8 Contour of the density gradient: $n = 4$ (*left*) and $n = 10$ (*right*)

case (8° of divergence angle), the value $n = 4$ is not adapted, whereas the value $n = 10$ allows to correctly simulate the sheet cavity dynamic (see Decaix and Goncalves 2013a).

The SST correction can be associated to other turbulence models. For example, we simulated the Venturi case using a KL-SST model. The value of the ratio $u'v'/k = c$ is a key point. The original value is $c = 0.3$. Yet, we have no idea of this value in cavitating flows. We performed three simulations by varying the parameter c from 0.3 to 0.1. Using the original value $c = 0.3$, we obtained a similar result in comparison with a computation without correction. The instantaneous density gradient contours are plotted in Fig. 9 for $c = 0.2$ (on the left) and $c = 0.1$ (on the right). The intermediate value $c = 0.2$ allows to simulate a sheet cavity in correct agreement with the experimental data. Using the smallest value $c = 0.1$, a large shedding of two-phase structures is computed, which is not observed in the experiments.

The time-averaged velocity and void ratio profiles at station 3 are plotted in Fig. 10. The three simulations with the KL-SST model are presented and also a KL realizable computation with $c = 0.2$. Using the value $c = 0.3$, the re-entrant jet is not computed at station 3 and the void ratio is close to 1. The decrease of the constant allows the development of the re-entrant jet. More comparisons can be found in Decaix and Goncalves (2012).

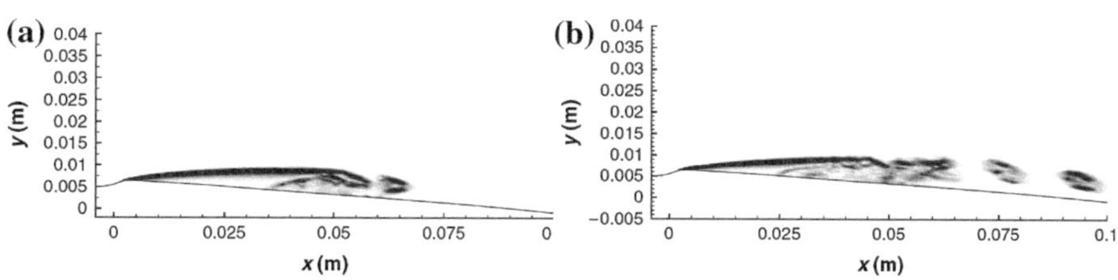

Fig. 9 Contour of density gradients: $c = 0.2$ (*left*) and $c = 0.1$ (*right*)

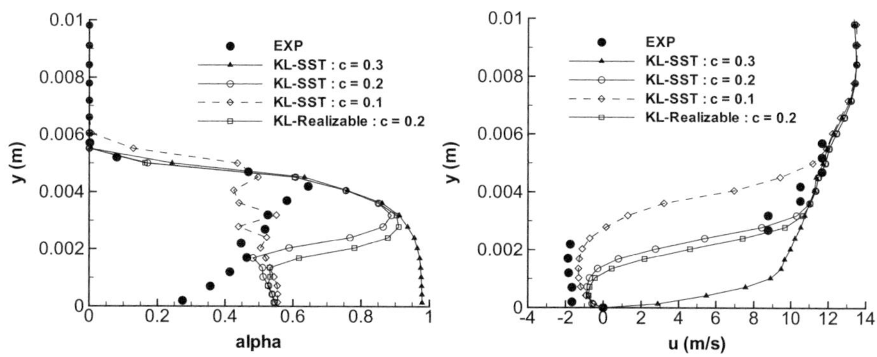

Fig. 10 Time-averaged void ratio (*left*) and velocity (*right*) profiles, KL-SST model

8.2 Compressibility Effects

We consider here the compressible formulation of the $k - \varepsilon$ model including the pressure-dilation term Π, the extra-dissipation ε_d and the mass flux term M. All these terms appear in the mixture TKE equation due to the fact that the divergence of the fluctuating velocity field is not null in two-phase flows. Unfortunately, closure relations in cavitating flows do not exist. In the following, we show numerical results obtained using the high-speed aerodynamic closure relations proposed by Sarkar.

The contours of density gradients at two instants are illustrated in Fig. 11 using the $k - \varepsilon$ model with all compressible terms Π, M and ϵ_d and with the original Sarkar values. We clearly observed the large shedding of two-phase structures and a very small attached cavity, which is absolutely not in agreement with the experiments.

We tried to evaluate separately the influence of the additional terms. First, we computed the Venturi case with the $k - \varepsilon$ model including only the mass flux term M. The closure relation involves a turbulent Schmidt number σ_p close to 1. Different values were tested for this parameter. The contours of the density gradient modulus are illustrated in Fig. 12 computed with $\sigma_p = 1$ (left) and $\sigma_p = 0.0001$ (right). Both results simulations exhibit a large attached cavity and a small recirculation on the closure part, which is similar to those obtained with the standard model. The influence of the mass flux term seems to be weak in the mixture TKE equation.

A computation using the $k - \varepsilon$ model with only the pressure-dilation term Π is presented in Fig. 13 where the density gradients are plotted. The influence of the parameter α_3 for the Sarkar correction is shown, the value of the constant α_2 being

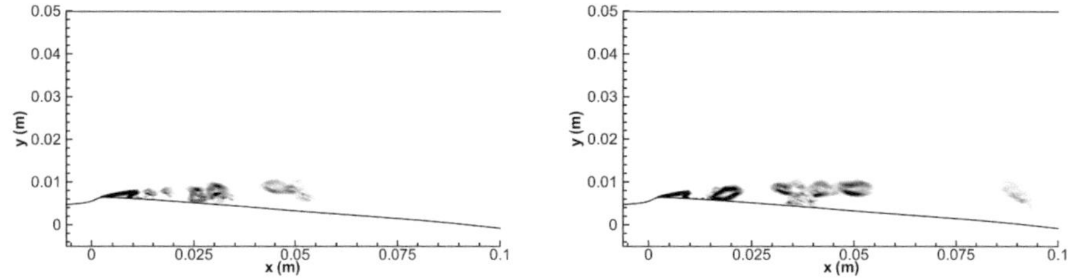

Fig. 11 Contours of the density gradient, $k - \epsilon + \Pi + M + \epsilon_d$ using Sarkar values, at 2 instants

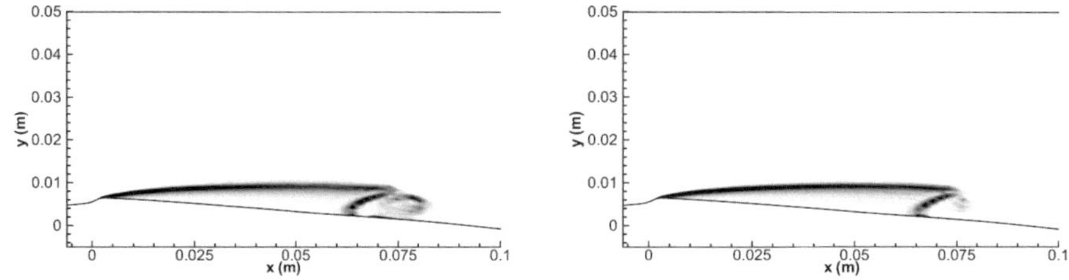

Fig. 12 $k - \epsilon +$ mass flux term, $\sigma_p = 1$ (*left*) and $\sigma_p = 0.0001$ (*right*)

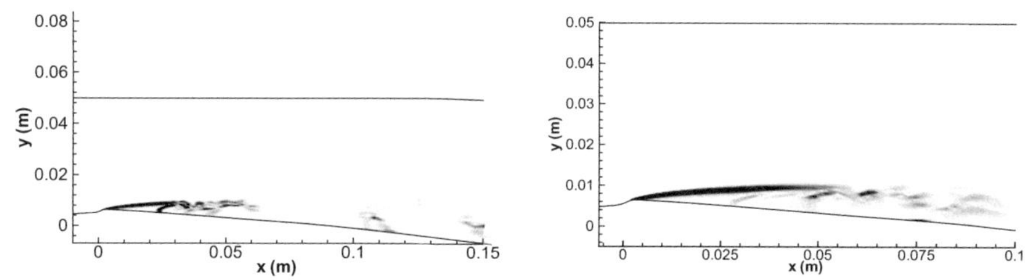

Fig. 13 Contours of the density gradient, $k - \varepsilon + \Pi$, α_2=0.15, $\alpha_3 = 0.001$ (*left*) and $\alpha_3 = 0.025$ (*right*)

set to 0.15. A weak value of α_3 leads to the large shedding of two-phase structures. By increasing α_3, it is possible to simulate a sheet cavity in relatively correct agreement with the experiments. Unfortunately, the simulation of the second Venturi case using this value $\alpha_3 = 0.025$ did not provide a correct result. A similar conclusion was done for the calibration of the extra-dissipation term. As a consequence, it was impossible to calibrate these compressibility corrections for both Venturi cases (see Decaix and Goncalves 2013a).

8.3 *Wall Models and Near-Wall Mesh*

This numerical work investigates the mesh influence in the near-wall area when a partial cavity develops and interacts with the turbulent boundary layer. All meshes contain 251 nodes in the flow direction and vary by the first cell size and the distribution in the orthogonal direction. The y^+ values obtained from a non cavitating simulation vary between 9 and 60 near the throat (Fig. 14). All meshes are computed using a two-layer wall model.

Comparisons are focused on the velocity profiles at station 3 and the development of the re-entrant jet. Four meshes and four turbulence models (coupled with the Reboud correction except the SST model) are compared in Fig. 15. According to

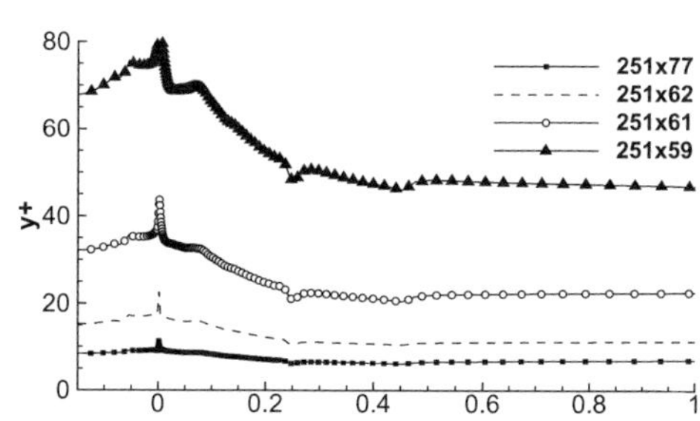

Fig. 14 Evolution of y^+ values near the throat, 4° Venturi

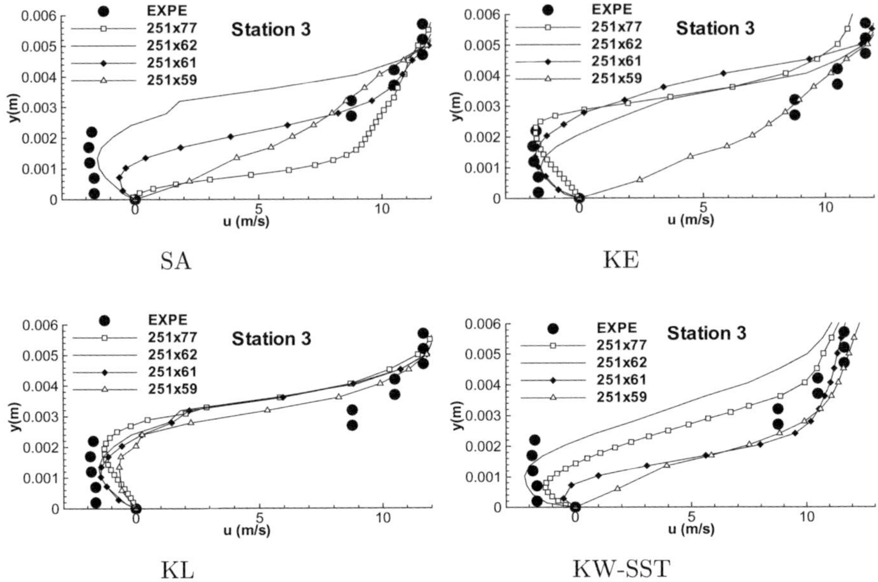

Fig. 15 Time-averaged velocity profiles at station 3, mesh and turbulence model comparisons

the turbulence model, different behaviour can be highlighted. For the $k - \ell$ model, results provided by all meshes are similar and in very close agreement with the experimental values. For the $k - \varepsilon$ model, the re-entrant jet is not predicted using the coarsest mesh 251×59. With the other meshes, results are relatively similar and in close agreement with the experiments. For the Spalart–Allmaras model, large discrepancies between meshes appear. Using the coarsest and finest meshes, the re-entrant jet is not predicted. For the intermediate mesh, the recirculating area obtained with the mesh 251×62 is in better agreement in comparison with the experimental data. The sensitivity to the near-wall mesh refinement seems important with this turbulence model. Using the Menter SST model, the best result is also obtained with the mesh 251×62. The finer mesh provides a less intense recirculating area and the solution is degraded when a coarser mesh is used. With the coarsest mesh, the re-entrant jet is not predicted. To conclude, the near-wall mesh can be determinant for the good simulation of the re-entrant jet.

Computations were performed to compare the two-layer wall functions and the TBLE model, associated with the Smith $k - \ell$ turbulence model and the mesh 251×62 nodes. Figure 16 shows the velocity and void ratio profiles at station 3. Results obtained with both wall treatments are similar and in very good agreement with the experimental data. A similar conclusion has been done for the computation of the second Venturi case (see Goncalves and Decaix (2012)). The two-layer single-phase wall functions seem to be a good approximation.

For the second Venturi case, the experimental frequency of the periodic shedding is around 45 Hz and the inlet cavitation parameter $\sigma_{inlet} \sim 2.15$. Similar to the pre-

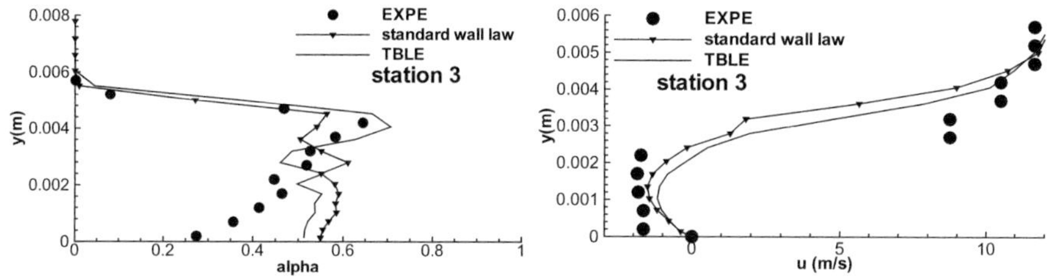

Fig. 16 Time-averaged velocity and void ratio profiles at station 3, two-layer model versus TBLE

Table 3 Frequency and CPU cost

Mesh	y^+	σ_{inlet}	Frequency (Hz)	CPU ratio
174 × 77	2	2.13	30	1
		2.18	No frequency	
174 × 62	4	2.13	35	0.817
		2.17	40	
174 × 60	8	2.13	35	0.80
		2.19	43.5	
174 × 59	12	2.14	44	0.776
174 × 57	20	2.145	46	0.747
174 × 56	30	2.14	46	0.736

vious study, different meshes varying by the first cell size are compared using the $k - \ell$ Reboud turbulence model coupled with a two-layer wall model. The simulation time is around 2 s. The numerical frequency and the cavitation parameter are given in Table 3.

The ratio between CPU time for a given case and the fine mesh computation is also given. With the coarsest mesh, the gain is around 25%. About the frequency prediction, with the finest mesh 174 × 77 computed without wall functions, it was not possible to capture a periodic self-oscillating cavity with a frequency close to 45 Hz. For values of the inlet cavitation parameter near 2.15, the obtained frequency is around 30 Hz. For higher values of σ_{inlet}, no frequency appears. For all other meshes computed with wall functions, a correct frequency can be obtained. According to the mesh refinement, the σ_{inlet} increase allows to reach the experimental value. For the coarsest meshes, both the frequency and the σ_{inlet} value are close to the experimental data. Moreover, the local velocity profiles obtained with all meshes are similar (not presented here).

8.4 Scale-Adaptive Simulation

We compare unsteady simulations performed on the 4° Venturi case using the $k - l$ turbulence model coupled with the Reboud correction and the scale-adaptive term. This term is activated in the mixture area and allows to reduce the eddy viscosity. A qualitative description of the computed sheet cavities is presented in Fig. 17 with the plot of the the density gradient contours. The attached cavity and downstream the recirculating area are well predicted by both models. The $k - \ell$ SAS model simulates a more homogeneous cavity in comparison with the $k - \ell$ Reboud model.

About the velocity and void ratio profiles, both models provided similar results close to the experimental data (see Decaix and Goncalves 2012). The dimensionless wall pressure distribution $(P - P_{vap})/P_{vap}$ is plotted in Fig. 18 (on the left) along the wall. The first five data are located inside the cavity (where the void ratio and velocity profiles are measured). All models provide a pressure distribution similar to the experimental measurements. The Root Mean Square (RMS) wall pressure fluctuations are plotted in the right part. The pressure fluctuation is divided by the time-averaged pressure Pav. Experimental data indicate an augmentation of pressure fluctuations at the end of the sheet cavity, with a peak located at the fifth station. The standard model (without any corrections) yields a pressure fluctuation profile close to the experimental profile even if the peak is slightly over-estimated. The pressure

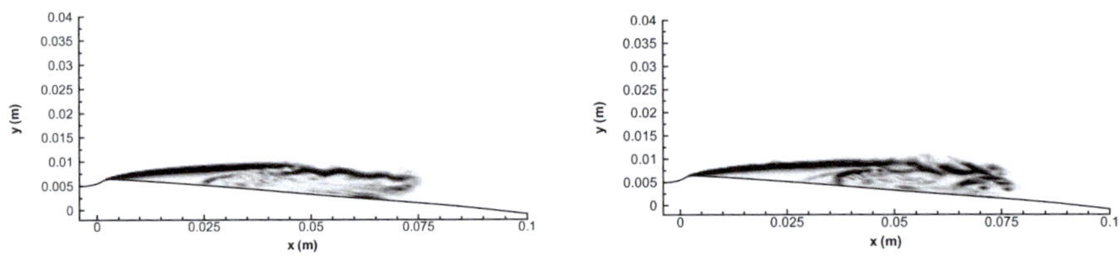

Fig. 17 Contours of density gradient: $k - \ell$ SAS (*left*) versus $k - \ell$ Reboud (*right*)

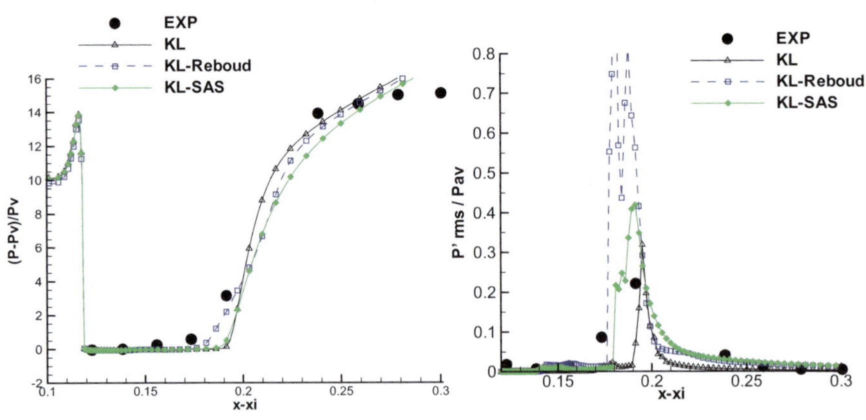

Fig. 18 Time-averaged wall pressure evolution (*left*) and RMS wall pressure fluctuations (*right*)

fluctuation computed by the $k - \ell$ Reboud model presents a different behaviour with two peaks more than three times higher than the experimental peak. Finally, the profile simulated by the $k - \ell$ SAS model is in good agreement with the experimental data even if the peak is over-estimated.

A three-dimensional study of the same Venturi case was performed. The square-section of the tunnel was meshed and a similar distribution was used in the span direction z and the y direction leading to $251 \times 62 \times 62$ nodes. The three-dimensional sheet cavity topology and the interaction with the side walls can be observed using an iso-surface of the void fraction. Figure 19 displays such an iso-surface for a value of the void fraction equal to 60%, for a $k - \ell$ SAS computation, at two different instants.

Comparisons between 2D and 3D simulations concern time-averaged quantities extracted on the mid-span plane. Figure 20 shows the longitudinal velocity and void ratio profiles at station 3 obtained with 2D and 3D SAS computations ($k - \ell$ and Spalart-Allmaras models). The void fraction predicted by the 3D computations is lower than in the 2D computations. For the velocity profiles, all simulations provide similar results in good agreement with the experimental data.

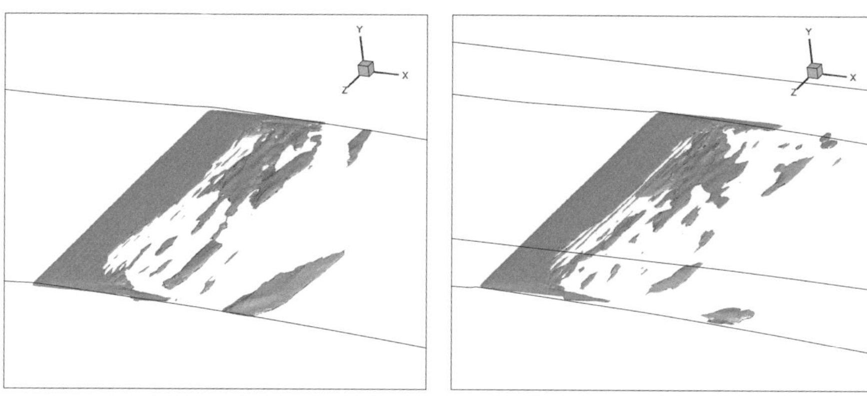

Fig. 19 Iso-surface of the void fraction for the value of 60% at two different instants, $k - \ell$ SAS model

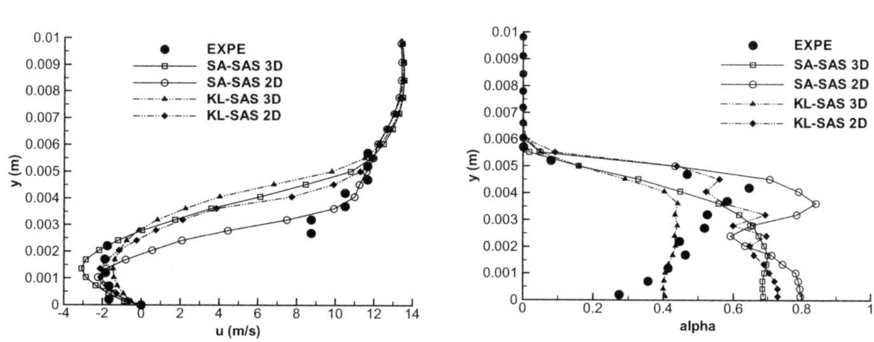

Fig. 20 Time-averaged velocity (*right*) and void ratio (*left*) profiles at station 3, mid-span plane

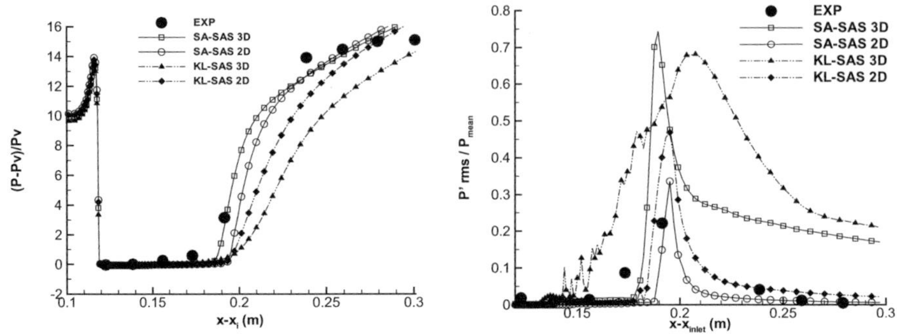

Fig. 21 Time-averaged wall pressure (*left*) and RMS fluctuations (*right*), mid-span plane

The time-averaged wall pressure, plotted in Fig. 21, does not evolve in the same way for the 3D computations. For the Spalart–Allmaras model, the 3D computation leads to a stronger re-compression area in comparison with the 2D computation. It is the inverse situation for the $k - \ell$ model. The RMS wall pressure fluctuations (on the right) put in evidence large discrepancies between 2D and 3D computations. For the 2D computations, the peak of fluctuations is correctly located with an amplitude in correct agreement with the experimental measurements. Moreover, the decrease of the fluctuations level in the wake of the cavitation sheet is well captured. This is not the case for the 3D computations since the level of pressure fluctuations is largely over-estimated in the wake. More details concerning the investigation of 3D computations can be found in Decaix and Goncalves (2013b).

8.5 Concluding Remarks

The turbulence modelling in cavitating flows is confronted with numerous questions such as the compressibility effects, the small scale effects, the near-wall modelling, the 3D phenomenon, the anisotropy of the Reynolds tensor, etc.

Hybrid RANS-LES models is a way to improve the level of resolved scales. We expect that large-eddy simulation of cavitating flows will bring us information on the flow dynamics, the turbulence structure and the interaction with cavitation.

9 Comparison of Cavitation Models

9.1 Inviscid Tube Cases

Different simulations were performed to compare the 4-equation sinus model and the 4-equation Hosangadi model by varying the initial discontinuity velocity u_{init}

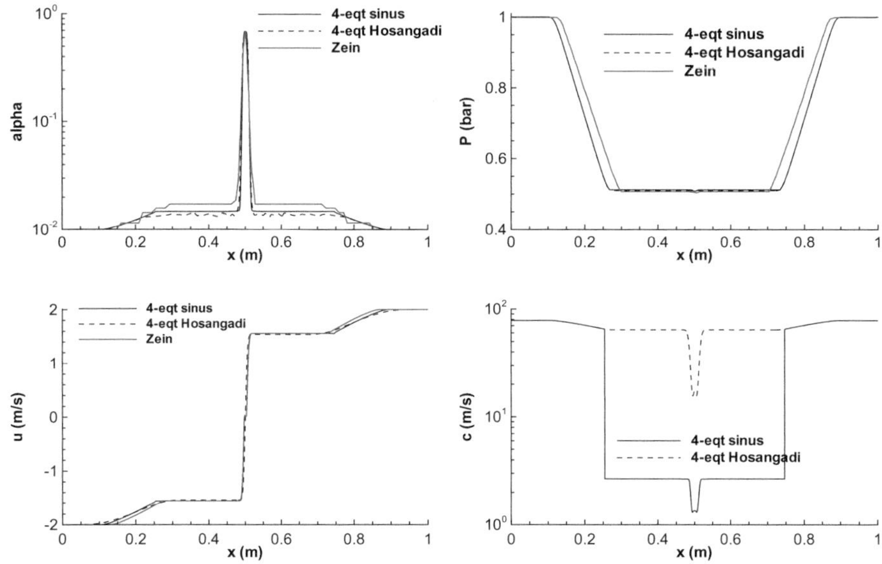

Fig. 22 Water–gas double rarefaction with cavitation $u = 2$ m/s, $\alpha_{init} = 10^{-2}$, mesh 5000 cells, t = 3.2 ms. Void ratio, pressure, velocity, and speed of sound profiles

and the initial value of the void fraction α_{init}. For these cases, only the evaporation process is present and only the C_{prod} parameter is studied. For the sinus model, the parameter c_{baro} is set to 1.31 m/s for all computations.

First, we computed the case $u_{init} = 2$ m/s and $\alpha_{init} = 10^{-2}$. Solutions are compared in Fig. 22 at time $t = 3.2$ ms with the two-fluid solution of Zein et al. (2010). The C_{prod} parameter was calibrated using the two-fluid solution and set to 1. Results obtained with both models are in close agreement with the 2-fluid solution for the void ratio and velocity profiles. For the pressure evolution, a small discrepancy appears for the location of the rarefaction fronts, certainly due to a small difference on the mixture speed of sound. An important difference between 4-equation models concerns the evaporation fronts, which are not simulated using the Hosangadi model. A part of the physics is missing with this model. Finally, large discrepancies are noticed about the mixture speed of sound profiles (plotted in log scale). It is due to the fact that the speed of sound associated to the Hosangadi model is the Wallis one.

Second, we computed the case $u_{init} = 100$ m/s and $\alpha_{init} = 10^{-2}$. Solutions are compared in Fig. 23 at time $t = 1.5$ ms with the two-fluid solution of Zein et al. (2010). The C_{prod} parameter was calibrated using the two-fluid solution and set to 100. Results obtained with both models are in good agreement with the 2-fluid solution for the void ratio and the velocity. In this case, evaporation is much more intense resulting in a large cavitation pocket where the gas volume fraction is close to 1. However, this pocket does not contain pure gas but a mixture at thermodynamic equilibrium. Moreover, thermal effects are not negligible and a large temperature drop

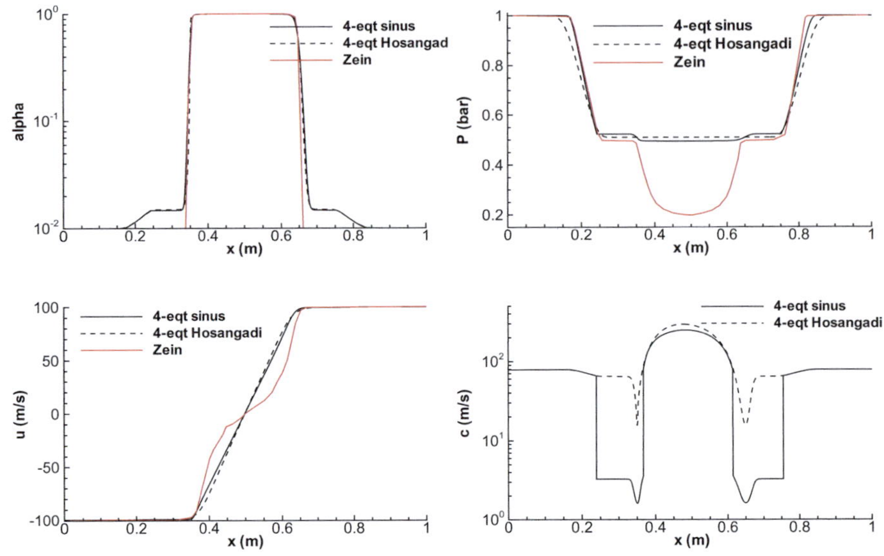

Fig. 23 Water–gas double rarefaction with cavitation $u = 100$ m/s, $\alpha_{init} = 10^{-2}$, mesh 5000 cells, t = 1.5 ms. Void ratio, pressure, velocity and speed of sound profiles

appears leading to a pressure drop in the cavitation area. This phenomenon is not taken into account by both 4-equation models, which explains the large difference with the 2-fluid solution (see the next section for the use of non isothermal models). As previously observed, evaporation fronts are not simulated by the Hosangadi model and large differences are put in evidence for the mixture speed of sound.

Finally, we computed the case $u_{init} = 100$ m/s with a small initial value of gas $\alpha_{init} = 10^{-10}$. For this case, no reference solutions are available. Various C_{prod} values were tested. From the value 10^6, the numerical solution did not move any more and we set the parameter to this value. Solutions are compared in Fig. 24 at time $t =$ 0.2 ms. The rarefaction fronts are stiffer in comparison with the previous case. The cavitation area and the evaporation fronts can be clearly observed with the 4-equation sinus solution, that is not the case for the other model. The mixture speed of sound is plotted with a logarithmic scale for both 4-equation models. Large variations are highlighted, from the pure liquid high value to small values in the cavitation pocket. At the middle of the tube where the void ratio is close to 1 the speed of sound is close to the vapour value. For the sinus speed of sound, the four waves can be observed and the ratio between extremal sound velocities is higher than 1000. These variations make the numerical solving of the system very stiff.

For the considered test cases, the domain of evolution of C_{prod} is $[1, 10^6]$ that is really questionable. Other comparisons can be found in Goncalves (2013).

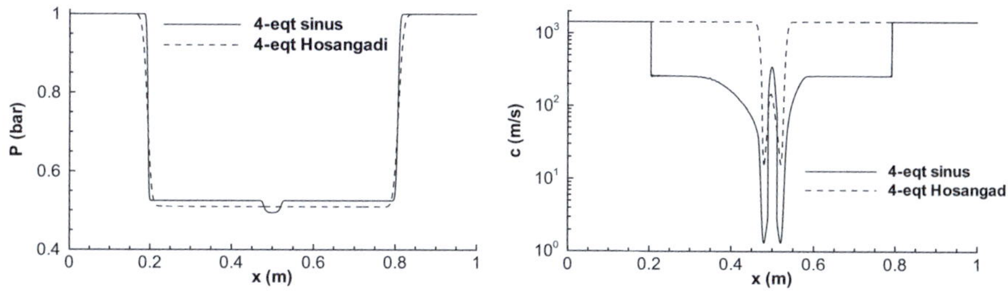

Fig. 24 Water–gas double rarefaction with cavitation $u = 100$ m/s, $\alpha_{init} = 10^{-10}$, mesh 5000 cells, t = 0.2 ms. Pressure and speed of sound profiles

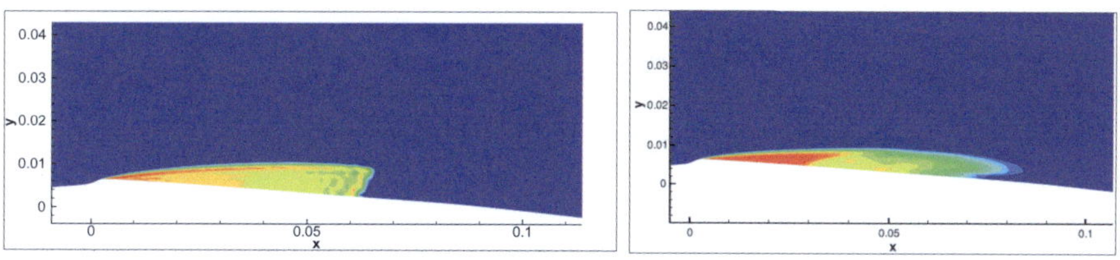

Fig. 25 Time-averaged void fraction downstream the Venturi throat

9.2 4° Venturi Case

Various unsteady simulations were performed in order to compare cavitation models on the 4° Venturi case. The same mesh 251 × 62 is used for all simulations. The simulation time is around 2 s. The considered turbulence model is the Spalart–Allmaras one coupled with the Reboud correction.

3-equation models We compared 3-equation models for which the mixture EOS is different: a mixture of stiffened gas versus a sinus law. A qualitative view of the sheet cavities is presented in Fig. 25 in which are plotted the time-averaged void ratio fields downstream the Venturi throat. Using the mixture of stiffened gas EOS, the re-entrant jet is not predicted and the repartition of void ratio is not in agreement with the experiments. On the contrary, using the sinus EOS, we can clearly observed the attached cavity with large values of void ratio (in red).

Figure 26 shows the time-averaged velocity and void ratio profiles at station 3. Except for the cavity thickness, similar results are obtained for the void ratio profiles. The velocity profiles clearly illustrates the absence of re-entrant jet for the mixture of stiffened gas EOS solution. This model seems not adaptable for such partial cavities.

3-equation versus 4-equation models We compared the 3-equation and 4-equation models coupled with the sinus EOS. A qualitative description of the sheet cavities is illustrated in Fig. 27 with the density gradient contours. Both simulations provide a solution in good agreement with the experimental visualizations.

In Fig. 28 are plotted the time-averaged velocity and void ratio profiles at station 3. Results obtained with both models are quite similar. For this test case, the use of a

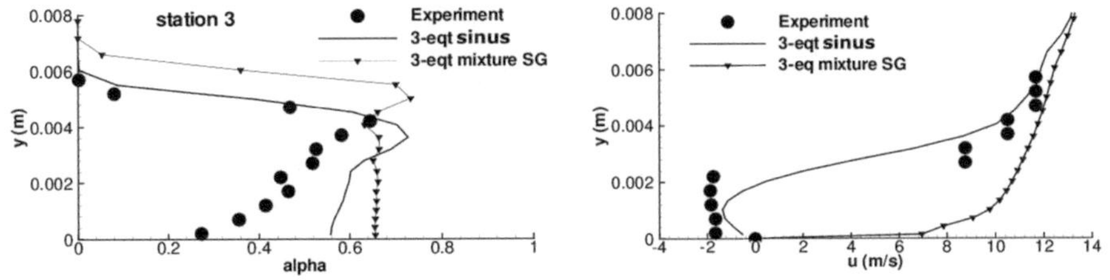

Fig. 26 Void ratio (*left*) and velocity (*right*) profiles at station 3

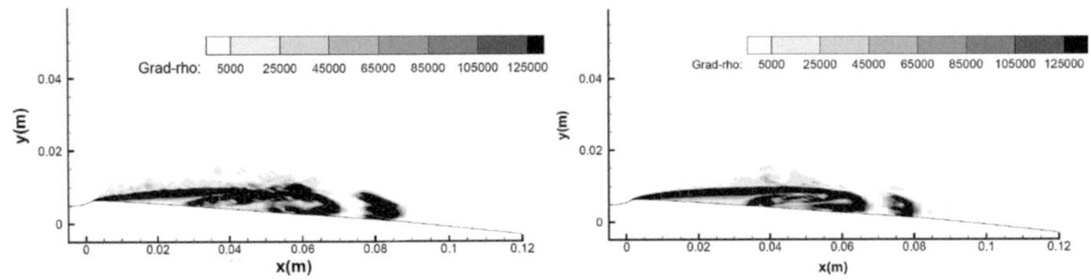

Fig. 27 Contours of the density gradient modulus: 3-equation model (*left*) and 4-equation model (*right*)

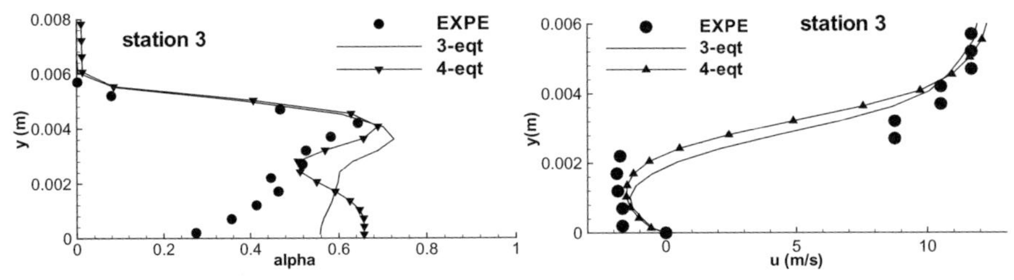

Fig. 28 Time-averaged void ratio and velocity profiles at station 3

transport-equation model did not improve the numerical results in comparison with the 3-equation sinus model (see Goncalves and Charriere 2014).

4-equation models Two formulations of 4-equation models have been previously presented in the theoretical part, based on either a mixture of stiffened gas EOS (called 4-equation SG model) or a sinus law (called 4-equation sinus model). The sheet cavities obtained with these models are compared in Fig. 29 with the plot of the instantaneous density gradient modulus. Both simulations put in evidence a quite similar dynamics involving an attached cavity (up to the abscissa $x = 0.04$ m) and a recirculating area with the generation of small vapour cloud shedding.

A local analysis is led by studying void ratio and velocity profiles, which are presented in Fig. 30 at station 3. Results obtained with both models are in good agreement with the experiments. The 4-equation SG cavitation model slightly provides a better prediction of the void ratio wall value. Moreover, this model provides a better

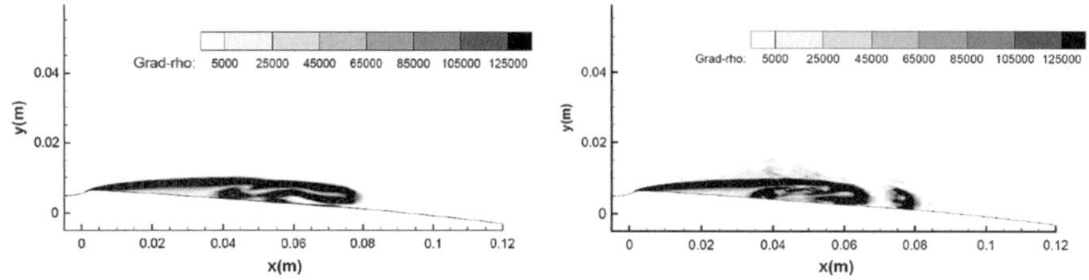

Fig. 29 Contours of the density gradient modulus: 4-equation SG model (*left*) and 4-equation sinus model (*right*)

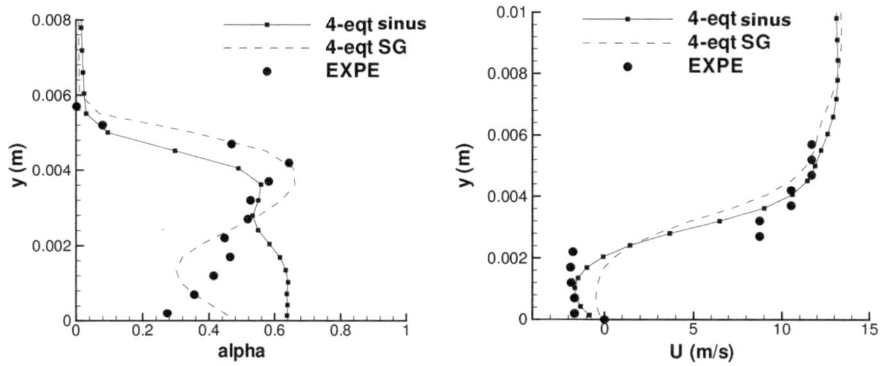

Fig. 30 Time-averaged void ratio and velocity profiles at station 3

estimation of the pressure fluctuations decrease in the re-compression area (Charriere et al. 2015).

Comparison with incompressible codes Based on homogeneous models, two numerical approaches using compressible and incompressible codes can be applied to capture large density variations and unsteady behaviours of cavitating flows. Both approaches seem to have strong and weak points. In this study, we compared the compressible CaviFlow solver with the incompressible IZ solver. This code solves a barotropic system coupled with transport-equation turbulence models. The numerical resolution is a pressure-correction method derived from the SIMPLE algorithm. All computations are time-dependent to take into account the variation of density. A sinus relation links the density to the pressure (Delannoy and Kueny 1990). More details can be found in Coutier-Delgosha et al. (2002).

Simulations were performed on the 4° Venturi case using the same mesh and a $k - \varepsilon$ turbulence model coupled with the Reboud correction. A qualitative view of the simulated sheet cavities is presented in Fig. 31 where the instantaneous void ratio is plotted. The numerical solutions obtained with both codes present some marked differences. The compressible code captured a sheet behaviour in good agreement with the experimental view. We can observe the attached cavity from the throat of the Venturi up to $x = 0.3$ m. Downstream, the recirculating area with two-phase structure

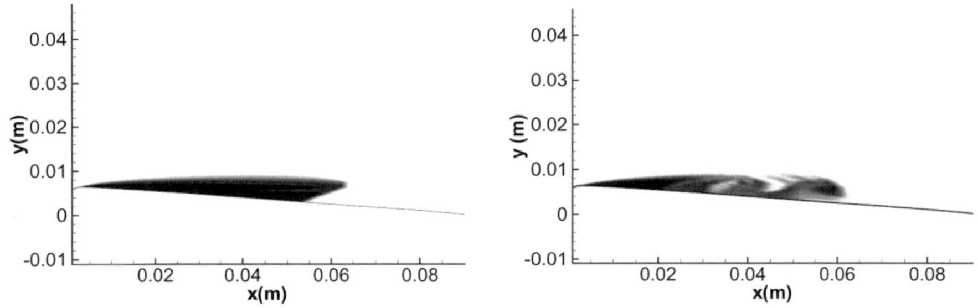

Fig. 31 Instantaneous contour of void fraction, incompressible code (*left*) and compressible code (*right*)

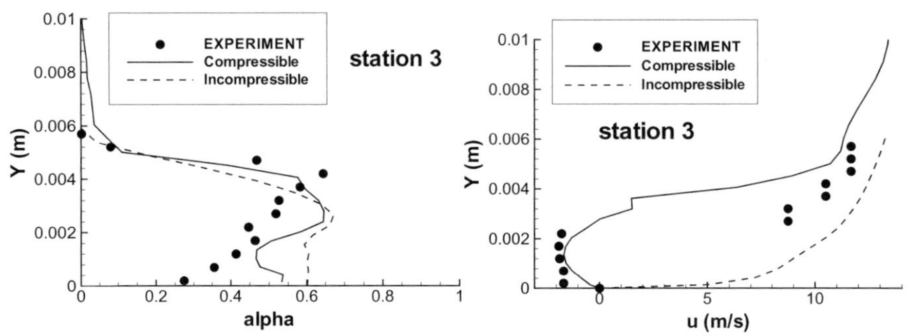

Fig. 32 Void ratio (*left*) and velocity (*right*) profiles at station 3

shedding is well illustrated. On the other hand, the incompressible code captured a low frequency periodic sheet dynamics without re-entrant jet (around 6 Hz).

Figure 32 shows the longitudinal velocity and void ratio profiles for experiments and numerical results at station 3. For the void ratio profiles, both codes provided a similar solution in good agreement with the experimental data. The recirculating behaviour with a re-entrant jet is well simulated by the compressible code. On the contrary, using the incompressible code the re-entrant jet is not captured. At stations 4 and 5, the re-entrant is largely under-estimated (see Goncalves et al. 2010a).

Another comparison is proposed with the incompressible code OpenFOAM. Simulations were performed on the 4° Venturi case using a similar mesh and the $k - \omega$ SST turbulence model. The considered cavitation models are the 4-equation SG model and the Kunz model for the CaviFlow and OpenFOAM codes, respectively (see Charriere et al. 2015).

Figure 33 shows the instantaneous void fraction field downstream the Venturi throat for both computations. Using the openFOAM solver, the attached cavity is extended with void ratio values close to 1 and a small liquid film can be observed along the wall from abscissa $x = 0.05$ m. On the contrary, the attached sheet cavity computed with the compressible CaviFlow solver is in good agreement with the experiments. Downstream, we can observe the recirculating area with two-phase structure shedding.

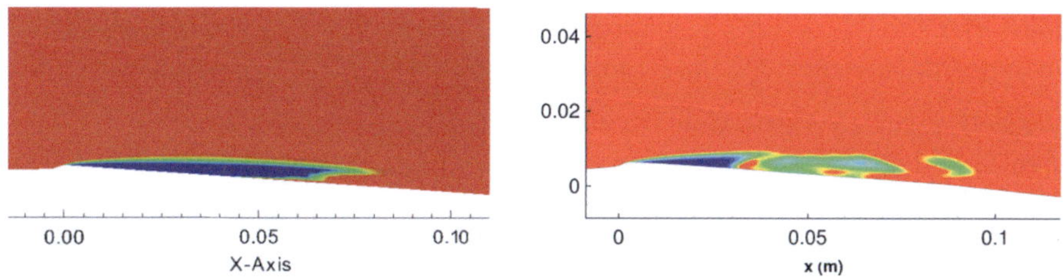

Fig. 33 Instantaneous contour of void fraction, incompressible code (*left*) and compressible code (*right*)

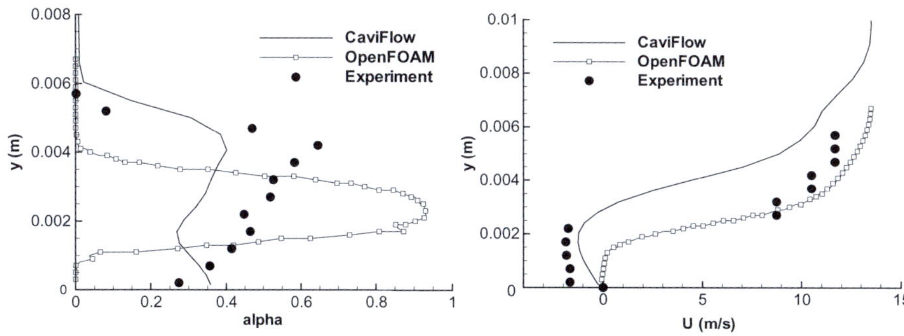

Fig. 34 Time-averaged void ratio (*left*) and velocity (*right*) profiles at station 3

In Fig. 34 are presented the time-averaged velocity and void ratio profiles at station 3. The compressible solver provides a better prediction of the void ratio profile due to the presence of a mixture in this area. The OpenFOAM computation over-predicts the vapour quantity inside the cavity, which can reach 95%. Moreover, the thickness of the cavity is under-estimated and a pure liquid phase is simulated close to the wall. As regard to the re-entrant jet, the OpenFOAM simulation badly reproduces the recirculation area. On the contrary, the recirculating behaviour with a re-entrant jet is correctly simulated by the compressible code.

To conclude, the topology of the cavitation pocket marks large discrepancies between compressible and incompressible codes, especially concerning the re-entrant jet development. Comparisons with measurements suggest that the compressible formulation provides an improved description of the cavitation dynamics compared with incompressible computations.

9.3 8° Venturi Case

Unsteady simulations were performed on the 8° Venturi case, which involves a self-sustained oscillating sheet cavity. The same mesh 174×56 is used for all simulations. The simulation time is around 2 s. The considered turbulence model is the Spalart–Allmaras one with the Reboud limiter.

3-equation versus 4-equation We compare the 3-equation and 4-equation models coupled with the sinus EOS. The obtained frequency is close to the experimental for both codes (around 45 Hz). A qualitative description of the sheet cavities is illustrated in Fig. 35 with the instantaneous density gradient contours. Using the 4-equation model, results provide a shedding phenomenon in better agreement with the experimental data. With the 3-equation model, we can observe an extended attached cavity, and downstream the fluctuating recirculation with two-phase structures shedding. These shedding are rapidly eliminated and under-estimated in comparison with the experimental visualizations and the 4-equation simulations.

Figure 36 presents the time-averaged void ratio and longitudinal velocity profiles at station 1. Near the throat, the void ratio is largely over-estimated using the 3-equation model. The attached cavity is too extended. Results are similar for the velocity profiles in good agreement with the experiments.

In Fig. 37 are plotted the time-averaged void ratio profiles at station 3 and the time-averaged dimensionless wall pressure distribution. Using the 3-equation model, the averaged value of the void ratio is null. As commented previously, the cloud shedding is not enough developed. On the contrary, it is more intense using the 4-equation model. For the average pressure, both models provide a result in correct agreement with the experimental data. More comparisons can be found in Goncalves and Charriere (2014).

To conclude between the 3 and 4-equation models and their ability to compute both Venturi cases:

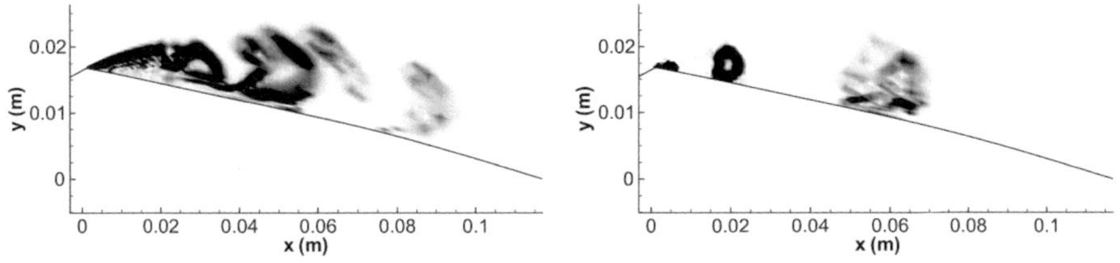

Fig. 35 Contours of the density gradient modulus: 3-equation model (*left*) and 4-equation model (*right*)

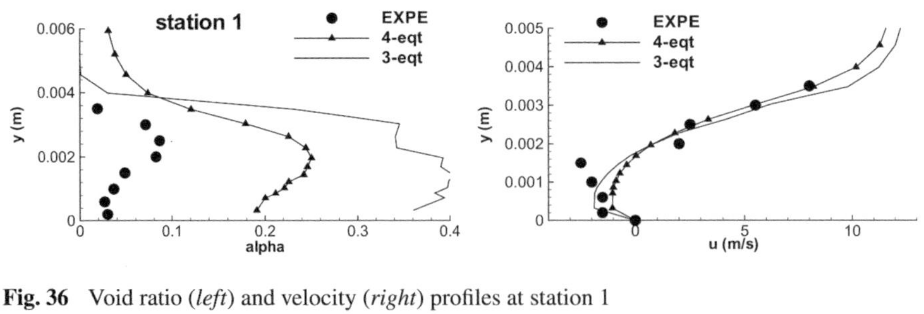

Fig. 36 Void ratio (*left*) and velocity (*right*) profiles at station 1

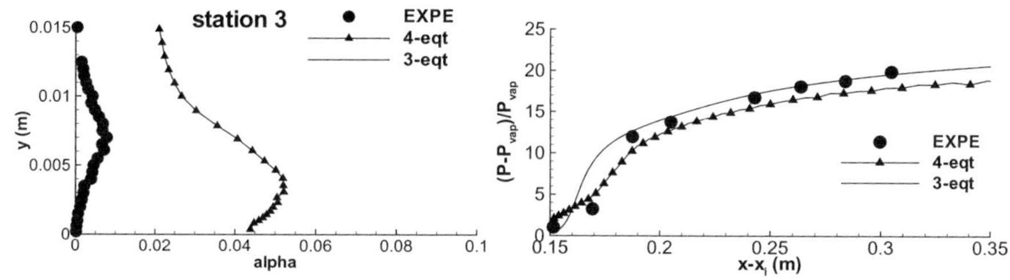

Fig. 37 Time-averaged void ratio profile at station 3 (*left*) and wall pressure evolution (*right*)

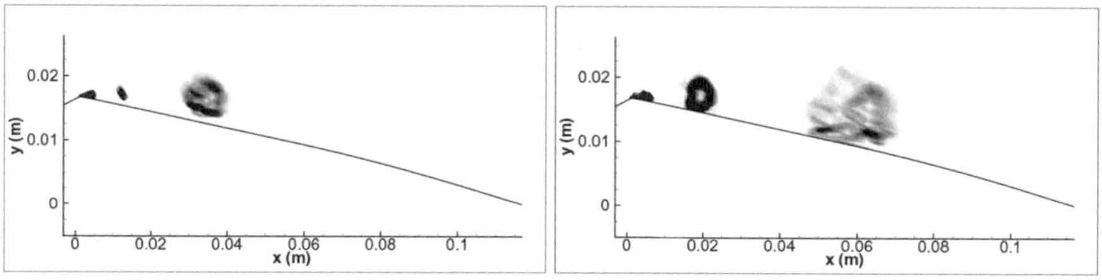

Fig. 38 Contours of the density gradient modulus: 4-equation SG model (*left*) and 4-equation sinus model (*right*)

- For the quasi-stable cavity, the void ratio transport-equation does not clearly improve the previous simulations obtained with a 3-equation model.
- For the periodic self-oscillating cavity, the two-phase structures shedding is more intense and in better agreement with the experimental data using the 4-equation model. The transport of void ratio seems to be a key point to correctly predict the sheet cavity dynamics.

4-equation models We compare the two formulations of 4-equation models based on a mixture of stiffened gas EOS and a sinus EOS. First, the obtained frequency using the 4-equation SG model is around 20 Hz instead of 45 Hz. Except the frequency, the overall agreement with the experimental visualizations is correct. The instantaneous density gradient contours are plotted in Fig. 38 for both models. The two-phase shedding is well captured by both models.

Figure 39 presents the time-averaged void ratio and longitudinal velocity profiles at station 1. For the velocity profile, both models provide a result in correct agreement with the experimental data. The void ratio profile is better predicted using the 4-equation SG model, especially the near-wall values.

Void ratio profiles at stations 2 and 3 are plotted in Fig. 40. As previously, a better agreement is observed using the 4-equation SG model, but the frequency of the phenomenon is not correct.

Comparison with an incompressible code Unsteady simulations obtained using the incompressible IZ and compressible CaviFlow codes are compared. The same mesh is used (251 $\times$ 61 nodes) and the turbulence model is a $k - \varepsilon$ with the Reboud

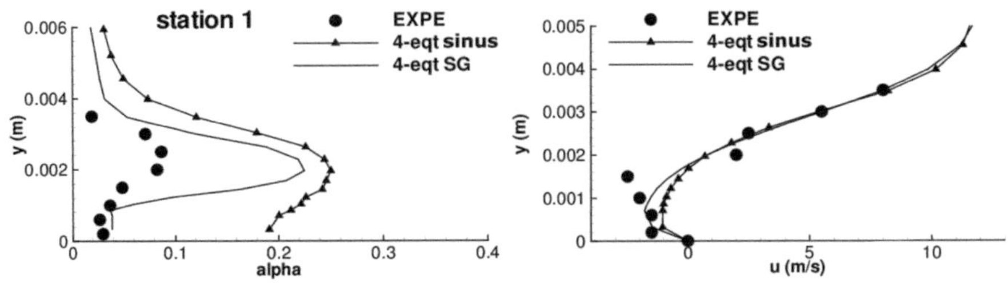

Fig. 39 Void ratio (*left*) and velocity (*right*) profiles at station 1

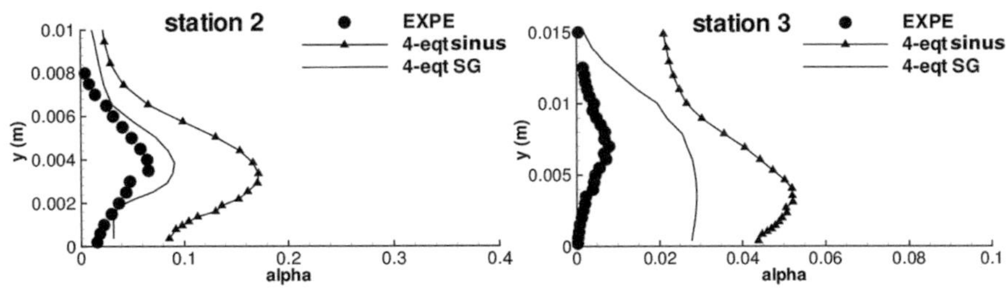

Fig. 40 Void ratio profiles at station 2 (*left*) and station 3 (*right*)

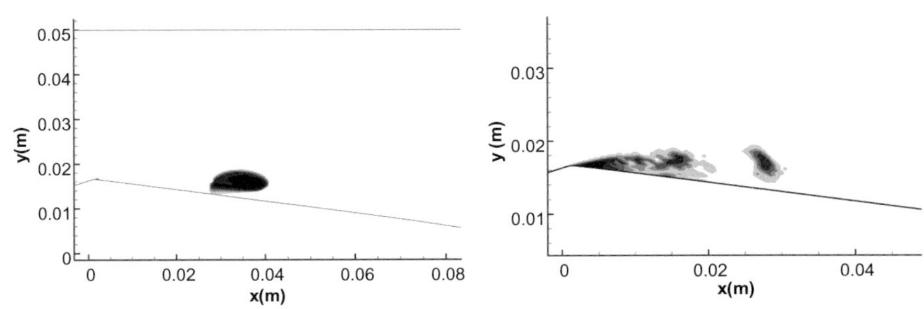

Fig. 41 Instantaneous void ratio field: incompressible code (*left*) and compressible code (*right*)

limiter. The 3-equation sinus model is considered for the compressible simulations and a barotropic sinus model for the incompressible ones. Both codes provided an oscillating phenomenon with the correct frequency in comparison with the experimental value.

An instantaneous view of the cavity (void ratio contours) obtained with the incompressible code is illustrated in the left part of Fig. 41. Large-scale two-phase structure shedding is clearly observed. The re-entrant jet is very intense and flows upstream to the throat of the Venturi, leading to the disappearance of the attached cavity. The solution obtained with the 3-equation compressible model presented in the right part of Fig. 41 is different. As previously commented, an extended attached cavity is predicted and two-phase structures convected in the flow are smaller and condense rapidly. Results are improved with the use of the 4-equation sinus model.

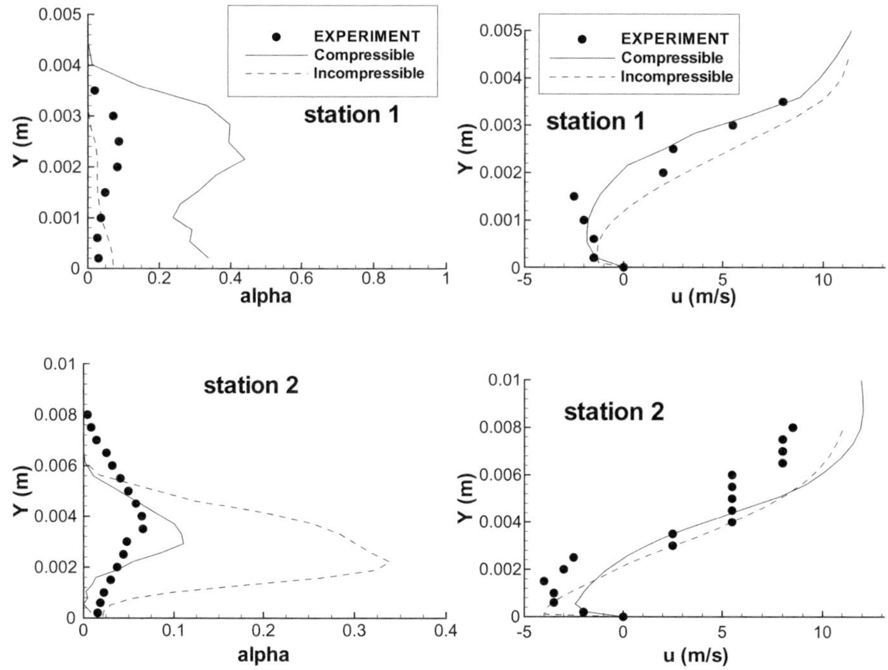

Fig. 42 Time-averaged void ratio (*left*) and velocity (*right*) profiles at stations 1 and 2

Figure 42 illustrates the time-averaged void ratio and velocity profiles at stations 1 and 2. At station 1, with the compressible code, the recirculating behaviour is very well captured, yet the void ratio is over-predicted. For the incompressible code, the void ratio is slightly under-estimated due to the disappearance of the attached cavity. At station 2, both codes under-evaluated the recirculating area thickness. For the void ratio profile, values obtained with the incompressible code are largely over-predicted, whereas values obtained by the compressible code are in better agreement with the experimental data.

To conclude these studies between compressible and incompressible codes:

- The incompressible IZ code seems to be unable to correctly simulate the quasi-stable Venturi case using a barotropic model and a Reboud limiter. Similarly, the openFOAM code using the Kunz's void ratio transport-equation model and the Menter SST model is unable to reproduce the re-entrant jet phenomenon. Is it due to a problem of waves propagation or compressibility effects?
- For the second Venturi case, using the incompressible IZ code a very intense re-entrant jet is captured, which flows upstream to the throat and completely eliminates the attached cavity. For the compressible CaviFlow code, large-scale cavitation structure shedding are under-predicted using the 3-equation model, whereas the phenomenon is correctly simulated using the void ratio transport-equation.

10 Cavitation Models and Thermodynamic Effects

10.1 Inviscid Tube Cases

As previously discussed, when the initial discontinuity velocity is high, thermal effects are no more negligible. We compare the two formulations of 4-equation models including thermal effects for the double rarefaction case for which the initial velocity is 100 m/s. For the sinus model, the parameter c_{baro} is set to 1.31 m/s, as previously. The value of dP_{vap}/dT, evaluated with a thermodynamic table, is set to 300 Pa/K (see Goncalves 2014).

The volume fraction, pressure, velocity and temperature evolutions are plotted in Fig. 43, at time t = 1.5 ms. No differences appear on the void ratio between models. For the pressure profiles, the pressure drop under $P_{vap}(T_{ref})$ is around 0.3 bar with both models, in close agreement with the two-fluid solution. Discrepancies appear for the velocity profile in comparison with the two-fluid solution: variations across the evaporation front are stiffer using the 4-equation models. The temperature drop inside the cavitation pocket due to the phase transition is very large. It reaches more than 70 K with the 4-equation SG model and 90 K with the 4-equation sinus model. Yet, we have no reference solution for the temperature profile.

A similar test case is based on the same conditions except regarding the initial discontinuity velocity which is set to $u = 500$ m/s. A large cavitation area is created. Inside the cavity, according to the solution presented in Zein et al. (2010), the

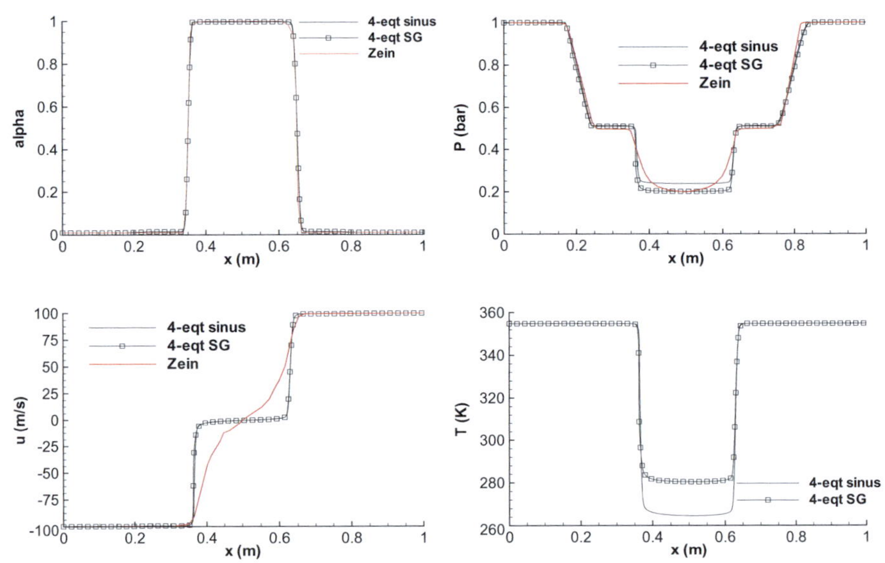

Fig. 43 Water–gas double rarefaction with cavitation $u = 100$ m/s, $\alpha_{init} = 10^{-2}$, mesh 5000 cells, t = 1.5 ms. Void ratio, pressure, velocity and temperature profiles

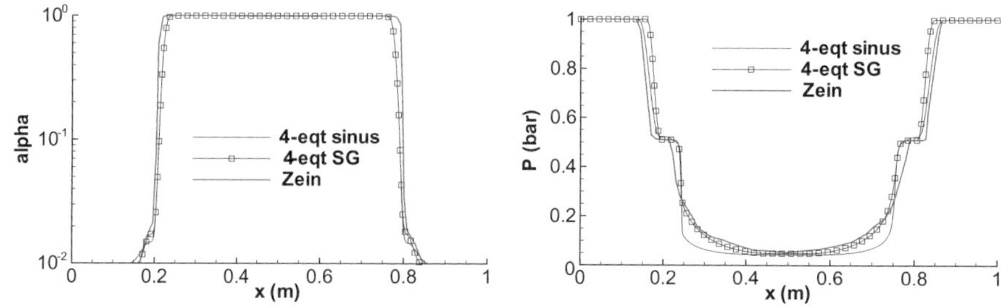

Fig. 44 Water–gas double rarefaction with cavitation $u = 500$ m/s, $\alpha_{init} = 10^{-2}$, mesh 5000 cells, t = 0.58 ms. Void ratio and pressure profiles

pressure drop is near 0.5 bar under the vapour pressure at the reference temperature. The value of dP_{vap}/dT is set to 130 Pa/K. In Fig. 44 are presented the void ratio and pressure profiles at time $t = 0.58$ ms. Results obtained with both 4-equation models are in good agreement with the two-fluid solution.

10.2 Venturi Case with Freon R-114

Steady simulations were performed on the freon R-114 Venturi case to investigate cavitation models. The considered mesh contained 251×77 nodes and the Spalart-Allmaras turbulence model was used for all simulations.

Isothermal and non isothermal 3-equation models Comparisons between the isothermal and non isothermal formulations of the 3-equation sinus model are proposed. The value of the constant dP_{vap}/dT is set to 5900 K/Pa and the parameter c_{baro} is set to 0.74 m/s. A study of the influence of this parameter is given in Goncalves and Patella (2010).

For both models, a steady pocket for witch the length is around 80 mm was simulated. Void ratio profiles at stations 1 to 3 are plotted in Fig. 45 with the experimental data. The cavity thickness is well predicted using the isothermal model but the maximal values of void ratio are too high (85% instead of 45% at station 1). Using the non isothermal model, the thickness is increased and the maximum values are decreased. The cooling effect during the phase transition is studied by plotting the temperature difference $T_{ref} - T$ at station 1 (Fig. 45). The measured wall temperature is 2.1 K. The wall value is correctly predicted by the non isothermal model. With the isothermal model, temperature variations are important inside the cavity and the peak of temperature deficit can reach 12 K.

Simulations were performed with other flow conditions: $T_{ref} = 303$ K and $U_{inlet} = 22$ m/s. Two inlet cavitation parameters were considered: $\sigma_{inlet} = 0.55$ and $\sigma_{inlet} = 0.59$. The goal was to obtain a stable cavitation sheet with a length close to the experimental one. Results provided by the 3-equation non isothermal model are compared with the experimental data. The wall temperature depression is plotted in Fig. 46

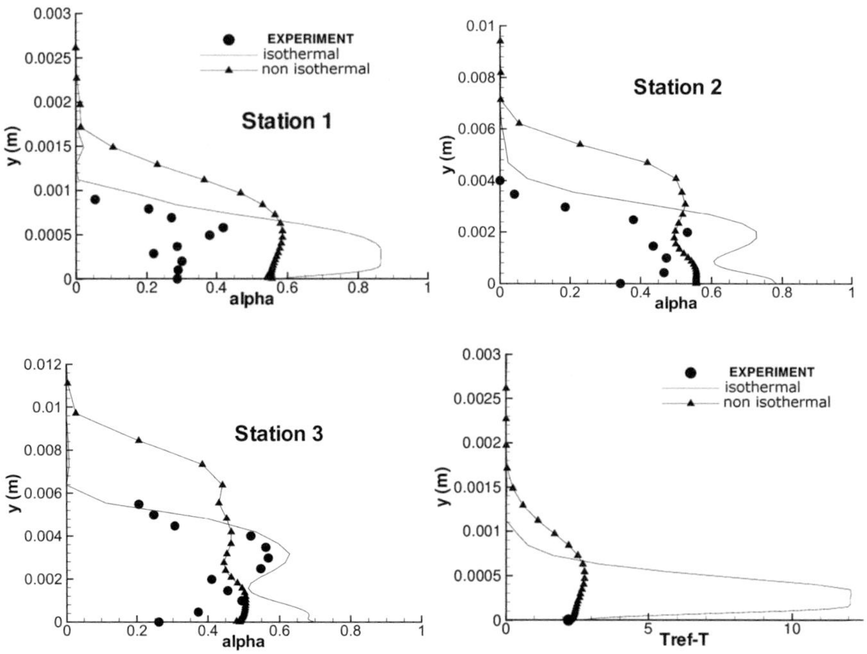

Fig. 45 Void ratio profiles and temperature difference at station 1

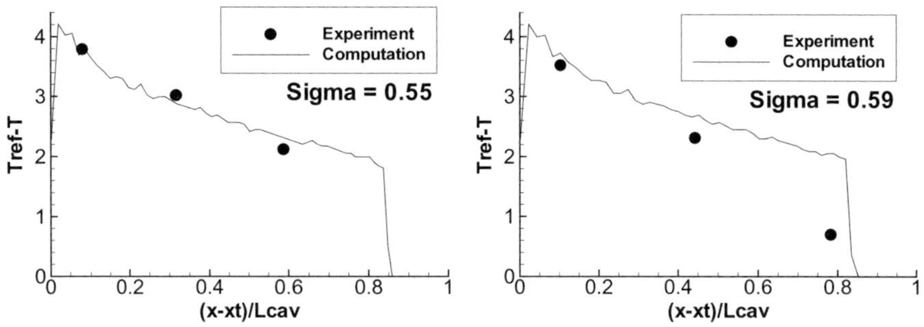

Fig. 46 Wall cooling $T_{ref} - T$ (K) at different inlet cavitation parameter

versus the distance $(x - x_{throat})/L_{cav}$ for the two inlet cavitation numbers. The temperature deficit is well predicted by computations and results are in good agreement with the experimental data. Yet, for higher inlet velocities, the model fails to predict the temperature depression (in comparison with the experimental data, the error is around 50%, see Goncalves and Patella 2010).

3-equation versus 4-equation models Both non isothermal sinus formulations have been compared on the Venturi case. The void ratio follows an analytical relation for the 3-equation model, whereas it follows a transport-equation for the 4-equation model.

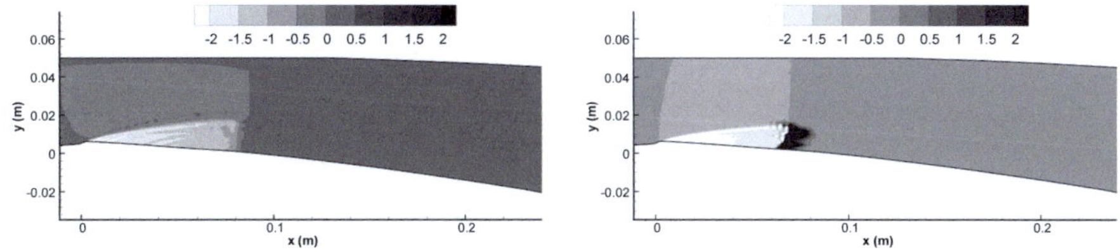

Fig. 47 Temperature difference $T - T_{ref}$ (K) downstream the Venturi throat: 3-equation model (*left*) versus 4-equation model

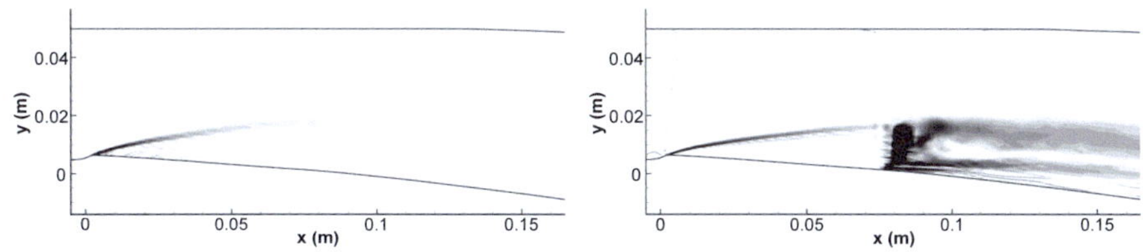

Fig. 48 Gradient of temperature (K/m) downstream the Venturi throat: 3-equation model (*left*) versus 4-equation model

Figure 47 shows the temperature deficit $T - T_{ref}$ inside the Venturi divergent. The cooling effect due to the vaporisation process is clearly observed for all simulations (negative values). A large discrepancy appears between the 3- and 4-equation models downstream the cavity in the re-compression area. Using the 3-equation model, the temperature deficit is close to zero downstream the cavitation pocket, that is the temperature goes back to the freestream temperature. On the contrary, we observe a warming effect downstream the cavity using the 4-equations model. Locally, the temperature exceeds the freestream temperature and values reach more than 4 K.

The temperature gradient is illustrated in Fig. 48 for both models. The grey scale is the same for all computations. We clearly observe differences on the behaviour of models in the closure part of the cavitation pocket. Using the 3-equation model, there is no temperature gradient at the end of the pocket. We just see a small gradient at the interface in the vicinity of the throat. With the 4-equation model, strong gradients are put in evidence in the closure part of the pocket and downstream in the re-compression area. Such a phenomenon was depicted in Petkovsek and Dular (2013) on a hot water Venturi flow due to the collapse of bubbles. Using high-speed infra-red thermography, authors measured a temperature depression of approximately 0.4 K in the vicinity of the throat and a temperature rise up to 1.4 K was recorded in the region of pressure recuperation.

Figure 49 illustrates the numerical void ratio obtained with three models (3-equation sinus, 4-equation sinus and 4-equation SG models) in comparison with the measurements at stations 2 and 3. The cavity thickness is clearly better predicted by the 4-equation sinus model. Using the 3-equation model, the thickness is overestimated with a factor 2. An intermediate result is provided by the 4-equation SG model. Other comparisons can be found in Goncalves (2014).

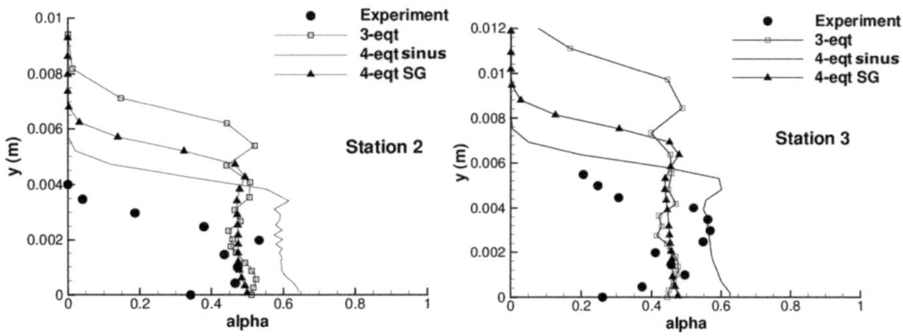

Fig. 49 Void ratio profiles at stations 2 and 3

11 Conclusion and Perspectives

The simulation of cavitating flows using 1-fluid RANS tools is confronted to open questions as regard to the turbulence modelling, the thermodynamic path during the phase transition, the mass transfer modelling and the turbulence-cavitation interaction. In this paper, various turbulence and cavitation models have been compared and tested in their ability to correctly simulate cavitation pockets which can occur in a variety of practical cases. Models has been implemented in a compressible RANS solver including a low Mach number preconditioning algorithm and has been applied for the simulation of two different partial cavities developing along Venturi geometries. Furthermore, a comparison with incompressible pressure-based solvers were proposed. Results lead to different concluding remarks:

(i) For the quasi-stable cavity, it was shown that both tested incompressible codes were unable to predict such sheet dynamics. The presence of wave propagation involved in the sheet dynamics is a possible explanation. Moreover, it seems that a void ratio transport-equation is not necessary to correctly simulate this case using the compressible code.

(ii) For the periodic self-oscillating cavity, the two-phase structures shedding was more intense and in better agreement with the experimental data using the 4-equation model in comparison with the 3-equation model. Using the incompressible code, the re-entrant jet phenomenon was over-estimated leading to the disappearance of the attached cavity near the Venturi throat.

On the other hand, the turbulence models comparison put in evidence the importance of the eddy viscosity limitation (Reboud limiter, SST correction, realizability constrains, scale-adaptive term) and the near-wall boundary-layer treatment for both Venturi cases. Moreover, it was impossible to calibrate the compressible or dilatational terms appearing in the mixture TKE equation for both cases using aerodynamic closures. The modelling of theses terms in cavitating flows is still an open question. Given that it is extremely difficult to quantify this term, it seems obvious to report the modelling assumptions to the dissipative scales by developing a large-eddy simulation formalism adapted to cavitating flows. Yet, the 1-fluid filtered equations have not been clearly expressed and investigated.

Finally, a study of cavitation models in the case of thermo-sensitive fluid was presented. A 4-equation model was developed to study thermal effects based on a void ratio transport equation and assuming a linear variation of the vapour pressure with the mixture temperature. First validations on inviscid one-dimensional cases showed the ability of 4-equation non isothermal models to simulate the cavitation development in which the running fluid is hot water. Secondly, RANS simulations were performed to study a stable cavitation pocket developing along a Venturi geometry in which the working fluid is freon R-114. In comparison with 3-equation simulations, the cavity thickness was clearly improved. Moreover, a warming effect was exhibited downstream the cavity in the re-compression area. The intensity of this warming effect is higher than the cooling effect observed near the Venturi throat. This rise of temperature is due to the collapse of bubbles in the closure part of the pocket. This new 4-equation model is therefore very attractive to study thermodynamic effects and cryogenic cavitation.

Acknowledgements The author is grateful to Prof. Salvetti and Prof. d'Agostino for their invitation to the Udine summer school. The author expresses his gratitude to the French Space Agency CNES and to the SNECMA Space Engines Division to support these studies. The author thanks the IDRIS–CNRS supercomputing centre for providing us their computing resources.

References

Aeschlimann, V., & Barre, S. (2009). PIV-LIF determination of mean velocity field and Reynolds stress tensor in a cavitatiting mixing layer. In *7th International Symposium on Cavitation CAV2009, Ann Arbor, USA*.

Aeschlimann, V., Barre, S., & Djeridi, H. (2011). Velocity field analysis in an experimental cavitating mixing layer. *Physics of Fluids*, *23*, 055105.

Aeschlimann, V., Barre, S., & Djeridi, H. (2013). Unsteady cavitation analysis using phase averaging and conditional approaches in a 2D Venturi. *Open Journal of Fluid Dynamics*, *3*, 171–183.

Ahn, S. J., & Kwon, O. J. (2013). Numerical investigation of cavitating flows for marine propulsors using an untructured mesh technique. *International Journal of Heat and Fluid Flow*, *43*, 259–267.

Barre, S., Rolland, J., Boitel, G., Goncalves, E., & Patella, R. F. (2009). Experiments and modelling of cavitating flows in Venturi: Attached sheet cavitation. *European Journal of Mechanics B/Fluids*, *28*, 444–464.

Barret, M., Faucher, E., & Herard, J. M. (2002). Schemes to compute unsteady flashing flows. *AIAA Journal*, *40*(5), 905–913.

Besnard, D. C., & Harlow, F. H. (1988). Turbulence in multiphase flow. *International Journal of Multiphase Flow*, *14*(6), 679–699.

Bilicki, Z., Kwidzinski, R., & Mohammadein, S. A. (1996). Evaluation of the relaxation time of heat and mass exchange in the liquid–vapour bubble flow. *International Journal of Heat and Mass Transfer*, *39*(4), 753–759.

Bois, G., Jamet, D., & Lebaigue, O. (2010). Towards large eddy simulation of two-phase flow with phase change: Direct numerical simulation of a pseudo-turbulent two-phase condensing flows. In *7th International Conference on Multiphase Flow, Tampa, USA*.

Charriere, B., Decaix, J., & Goncalves, E. (2015). A comparative study of cavitation models in a venturi flow. *European Journal of Mechanics B/Fluids*, *49*, 287–297.

Chen, Y., Lu, C. J., & Wu, L. (2006). Modelling and computation of unsteady turbulent cavitation flows. *Journal of Hydrodynamics*, *18*(5), 559–566.

Choi, Y. H., & Merkle, C. L. (1993). The application of preconditioning to viscous flows. *Journal of Computational Physics*, *105*(2), 207–223.

Clerc, S. (2000). Numerical simulation of the homogeneous equilibrium model for two-phase flows. *Journal of Computational Physics*, *161*(1), 354–375.

Coutier-Delgosha, O., Fortes-Patella, R., & Reboud, J. L. (2002). Simulation of unsteady cavitation with a two-equation turbulence model including compressibility effects. *Journal of Turbulence*, *3*(58).

Davidson, L. (2006). Evaluation of the SST-SAS model: Channel flow, asymmetric diffuser and axi-symmetric hill. In *Proceedings of ECCOMAS CFD 2006, Egmond aan Zee, The Netherlands*.

Decaix, J., & Goncalves, E. (2012). Time-dependent simulation of cavitating flow with $k - \ell$ turbulence models. *International Journal for Numerical Methods in Fluids*, *68*, 1053–1072.

Decaix, J., & Goncalves, E. (2013a). Compressible effects modelling in turbulent cavitating flows. *European Journal of Mechanics B/Fluids*, *39*, 11–31.

Decaix, J., & Goncalves, E. (2013b). Investigation of three-dimensional effects on a cavitating venturi flow. *International Journal of Heat and Fluid Flow*, *44*, 576–595.

Delannoy, Y., & Kueny, J. L. (1990). Two phase flow approach in unsteady cavitation modelling. *Cavitation and Multiphase Flow Forum, ASME-FED*, *98*, 153–158.

Downar-Zapolski, P., Bilicki, Z., Bolle, L., & Franco, J. (1996). The non-equilibrium relaxation model for one-dimensional flashing liquid flow. *International Journal of Multiphase Flow*, *22*(3), 473–483.

Durbin, P. A. (2009). Limiters and wall treatments in applied turbulence modeling. *Fluid Dynamics Research*, *41*(1), 012203.

Edwards, J. R., & Franklin, R. K. (2000). Low-diffusion flux splitting methods for real fluid flows with phase transition. *AIAA Journal*, *38*(9), 1624–1633.

Elghobashi, S. E., & Abou-Arab, T. W. (1983). A two-equation turbulence model for two-phase flows. *Physics of Fluids*, *26*(4), 931–938.

Fruman, D. H., Reboud, J. L., & Stutz, B. (1999). Estimation of thermal effects in cavitation of thermosensible liquids. *International Journal of Heat and Mass Transfer*, *42*, 3195–3204.

Goncalves, E. (2011). Numerical study of unsteady turbulent cavitating flows. *European Journal of Mechanics B/Fluids*, *30*(1), 26–40.

Goncalves, E. (2013). Numerical study of expansion tube problems: Toward the simulation of cavitation. *Computers & Fluids*, *72*, 1–19.

Goncalves, E. (2014). Modeling for non isothermal cavitation using 4-equation model. *International Journal of Heat and Mass Transfer*, *76*, 247–262.

Goncalves, E., & Charriere, B. (2014). Modelling for isothermal cavitation with a four-equation model. *International Journal of Multiphase Flow*, *59*, 54–72.

Goncalves, E., & Decaix, J. (2012). Wall model and mesh influence study for partial cavities. *European Journal of Mechanics B/Fluids*, *31*(1), 12–29.

Goncalves, E., & Fortes, R. (2009). Patella. Numerical simulation of cavitating flows with homogeneous models. *Computers & Fluids*, *38*(9), 1682–1696.

Goncalves, E., & Patella, R. F. (2010). Numerical study of cavitating flows with thermodynamic effect. *Computers & Fluids*, *39*(1), 99–113.

Goncalves, E., & Patella, R. F. (2011). Constraints on equation of state for cavitating flows with thermodynamic effects. *Applied Mathematics and Computation*, *217*, 5095–5102.

Goncalves, E., Champagnac, M., & Fortes, R. (2010a). Patella. Comparison of numerical solvers for cavitating flows. *International Journal of Computational Fluid Dynamics*, *24*(6), 201–216.

Goncalves, E., & Fortes, R. (2010b). Patella, J. Rolland, B. Pouffary, and G. Challier. Thermodynamic effect on a cavitating inducer in liquid hydrogen. *Journal of Fluids Engineering*, *132*, 111305.

Guillard, H., & Viozat, C. (1999). On the behaviour of upwind schemes in the low Mach number limit. *Computers & Fluids*, *28*(1), 63–86.

Hosangadi, A., & Ahuja, V. (2005). Numerical study of cavitation in cryogenic fluids. *Journal of Fluids Engineering*, *127*(2), 267–281.

Hu, Z. M., Dou, H. S., & Khoo, B. C. (2011). On the modified dispersion-controlled dissipative (DCD) scheme for computation of flow supercavitation. *Computers & Fluids*, *40*(1), 315–323.

Ihm, S. W., & Kim, C. (2008). Computations of homogeneous-equilibrium two-phase flows with accurate and efficient shock-stable schemes. *AIAA Journal*, *46*(12), 3012–3037.

Ishii, C. M., & Hibiki, T. (2006). *Thermo-fluid dynamics of two-phase flow*. Springer.

Jameson, A., Schmidt, W., & Turkel, E. (1981). Numerical solution of the Euler equations by finite volume methods using Runge-Kutta time stepping schemes. *In AIAA Paper*, 81–1259.

Ji, B., Luo, X., Wu, Y., Peng, X., & Duan, Y. (2013). Numerical analysis of unsteady cavitating turbulent flow and sheding horse-shoe vortex structure around a twisted hydrofoil. *Interational Journal of Multiphase Flow*, *51*, 33–43.

Jones, W. P. (1979). Prediction methods for turbulent flows. In *Von Karman Institute*. Lecture Series 1979–2.

Jones, W. P., & Launder, B. E. (1972). The prediction of laminarization with a two-equation model of turbulence. *International Journal of Heat and Mass Transfer*, *15*, 301–314.

Kapila, A. K., Menikoff, R., Bdzil, J. B., Son, S. F., & Stewart, D. S. (2001). Two-phase modeling of deflagration-to-detonation transition in granular materials: Reduced equations. *Physics of Fluids*, *13*(10), 3002–3024.

Kataoka, I., & Serizawa, A. (1989). Basic equations of turbulence in gas-liquid two-phase flow. *International Journal of Multiphase Flow*, *15*(5), 843–855.

Kunz, R. F. Boger, D. A., Stinebring, D. R., Chyczewski, T. S., Lindau, J. W., Gibeling, H. J., Venkateswaran, S., & Govindan, T. R. (2000). A preconditioned Navier-Stokes method for two-phase flows with application to cavitation prediction. *Computers & Fluids*, *29*(8), 849–875.

Labourasse, E., Lacanette, D., Toutant, A., Vincent, S., Lubin, P., Lebaigue, O., Caltagirone, J.-P., & Sagaut, P. (2007). Towards large eddy simulation of isothermal two-phase flows: Governing equations and a priori test. *The International Journal of Multiphase Flow*, *33*(1), 1–39.

Lance, M., Marie, J. L., & Bataille, J. (1984). Construction of a model for liquid phase turbulence in bubble flow. *La Houille Blanche*, *3–4*, 255–260.

Larocque, J., Vincent, S., Lacanette, D., Lubin, P., & Caltagirone, J.-P. (2010). Parametric study of LES subgrid terms in a turbulent phase separation. *International Journal of Heat and Fluid Flow*, *31*, 536–544.

Liovic, P., & Lakehal, D. (2012). Subgrid-scale modelling of surface tension within interface tracking-based Large Eddy and Interface Simulation of 3D interfacial flows. *Computers & Fluids*, *63*, 27–46.

Liu, T. G., Khoo, B. C., & Xie, W. F. (2004). Isentropic one-fluid modelling of unsteady cavitating flow. *Journal of Computational Physics*, *201*(1), 80–108.

Menter, F. R. (2010a). The scale-adaptive simulation mathod for unsteady turbulent flow predictions. Part 1: Theory and model description. *Flow, Turbulence and Combustion*, *85*(1), 113–138.

Menter, F. R. (2010b). The scale-adaptive simulation method for unsteady turbulent flow predictions. Part 2: Application to complex flows. *Flow, Turbulence and Combustion*, *85*(1), 139–165.

Menter, F. R. (1994). Two-equation eddy-viscosity turbulence models for engineering applications. *AIAA Journal*, *32*(8), 1598–1605.

Menter, F. R. (1997). Eddy viscosity transport equations and their relation to the $k-\epsilon$ model. *Journal of Fluids Engineering*, *119*, 876–884.

Menter, F. R., & Egorov, Y. (2004). Re-visiting the turbulent scale equation. In *Proceeding IUTAM Symposium; One Hundred Years of Boundary Layer Research—Gottingen*.

Merci, B., Steelant, J., Vierendeels, J., Riemslagh, K., & Dick, E. (2000). Computational treatment of source terms in two-equation turbulence models. *AIAA Journal*, *38*(11), 2085–2093.

Merkle, C. L., Feng, J., & Buelow, P. E. (1998). Computation modeling of the dynamics of sheet cavitation. In *3rd International Symposium on Cavitation CAV1998, Grenoble, France*.

Le Metayer, O., Massoni, J., & Saurel, R. (2004). Elaborating equations of state of a liquid and its vapor for two-phase flow models. *International Journal of Thermal Sciences*, *43*, 265–276.

Petitpas, F., Massoni, J., Saurel, R., Lapedie, E., & Munier, L. (2009). Diffuse interface model for high speed cavitating underwater systems. *International Journal of Multiphase Flow*, *35*, 747–759.

Petkovsek, M., & Dular, M. (2013). IR measurements of the thermodynamic effects in cavitating flow. *International Journal of Heat and Fluid Flow*, *44*, 756–763.

Reboud, J.-L., Stutz, B., & Coutier, O. (1998). Two-phase flow structure of cavitation: Experiment and modelling of unsteady effects. In *3rd International Symposium on Cavitation CAV1998, Grenoble, France*.

Roe, P. L. (1981). Approximate Riemann solvers, parameters vectors, and difference schemes. *Journal of Computational Physics*, *43*, 357–372.

Sarkar, S. (1992). The pressure-dilatation correlation in compressible flows. *Physics of Fluids*, *4*(12), 2674–2682.

Saurel, R., Cocchi, J. P., & Barry, P. (1999). Butler. Numerical study of cavitation in the wake of a hypervelocity underwater projectile. *Journal of Propulsion and Power*, *15*(3), 513–522.

Saurel, R., Petitpas, F., & Abgrall, R. (2008). Modelling phase transition in metastable liquids: application to cavitating and flashing flows. *Journal of Fluid Mechanics*, *607*, 313–350.

Saurel, R., Petitpas, F., & Berry, A. (2009). Simple and efficient relaxation methods for interfaces separating compressible fluids, cavitating flows and shocks in multiphase mixtures. *Journal of Computational Physics*, *228*(5), 1678–1712.

Schmidt, D., Rutland, C., & Corradini, M. L. (1999). A fully compressible, two-dimensional model of small, high-speed, cavitating nozzles. *Atomization and Sprays*, *9*, 255–276.

Schmidt, S. J., Sezal, I. H., & Schnerr, G. H. (2006). Compressible simulation of high-speed hydrodynamics with phase change. In *European Conference on Computational Fluid Dynamics ECCOMAS 2006, Delft, The Netherlands*.

Senocak, I., & Shyy, W. (2002). A pressure-based method for turbulent cavitating flow computations. *Journal of Computational Physics*, *176*(2), 363–383.

Shin, B. R., Iwata, Y., & Ikohagi, T. (2003). Numerical simulation of unsteady cavitating flows using a homogeneous equilibrium model. *Computational Mechanics*, *30*, 388–395.

Singhal, A. K., Athavale, M. M., Li, H., & Jiang, Y. (2002). Mathematical basis and validation of the full cavitation model. *Journal of Fluids Engineering*, *124*(3), 617–624.

Sinibaldi, E., Beux, F., & Salvetti, M. V. (2006). A numerical method for 3D barotropic flows in turbomachinery. *Flow Turbulence Combustion*, *76*, 371–381.

Smith, B. R. (1994). A near wall model for the $k - l$ two equation turbulence model. In *AIAA 94–2386, 25th Fluid Dynamics Conference—Colorado Springs, Colorado*.

Spalart, P. R., & Allmaras, S. R. (1994). A one-equation turbulence model for aerodynamic flows. *La Recherche Aérospatiale*, *1*, 5–21.

Srinivasan, V., Salazar, A. J., & Saito, K. (2009). Numerical simulation of cavitation dynamics using a cavitation-induced-momentum-defect (CIMD) correction approach. *Applied Mathematical Modelling*, *33*, 1529–1559.

Stutz, B. (1996). *Analyse de la structure diphasique et instationnaire de poches de cavitation*. Ph.D. thesis, Institut National Polytechnique de Grenoble.

Toutant, A., Chandesris, M., Jamet, D., & Lebaigue, O. (2009). Jump conditions for filtered quantities at under-resolved discontinuous interface. Part 1: Theoretical development. *International Journal of Multiphase Flow*, *35*, 1100–1118.

Turkel, E. (1987). Preconditioned methods for solving the incompressible and low speed compressible equations. *Journal of Computational Physics*, *172*(2), 277–298.

Utturkar, Y., Wu, J., Wang, G., & Shyy, W. (2005). Recent progress in modelling of cryogenic cavitation for liquid rocket propulsion. *Progress in Aerospace Sciences*, *41*, 558–608.

Venkateswaran, S., Lindau, J. W., Kunz, R. F., & Merkle, C. L. (2002). Computation of multiphase mixture flows with compressibility effects. *Journal of Computational Physics*, *180*(1), 54–77.

Ventikos, Y., & Tzabiras, G. (2000). A numerical method for the simulation of steady and unsteady cavitating flows. *Computers & Fluids*, *29*(1), 63–88.

Viegas, J. R., & Rubesin, M. W. (1983). Wall–function boundary conditions in the solu tion of the Navier–Stokes equations for complex compressible flows. In *AIAA 83–1694, 16th Fluid and Plasma Dynamics Conference*, Danver, Massachussetts, 12–14 July 1983.

Vincent, S., Larocque, J., Lacanette, D., Toutant, A., Lubin, P., & Sagaut, P. (2008). *Computers & Fluids*, *37*, 898–906.

Vortmann, C., Schnerr, G. H., & Seelecke, S. (2003). Thermodynamic modeling and simulation of cavitating nozzle flow. *International Journal of Heat and Fluid Flow*, *24*, 774–783.

De Vuyst, F., Ghidaglia, J. M., & Le Coq, G. (2005). On the numerical simulation of multiphase water flows with changes of phase and strong gradients using the Homogeneous Equilibrium Model. *International Journal on Finite Volumes*, *2*(1), 1–36.

Wallis, G. (1967). *One-dimensional two-phase flow*. New York: McGraw-Hill.

Weller, H., Tabor, G., Jasak, H., & Fureby, C. (1998). A tensorial approach to computational continuum mechanics using objected-oriented techniques. *Computers in Physics*, *12*, 620–631.

Wilcox, D. C. (1988). Reassement of the scale-determining equation for advanced turbulence models. *AIAA Journal*, *26*(11), 1299–1310.

Wilcox, D. C. (1998). *Turbulence modeling for CFD*. DCW Industries Inc.

Wu, J., Wang, G., & Shyy, W. (2005). Time-dependent turbulent cavitating flow computations with interfacial transport and filter-based models. *International Journal for Numerical Methods in Fluids*, *49*(7), 739–761.

Zein, A., Hantke, M., & Warnecke, G. (2010). Modeling phase transition for compressible two-phase flows applied to metastable liquids. *Journal of Computational Physics*, *229*(8), 2964–2998.

Zhang, X. B., Qiu, L. M., Gao, Y., & Zhang, X. J. (2008). Computational fluid dynamic study on cavitation in liquid nitrogen. *Cryogenics*, *48*, 432–438.

Zhou, L., & Wang, Z. (2008). Numerical simulation of cavitation around a hydrofoil and evaluation of a RNG $k-\varepsilon$ model. *Journal of Fluid Engineering*, *130*(1), 011302.

Numerical Simulation of Cavitating Flows in Complex Geometries

Maria Vittoria Salvetti

Abstract The present contribution focusses on numerical simulation of cavitating flows in complex geometries. We consider compressible flows and cavitation models assuming a homogeneous barotropic flow behavior. Different numerical issues are analyzed and possible solutions are presented and validated. Finally, an application to the simulation of the flow in a real turbopump inducer designed for liquid-propelled rockets is presented.

1 Introduction

Cavitating flows occur in many engineering devices, as, for instance, rocket turbopumps, turbomachinery, hydrofoils, marine propellers, or nozzles. It is well known that cavitation has strong effects, usually negative, on performance and life of such devices. These kinds of engineering problems have some common features, which impact on physical modeling and numerical methodology adopted in numerical simulations of these flows. First, the fluid flows over solid walls having complex geometry and in many cases moving. Moreover, the characteristics of the inlet flow stream are only partially known; for instance, the size and distribution of active cavitation nuclei are usually not known. In general, for engineering applications, the aim is to obtain predictions of the global performance of the device and of the effects of cavitation rather than a detailed description of the flow or of cavitation phenomena. On the other hand, the various effects of cavitation may have different importance depending on the engineering application; for instance, the analysis of liquid-propellant fed turbopumps for space application put especial prize on the suction and dynamic performance of the machine rather than on its resistance to erosion and other long-term effects of cavitation, which are typically a major concern in other applications.

From a physical viewpoint, cavitating flows are characterized by complex phenomena interacting each other, such as change of phase, evolving vapor to liquid

M.V. Salvetti (✉)
DICI, University of Pisa, Pisa, Italy
e-mail: mv.salvetti@ing.unipi.it

L. d'Agostino and M.V. Salvetti (eds.), *Cavitation Instabilities and Rotordynamic Effects in Turbopumps and Hydroturbines*, CISM International Centre for Mechanical Sciences 575, DOI 10.1007/978-3-319-49719-8_9

interfaces and turbulence. Therefore, the modeling of cavitating flows is a very complex task. Several models exist in the literature of different levels of complexity. We focus here on the so-called *one-fluid or homogeneous-flow* models, in which the cavitating flow is described in terms of a single fluid or mixture, whose properties depend on the void fraction. We refer to the contributions by Goncalves and by Saurel et al. in this volume for more complex two-phase cavitation models. Homogeneous-flow cavitation models are based on the following assumptions: (i) local kinematic equilibrium, i.e., same local velocity for both liquid and vapor phases; (ii) local thermodynamic equilibrium, i.e., local temperature and pressure equality between phases; and (iii) instantaneous vaporization or condensation processes. Clearly, this kind of modeling is not suitable if strong non-equilibrium effects between the two phases are present. On the other hand, these models are attracting because of their simplicity and because they a-priori have the capability of describing the global effects of cavitation, which are the most interesting for practical applications. In the framework of the previous simplifying assumptions, the thermodynamic properties for the pure liquid, the pure vapor, and the mixture must be specified through an equation of state (EOS). Different ways have been followed in the literature, e.g., tabulated EOS and thermodynamic properties (e.g., Clerc 2000 or Ventikos and Tzabiras 2000), entropy maximization procedure (e.g., Barberon and Helluy 2005), partial mass and partial density properties (e.g., Meng and Yang 2003), and saturation equations and thermodynamics tables (e.g., Saurel et al. 1999). Homogeneous-flow cavitation models using the assumption of a barotropic behavior for pure liquid and vapor and for the liquid–vapor mixture (see e.g., Delannoy and Kueny 1990; d'Agostino et al. 2011; Qin et al. 2003; Liu et al. 2004) are considered herein.

In the homogenous-flow description, the physical properties of the flow change dramatically between the zones of pure liquid and the cavitating regions. In particular, in non-cavitating zones the flow is almost incompressible, while the cavitating mixture is described as a highly compressible fluid, characterized by speed of sound values of several orders of magnitude lower than those of the pure liquid. Moreover, an abrupt transition from the wetted (incompressible) to the cavitating (highly supersonic) regimes occurs. These features yield strong difficulties for numerical simulation, in spite of the simplifying assumptions made and the apparent simplicity of barotropic homogeneous-flow cavitation models. It is clear, on the basis of the previous considerations, that the numerical schemes must be designed in order to cope with nearly incompressible and highly compressible regions coexisting in the flow. Two opposite approaches can be found in the literature: adaptation to compressible flows of methods developed for incompressible flows, usually through pressure correction (e.g., Srinivasan et al. 2009; Senocak and Shyy 2002; Coutier-Delgosha et al. 2003) or adaptation to the incompressible limit, usually through ad-hoc preconditioning, of compressible flow solvers (e.g., Sinibaldi et al. 2006; Bilanceri et al. 2010; Goncalves and Patella 2009; Coutier-Delgosha et al. 2005; Kunz et al. 2000). Each one of these approaches has advantages and drawbacks. Incompressible solvers may lead to erroneous acoustic speeds in the cavitating regions, while compressible solvers have well-known accuracy and efficiency/robustness problems in the low Mach number regions (Sinibaldi et al. 2006; Guillard and Viozat 1999).

A mixed finite-volume/finite-element compressible flow solver on unstructured grids is considered herein. Unstructured grids are interesting when complex geometries are considered. Due to the previously mentioned difficulties, the numerical method should be robust and efficient. Two key ingredients are used in this work. First, a HLL flux function for the convective terms is used, in which an anti-diffusive term is introduced to counteract accuracy problems for viscous flows typical of this class of schemes. On the basis of an asymptotic analysis in power of Mach, a preconditioning of the same kind of that used in Guillard and Viozat (1999) for perfect gases was adapted to barotropic flows and introduced in the upwind part of the numerical flux, in order to counteract the accuracy problems encountered in the low Mach regime. Since the preconditioning matrix multiplies the upwind part of the flux only, consistency in time is preserved. The second key point is the linearization in time of the numerical flux function in a implicit time-advancing algorithm: a linearization taking into account the time variation of the flux upwind part is presented.

As an example of a real engineering application, all these numerical ingredients are applied to the simulation of the flow in a real inducer for space rocket turbopumps in non-cavitating and cavitating conditions.

2 Governing Equations and Cavitation Modeling

As previously said, a homogeneous-flow cavitation model is considered, together with a barotropic EOS. Therefore, the energy equation can be discarded since it is decoupled from the mass and momentum balances.

An additional modeling issue for cavitating flow simulations is turbulence and its interaction with cavitation phenomena. This aspect is not investigated in detail herein; we refer to the contribution by E. Goncalves in this volume. We simply consider the unsteady Reynolds-averaged Navier–Stokes (URANS) equations, together with the standard k-ε model (Launder and Spalding 1974). Although corrections aimed at limiting the turbulent viscosity have been proposed for cavitating flows (see again the contribution by E. Goncalves), the standard k-ε model is commonly applied to a wide range of cases and its limitations for non-cavitating flows are well known. As a consequence, this model seems a good starting point for a first appraisal of turbulence effects.

Thus, considering a reference frame rotating with constant angular velocity $\boldsymbol{\omega}$, the following system of equations is obtained:

$$\frac{\partial \mathbf{W}}{\partial t} + \frac{\partial}{\partial x_j}\mathbf{F}_j(\mathbf{W}) + \frac{\partial}{\partial x_j}\widetilde{\mathbf{F}}_j(\mathbf{W}) - \frac{\partial}{\partial x_j}\mu \mathbf{V}_j(\mathbf{W}, \nabla \mathbf{W}) - \frac{\partial}{\partial x_j}\mu_t \widetilde{\mathbf{V}}_j(\mathbf{W}, \nabla \mathbf{W}) = \boldsymbol{\Omega}(\mathbf{W}) + \mathbf{S}\,(\omega, \mathbf{x}, \mathbf{W}) . \quad (1)$$

In Eq. (1) the Einstein notation is used and the vector of unknowns **W** is defined as follows:

$$\mathbf{W} = (\rho,\ \rho u,\ \rho v,\ \rho w,\ \rho k,\ \rho \varepsilon)^T, \tag{2}$$

where ρ is the density; u, v, and w are the velocity components in the x, y, and z directions; k is the turbulent kinetic energy; and ε is the turbulent dissipation. The other terms appearing in Eq. (1) are the convective fluxes F_j, the turbulence contribution to convective fluxes $\widetilde{F}_j$, the viscous laminar fluxes V_j, the viscous turbulent fluxes $\widetilde{V}_j$, and the source term Ω related to the k-ε model. Their expressions are the following:

$$\begin{cases} \mathbf{F}_1 = \left(\rho u,\ \rho u^2 + p,\ \rho uv,\ \rho uw,\ \rho uk,\ \rho u\varepsilon\right)^T \\ \mathbf{F}_2 = \left(\rho v,\ \rho vu,\ \rho v^2 + p,\ \rho vw,\ \rho vk,\ \rho v\varepsilon\right)^T \\ \mathbf{F}_3 = \left(\rho w,\ \rho wu,\ \rho wv,\ \rho w^2 + p,\ \rho wk,\ \rho w\varepsilon\right)^T \end{cases} \tag{3}$$

$$\begin{cases} \widetilde{\mathbf{F}}_1 = \left(0,\ \tfrac{2}{3}\rho k,\ 0,\ 0,\ 0, 0\right)^T \\ \widetilde{\mathbf{F}}_2 = \left(0,\ 0,\ \tfrac{2}{3}\rho k,\ 0,\ 0, 0\right)^T \\ \widetilde{\mathbf{F}}_3 = \left(0,\ 0,\ 0,\ \tfrac{2}{3}\rho k, 0, 0\right)^T \end{cases} \tag{4}$$

$$\begin{cases} \mathbf{V}_1 = \left(0,\ \sigma_{11},\ \sigma_{12},\ \sigma_{13},\ \dfrac{\partial k}{\partial x_1},\ \dfrac{\partial \varepsilon}{\partial x_1}\right)^T \\ \mathbf{V}_2 = \left(0,\ \sigma_{21},\ \sigma_{22},\ \sigma_{23},\ \dfrac{\partial k}{\partial x_2},\ \dfrac{\partial \varepsilon}{\partial x_2}\right)^T \\ \mathbf{V}_3 = \left(0,\ \sigma_{31},\ \sigma_{32},\ \sigma_{33},\ \dfrac{\partial k}{\partial x_3},\ \dfrac{\partial \varepsilon}{\partial x_3}\right)^T \end{cases} \tag{5}$$

$$\begin{cases} \widetilde{\mathbf{V}}_1 = \left(0,\ \sigma_{11},\ \sigma_{12},\ \sigma_{13},\ \dfrac{1}{\sigma_k}\dfrac{\partial k}{\partial x_1},\ \dfrac{1}{\sigma_\varepsilon}\dfrac{\partial \varepsilon}{\partial x_1}\right)^T \\ \widetilde{\mathbf{V}}_2 = \left(0,\ \sigma_{21},\ \sigma_{22},\ \sigma_{23},\ \dfrac{1}{\sigma_k}\dfrac{\partial k}{\partial x_2},\ \dfrac{1}{\sigma_\varepsilon}\dfrac{\partial \varepsilon}{\partial x_2}\right)^T \\ \widetilde{\mathbf{V}}_3 = \left(0,\ \sigma_{31},\ \sigma_{32},\ \sigma_{33},\ \dfrac{1}{\sigma_k}\dfrac{\partial k}{\partial x_3},\ \dfrac{1}{\sigma_\varepsilon}\dfrac{\partial \varepsilon}{\partial x_3}\right)^T, \end{cases} \tag{6}$$

in which p is the pressure, σ_{ij} are the components of the viscous stress tensor, while σ_ε and σ_k are empirical parameters.

The formulation of the source term $\mathbf{\Omega}\,(\mathbf{W})$ is:

$$
\begin{cases}
\mathcal{P} = -\left(\frac{2}{3}\rho k \delta_{ij} - \mu_t \left(\frac{\partial u_i}{\partial x_j} + \frac{\partial u_j}{\partial x_i} - \frac{2}{3}\frac{\partial u_k}{\partial x_k}\delta_{ij}\right)\right)\frac{\partial u_i}{\partial x_j} \\
\omega_k = -\rho\varepsilon + \mathcal{P} \\
\omega_\varepsilon = c_{\varepsilon 1}\frac{k}{\varepsilon}\mathcal{P} - c_{\varepsilon 2}\frac{\rho\varepsilon^2}{k} \\
\mathbf{\Omega}(\mathbf{W}) = \left(0\;0\;0\;0\;\omega_k\;\omega_\varepsilon\right)^T,
\end{cases}
\tag{7}
$$

$\mathcal{P}$ being the production term of the turbulent kinetic energy and $c_{\varepsilon 1}$ and $c_{\varepsilon 2}$ are empirical parameters.

The molecular and turbulent viscosities are, respectively, μ, defined in the following (Eq. (12)) and μ_t defined as follows:

$$
\mu_t = C_\mu \frac{\rho k^2}{\varepsilon},
\tag{8}
$$

C_μ being an additional empirical parameter. For all the empirical parameters contained in the k-ε model, the standard values proposed in Launder and Spalding (1974) are used herein.

Finally, $\mathbf{S}$ is the source term due to reference frame rotation:

$$
\begin{cases}
\overline{\mathbf{S}} = -\left(2\boldsymbol{\omega} \wedge \rho\mathbf{u} + \rho\boldsymbol{\omega} \wedge (\boldsymbol{\omega} \wedge \mathbf{x})\right) \\
\mathbf{S}(\boldsymbol{\omega}, \mathbf{x}, \mathbf{W}) = \left(0,\; \overline{\mathbf{S}}^T,\; 0,\; 0\right)^T,
\end{cases}
\tag{9}
$$

$\mathbf{x}$ and $\mathbf{u}$ being the position and velocity vectors, respectively.

System (1) is completely defined once a suitable constitutive equation $p = p(\rho)$ is introduced. In this work a weakly compressible liquid at constant temperature T_L is considered as working fluid. The liquid density ρ is allowed to locally fall below the saturation limit $\rho_{Lsat} = \rho_{Lsat}\left(T_L\right)$ thus originating cavitation phenomena. A regime-dependent (wetted/cavitating) constitutive relation is therefore adopted. As for the wetted regime ($\rho \geq \rho_{Lsat}$), a barotropic model of the form

$$
p = p_{sat} + \frac{1}{\beta_{sL}} \ln\left(\frac{\rho}{\rho_{Lsat}}\right)
\tag{10}
$$

is adopted, $p_{sat} = p_{sat}\left(T_L\right)$ and $\beta_{sL} = \beta_{sL}\left(T_L\right)$ being the saturation pressure and the liquid isentropic compressibility, respectively. As for the cavitating regime ($\rho < \rho_{Lsat}$), a homogeneous-flow model explicitly accounting for thermal cavitation effects and for the concentration of the active cavitation nuclei in the pure liquid has been adopted (d'Agostino et al. 2011):

$$
\frac{p}{\rho}\frac{d\rho}{dp} = (1-\alpha)\left[(1-\varepsilon_L)\frac{p}{\rho_{Lsat}a_{Lsat}^2} + \varepsilon_L g^\star \left(\frac{p_c}{p}\right)^\eta\right] + \frac{\alpha}{\gamma_V},
\tag{11}
$$

where $g^\star$, η, γ_V, and p_c are the liquid parameters, a_{Lsat} is the liquid sound speed at saturation, $\alpha = 1 - \rho/\rho_{Lsat}$, and $\varepsilon_L = \varepsilon_L(\alpha, \zeta)$ is a given function (see d'Agostino et al. 2011 for its physical interpretation and for more details). The resulting unified barotropic state law for the liquid and for the cavitating mixture only depends on the two parameters T_L and ζ. For instance, for water at $T_L = 293.16\,\mathrm{K}$, the other parameters involved in (10) and (11) are $p_{sat} = 2806.82$ Pa, $\rho_{Lsat} = 997.29$ kg/m^3, $\beta_{sL} = 5\,10^{-10}$ Pa^{-1}, $g^* = 1.67$, $\eta = 0.73$, $\gamma_V = 1.28$, $p_c = 2.21\,10^7$ Pa and $a_{Lsat} = 1415$ m/s (Torre et al. 2010). Note that, despite the model simplifications leading to a unified barotropic state law, the transition between wetted and cavitating regimes is extremely abrupt. Indeed, the sound speed falls from values of order 10^3 m/s in the pure liquid down to values of order 0.1 or 1 m/s in the mixture. The corresponding Mach number variation renders this state law very stiff from a numerical viewpoint.

As for the definition of the molecular viscosity, a simple model, which is linear in the cavitating regime, is considered:

$$\mu(\rho) = \begin{cases} \mu_L & \text{if} \quad \rho \geq \rho_{Lsat} \\ \mu_v & \text{if} \quad \rho \leq \rho_v \\ \alpha\mu_v + (1-\alpha)\mu_L & \text{otherwise,} \end{cases} \tag{12}$$

in which μ_v and μ_L are the molecular viscosity of the vapor and liquid, respectively, which, consistently with the assumptions made in the adopted cavitation model, are considered constant and computed at $T = T_L$.

3 Numerical Methodology

The spatial discretization of the governing equations is based on a mixed finite-element/finite-volume formulation on unstructured grids. Starting from an unstructured tetrahedral grid, a dual finite-volume mesh is obtained by the rule of medians: a cell C_i is built around each vertex i, and boundaries between cells are made of triangular interface facets. Let us consider the integral formulation of system (1) for the case in which the control volume $\mathcal{V}$ is the finite-volume cell C_i:

$$\begin{aligned} &\int_{C_i} \frac{\partial \mathbf{W}}{\partial t} \mathrm{d}\mathcal{V} + \int_{\partial C_i} n_k \left(\mathbf{F}_k + \widetilde{\mathbf{F}}_k \right) \mathrm{d}S - \int_{C_i} \frac{\partial}{\partial x_k} \left(\mu \mathbf{V}_k + \mu_t \widetilde{\mathbf{V}}_k \right) \mathrm{d}\mathcal{V} = \\ &\int_{C_i} (\mathbf{\Omega} + \mathbf{S}) \mathrm{d}\mathcal{V}, \end{aligned} \tag{13}$$

where ∂C_i is the boundary of C_i and n_k are the components of the unit vector normal to ∂C_i. It is possible to reformulate (13) in the following semi-discrete form:

$$\mathcal{V}_i \frac{\mathrm{d}\mathbf{W}_i}{\mathrm{d}t} + \sum_{j \in N(i)} \mathbf{\Phi}_{ij} + \mathbf{\Upsilon}_i = \mathbf{\Omega}_i + \mathbf{S}_i, \tag{14}$$

where $\mathbf{W}_i$ is the semi-discrete unknown associated with C_i, defined as follows:

$$\mathbf{W}_i = \frac{1}{\mathcal{V}_i} \int_{C_i} \mathbf{W} \mathrm{d}\mathcal{V}, \tag{15}$$

where $\mathcal{V}_i$ is the cell volume, and $N(i)$ represents the set of neighbors of the ith cell. The numerical discretization of the convective flux crossing the boundary ∂C_{ij} shared by C_i and C_j (positive towards C_j) is denoted $\mathbf{\Phi}_{ij}$, while $\mathbf{\Upsilon}_i$, $\mathbf{\Omega}_i$, and $\mathbf{S}_i$ are the numerical discretizations for, respectively, the viscous fluxes and the source terms.

3.1 Numerical Discretization for a 1D Inviscid Flow

Since the key ingredients of the proposed numerical discretization are the numerical approximation of the convective fluxes and time advancing, the 1D inviscid case is used as a first step for the definition of the numerical convective fluxes:

$$\frac{\partial \mathbf{W}}{\partial t} + \frac{\partial \mathbf{F}(\mathbf{W})}{\partial x} = 0 \tag{16}$$

where $\mathbf{W} = (\rho, \rho u, \rho\xi)^T$ and $\mathbf{F}(\mathbf{W}) = \left(\rho u, \rho u^2 + p, \rho u \xi\right)^T$, ξ denoting a passive scalar.

The associated semi-discrete problem is the following:

$$\delta x_i \frac{\mathrm{d}\mathbf{W}_i}{\mathrm{d}t} + \mathbf{\Phi}_{i,i+1} - \mathbf{\Phi}_{i-1,i} = 0, \tag{17}$$

δx_i being the width of the finite-volume cell centered around node i.

3.1.1 First-Order Numerical Approximation of Convective Fluxes

The numerical scheme herein considered belongs to the so-called approximate Riemann solvers, and, more particularly, to the HLL schemes (Harten et al. 1983), which assume one intermediate state between two acoustic waves in the approximate Riemann problem. The simplest HLL scheme is the so-called Rusanov scheme (Rusanov 1961; Park and Kwon 2003):

$$\mathbf{\Phi}_{i,j} = \frac{\mathbf{F}(\mathbf{W}_i) + \mathbf{F}(\mathbf{W}_j)}{2} - \frac{\lambda_{ij}}{2}(\mathbf{W}_j - \mathbf{W}_i), \tag{18}$$

where the first term of the right-hand side is a centered approximation of the fluxes and the second one the upwind part, introducing numerical viscosity, with $\lambda_{ij} = s_R$ being an upper bound for the fastest wave speed. A classical choice for λ_{ij} is the largest absolute value of the Roe matrix eigenvalues (Roe 1981). A suitable Roe matrix for a generic barotropic EOS was defined in Sinibaldi et al. (2006); its eigenvalues are the following:

$$\tilde{\lambda}_1 = \tilde{u}_{ij} + \tilde{a}_{ij}\,,\ \tilde{\lambda}_2 = \tilde{u}_{ij} - \tilde{a}_{ij} \text{ and } \tilde{\lambda}_3 = \tilde{u}_{ij},$$

where $\tilde{u}_{ij}$ is the classical 'Roe average' of u:

$$\tilde{u}_{ij} = \frac{\sqrt{\rho_j}u_j + \sqrt{\rho_i}u_i}{\sqrt{\rho_j} + \sqrt{\rho_i}}, \tag{19}$$

whereas $\tilde{a}_{ij}$, which can be considered as a Roe average for the sound speed, is defined as:

$$\tilde{a}_{ij} = \begin{cases} \sqrt{\dfrac{p(\rho_j) - p(\rho_i)}{\rho_j - \rho_i}} & \text{if } \rho_j \neq \rho_i \\ a\left(\rho_\star, p(\rho_\star)\right) & \text{if } \rho_i = \rho_j = \rho_\star. \end{cases} \tag{20}$$

Therefore, we obtain

$$\lambda_{ij} = \max_p(|\tilde{\lambda}_p|) = |\tilde{u}_{ij}| + \tilde{a}_{ij}.$$

The Rusanov scheme, as all the HLL schemes, is known to introduce excessive numerical diffusion in the presence of contact discontinuities. Indeed, due to the assumption of a two-wave configuration with only one intermediate state, the two intermediate states of the exact Riemann solution are averaged, smearing, in this way, the contact discontinuity. In the barotropic case, since density, velocity and pressure are continuous across the contact discontinuity, the presence of two different intermediate states in the Riemann problem is only due to the passive scalar (see Sinibaldi 2006). Moreover, only one eigenvalue is associated with the contact discontinuity. Thus, a possible way to counteract the excessive diffusion of the Rusanov scheme is to introduce an anti-diffusive term in the discretization of the third equation of system (16). Based on this idea, the following LD-HLL (Low-Diffusive HLL) was proposed in Bilanceri et al. (2010):

$$\boldsymbol{\Phi}_{i,j} = \frac{\mathbf{F}(\mathbf{W}_i) + \mathbf{F}(\mathbf{W}_j)}{2} - \frac{1}{2}\underbrace{\begin{pmatrix} \lambda_{ij} & 0 & 0 \\ 0 & \lambda_{ij} & 0 \\ 0 & 0 & |\tilde{u}_{ij}| \end{pmatrix}}_{Q_{ij}}(\mathbf{W}_j - \mathbf{W}_i). \tag{21}$$

Compared with the original Rusanov scheme, the first two equations, which are related to the acoustic waves, are unchanged, while for the third equation, i.e., the one directly related to the contact discontinuity, the diffusive part of the scheme has been reduced (see Bilanceri et al. 2010 for more details).

As mentioned in the Introduction, compressible flow solvers suffer of accuracy and efficiency problems for nearly incompressible flows, as encountered in the non-cavitating flow regions. Following the approach proposed by Guillard and Viozat (1999), it was shown in Bilanceri (2011) that, as expected, also the LD-HLL scheme (21) suffers of accuracy problems in the low Mach number limit. Therefore, suitable preconditioning is needed. It was also shown in Bilanceri (2011) that the correct asymptotic behavior can be recovered by acting on the acoustic terms of the matrix Q_{ij}. This can be done by multiplying Q_{ij} by the following preconditioning matrix:

$$P_{ij} = \begin{pmatrix} \theta^{-1} & 0 & 0 \\ 0 & \theta & 0 \\ 0 & 0 & 1 \end{pmatrix}, \tag{22}$$

in which

$$\theta = \theta(M) = \begin{cases} 10^{-6} & \text{if } M \leq 10^{-6} \\ \min(M, 1) & \text{otherwise} \end{cases} \quad . \quad M = \frac{|\tilde{u}_{ij}|}{\tilde{a}_{ij}}. \tag{23}$$

This leads to the following preconditioned LD-HLL scheme:

$$\mathbf{\Phi}_{i,j} = \frac{\mathbf{F}(\mathbf{W}_i) + \mathbf{F}(\mathbf{W}_j)}{2} - \frac{1}{2} \underbrace{\begin{pmatrix} \lambda_{ij}\theta^{-1} & 0 & 0 \\ 0 & \lambda_{ij}\theta & 0 \\ 0 & 0 & |\tilde{u}_{ij}| \end{pmatrix}}_{Q_{ij}} (\mathbf{W}_j - \mathbf{W}_i). \tag{24}$$

Note that since preconditioning acts only on the upwind part of the scheme, consistency in time is preserved. Moreover, preconditioning does not affect the low-diffusion correction introduced for the passive scalar equation.

3.1.2 Linearized Implicit Time Advancing

Let us consider an implicit backward Euler method applied to the semi-discrete problem (17):

$$\frac{\delta x_i}{\Delta t} \Delta^n \mathbf{W}_i + \Delta^n \mathbf{\Phi}_{i,i+1} - \Delta^n \mathbf{\Phi}_{i-1,i} = -\left(\mathbf{\Phi}^n_{i,i+1} - \mathbf{\Phi}^n_{i-1,i}\right), \tag{25}$$

where $\Delta^n(\cdot) = (\cdot)^{n+1} - (\cdot)^n$. To avoid the solution of a nonlinear system at each time step, a linearization of $\Delta^n \Phi_{ij}$ is usually adopted. A way to obtain such a linearization is to find two matrices D_1 and D_2 such that

$$\Delta^n \mathbf{\Phi}_{ij} \simeq D_1(\mathbf{W}_i^n, \mathbf{W}_j^n)\, \Delta^n \mathbf{W}_i + D_2(\mathbf{W}_i^n, \mathbf{W}_j^n)\, \Delta^n \mathbf{W}_j \tag{26}$$

In this case (25) is reduced to the following block tridiagonal linear system:

$$B_{-1}^{i,n}\, \Delta^n \mathbf{W}_{i-1} + B_0^{i,n}\, \Delta^n \mathbf{W}_i + B_1^{i,n} \Delta^n \mathbf{W}_{i+1} = -\left(\mathbf{\Phi}_{i,i+1}^n - \mathbf{\Phi}_{i-1,i}^n\right), \tag{27}$$

where

$$\begin{cases} B_{-1}^{i,n} = -D_1(\mathbf{W}_{i-1}^n, \mathbf{W}_i^n) \\ B_0^{i,n} = \dfrac{\delta x_i}{\Delta t} I + D_1(\mathbf{W}_i^n, \mathbf{W}_{i+1}^n) - D_2(\mathbf{W}_{i-1}^n, \mathbf{W}_i^n) \\ B_1^{i,n} = D_2(\mathbf{W}_i^n, \mathbf{W}_{i+1}^n). \end{cases} \tag{28}$$

A classical linearization of type (26) consists in applying a first-order Taylor expansion in time but with a complete differentiation only for the centered part of the numerical flux function while the matrix Q_{ij} in the upwind part is frozen at time t^n (see, e.g., Yee 1987; Delis et al. 2000; Luo et al. 2001). This results in the following approximation:

$$\begin{aligned} \Delta^n \mathbf{\Phi}_{ij} \simeq & \frac{1}{2}\left(A(\mathbf{W}_i^n)\Delta^n \mathbf{W}_i + A(\mathbf{W}_j^n)\Delta^n \mathbf{W}_j\right) \\ & - \frac{1}{2} Q_{ij}^n (\Delta^n \mathbf{W}_j - \Delta^n \mathbf{W}_i), \end{aligned} \tag{29}$$

in which A is the Jacobian matrix of $\mathbf{F}$. Note that this approach can be used independently of the differentiability or of the complexity in the differentiation of Q_{ij}.

This linearization can be reinterpreted by rewriting the time variation of the upwind term of $\mathbf{\Phi}_{ij}$ as follows:

$$\Delta^n \mathbf{\Phi}_{ij,u} = -\frac{Q_{ij}^n}{2}(\Delta^n \mathbf{W}_j - \Delta^n \mathbf{W}_i) - \mathbf{\Gamma}_{ij}^{n,n+1}, \tag{30}$$

in which $\mathbf{\Gamma}_{ij}^{n,n+1} \doteq \dfrac{\Delta^n Q_{ij}}{2}(\mathbf{W}_j^{n+1} - \mathbf{W}_i^{n+1})$.

Linearization (29) is obtained just by neglecting the term $\mathbf{\Gamma}_{ij}^{n,n+1}$. This term can be neglected as long as that the solution is regular enough to satisfy

$$\mathbf{W}_j^{n+1} - \mathbf{W}_i^{n+1} \propto O(\delta x) \text{ and } \Delta^n Q_{ij} \propto O(\Delta t). \tag{31}$$

Even if the assumption (31) is in general a reasonable one, there are situations of practical interest in which it is not satisfied. Indeed, if a discontinuity is present the magnitude of the term $\mathbf{W}_j^{n+1} - \mathbf{W}_i^{n+1}$ can be large independently of the size of Δx. Moreover, the term $\Delta^n Q_{ij}$ can also be large. This can happen again in the presence of large variations of the flow velocity but also when the speed of sound has a stiff change in magnitude. The latter is a typical situation occurring in the presence of cavitation. A more complete linearization was proposed by Bilanceri et al. (2010) for cavitating barotropic flows, which takes into account, at least in an approximate way, the term $\mathbf{\Gamma}_{ij}^{n,n+1}$. This linearization exploits the fact that Q_{ij} is a diagonal matrix such that any of its diagonal coefficients, q_k, can be written as a composite function of two variables, $\tilde{a}$ and $\tilde{u}$. Then, through differentiation, by neglecting terms of higher order and after some mathematical developments (see Bilanceri et al. 2010 for the detailed procedure), the following approximation of the previously neglected term is obtained:

$$\mathbf{\Gamma}_{ij}^{n,n+1} \simeq \frac{1}{2} K_{ij} \Delta^n \mathbf{W}_i - \frac{1}{2} K_{ji} \Delta^n \mathbf{W}_j, \tag{32}$$

in which the generic element of K_{ij} is defined by the following expression:

$$\left(K_{ij}\right)_{km} = \left(\frac{\partial q_k}{\partial \tilde{u}} \frac{\partial \tilde{u}}{\partial W_{i,m}} + \frac{\partial q_k}{\partial \tilde{a}} \frac{\partial \tilde{a}}{\partial W_{i,m}} \right) \left(W_{j,k}^n - W_{i,k}^n \right), \tag{33}$$

where $W_{i,m}$ denotes the m-th element of the vector $\mathbf{W}_i$. Finally, considering $\mathbf{\Gamma}_{ij}^{n,n+1}$ from (32) and (33) in the evaluation of the upwind part of the numerical flux time variation, the following more complete approximation is obtained instead of (29):

$$\begin{aligned} \Delta^n \mathbf{\Phi}_{ij} \simeq & \frac{1}{2}\left(A(\mathbf{W}_i^n) + Q_{ij}^n - K_{ij} \right) \Delta^n \mathbf{W}_i \\ & + \frac{1}{2}\left(A(\mathbf{W}_j^n) - Q_{ij}^n + K_{ji} \right) \Delta^n \mathbf{W}_j. \end{aligned} \tag{34}$$

Some simplifications can be made, based on physical considerations related to the particular kind of applications of interest. Since $\tilde{u}$ and $\tilde{a}$ are directly related to the flow velocity and speed of sound, respectively, we have that the derivatives of $\tilde{u}$ respect to the flow variables are much smaller than the ones of $\tilde{a}$, because, when a transition from vapor to liquid occurs, in the used cavitation model this leads to a step-like change of the speed of sound of a few order of magnitudes. Consequently, it seems reasonable to neglect the variation in $\tilde{u}$ in the computation of K_{ij} in (33), and then, to approximate the variation of the matrix Q_{ij} only through the variation in $\tilde{a}$. Under this assumption and using the fact that $\tilde{a}$ only depends on the density, i.e., $\tilde{a} = \tilde{a}(\rho_i, \rho_j)$, the matrix K_{ij} previously defined in (33) reduces here to

$$K_{ij} = \begin{pmatrix} \partial_{\rho_i} q_1 \left(\rho_j^n - \rho_i^n\right) 0\ 0 \\ \\ \partial_{\rho_i} q_2 \left((\rho u)_j^n - (\rho u)_i^n\right) 0\ 0 \\ \\ \partial_{\rho_i} q_3 \left((\rho \xi)_j^n - (\rho \xi)_i^n\right) 0\ 0 \end{pmatrix}, \tag{35}$$

in which $\partial_{\rho_i} q_k$ represents the partial derivative of q_k with respect to ρ_i, but considering only the variation in $\tilde{a}$, i.e.,

$$\partial_{\rho_i} q_k = \frac{\partial}{\partial \rho_i} \left(q_k \left(\tilde{u}, \tilde{a}(\rho_i, \rho_j)\right) \right).$$

These partial derivatives are numerically computed through centered finite differences (see Bilanceri 2011 for more details).

3.1.3 Extension to Second-Order Accuracy

The extension to second-order accuracy in space can be achieved using a classical MUSCL technique (van Leer 1979), in which instead of $\mathbf{\Phi}_{i,i+1} = \mathbf{\Phi}(\mathbf{W}_i, \mathbf{W}_{i+1})$, the numerical flux $\mathbf{\Phi}_{i+\frac{1}{2}} = \mathbf{\Phi}(\mathbf{W}^-_{i+\frac{1}{2}}, \mathbf{W}^+_{i+\frac{1}{2}})$ is considered at the cell interface $x_{i+\frac{1}{2}}$. The considered values $\mathbf{W}^{\pm}_{i+\frac{1}{2}}$ are defined by piecewise linear reconstruction of the solution. They can be expressed by the following β-scheme (Harten et al. 1983) including limiters (Leveque 1994):

$$\begin{cases} \mathbf{W}^-_{i+\frac{1}{2}} = \mathbf{W}_i + \dfrac{h_{i+1}}{2} \left((1-\beta)\, \Lambda^-_{i+\frac{1}{2}} \dfrac{\mathbf{W}_{i+1} - \mathbf{W}_i}{h_{i+1}} + \beta\, \Lambda^+_{i-\frac{1}{2}} \dfrac{\mathbf{W}_i - \mathbf{W}_{i-1}}{h_i} \right) \\ \\ \mathbf{W}^+_{i+\frac{1}{2}} = \mathbf{W}_{i+1} - \dfrac{h_{i+1}}{2} \left((1-\beta)\, \Lambda^+_{i+\frac{1}{2}} \dfrac{\mathbf{W}_{i+1} - \mathbf{W}^n_i}{h_{i+1}} + \beta\, \Lambda^-_{i+\frac{3}{2}} \dfrac{\mathbf{W}^n_{i+2} - \mathbf{W}_{i+1}}{h_{i+2}} \right), \end{cases} \tag{36}$$

in which β is a free parameter, $h_i = x_i - x_{i-1}$, and $\Lambda^{\pm}_l$ are diagonal matrices introducing the (nonlinear) limiter function. Several classical choices can be used to obtain a total variation diminishing scheme (Toro 1997). We adopted the Minmod limiter (see e.g., Toro 1997). An analogous procedure is obviously carried out for the cell interface $x_{i-\frac{1}{2}}$.

The second-order accuracy in time is then achieved through the use of a backward differentiation formula for the discretization of the time derivative; this leads to the following implicit formulation:

$$\delta x_i \frac{3\mathbf{W}_i^{n+1} - 4\mathbf{W}_i^n + \mathbf{W}_i^{n-1}}{2\Delta t} + \Delta^n \boldsymbol{\Phi}_{i+\frac{1}{2}} - \Delta^n \boldsymbol{\Phi}_{i-\frac{1}{2}} = -\left(\boldsymbol{\Phi}^n_{i+\frac{1}{2}} - \boldsymbol{\Phi}^n_{i-\frac{1}{2}}\right), \tag{37}$$

where $\boldsymbol{\Phi}_{i\pm\frac{1}{2}} = \boldsymbol{\Phi}(\mathbf{W}^-_{i\pm\frac{1}{2}}, \mathbf{W}^+_{i\pm\frac{1}{2}})$ are the second-order accurate numerical fluxes computed using extrapolated variable values at the cell interface (36).

Similar to the first-order case, a linearization of $\Delta^n \boldsymbol{\Phi}_{i\pm\frac{1}{2}}$ must be carried out in order to avoid the solution of a nonlinear system at each time step. However, the linearization for the second-order accurate fluxes and the solution of the resulting linear system implies significant computational costs and memory requirements. Thus, a defect correction technique (DeC see e.g., Martin and Guillard 1996) is used here, which consists in iteratively solving simpler problems obtained by considering the same linearization as used for the first-order scheme:

$$\begin{cases} \mathcal{W}^0 = \mathbf{W}^n \\ \mathcal{B}^{i,s}_{-1} \Delta^s \mathcal{W}_{i-1} + \mathcal{B}^{i,s}_0 \Delta^s \mathcal{W}_i + \mathcal{B}^{i,s}_1 \Delta^s \mathcal{W}_{i+1} = \mathcal{S}^s_i \quad s = 0, \dots, m-1 \\ \mathbf{W}^{n+1} = \mathcal{W}^m, \end{cases}$$

in which

$$\begin{cases} \mathcal{B}^{i,s}_{-1} = -D_1(\mathcal{W}^s_{i-1}, \mathcal{W}^s_i) \\ \mathcal{B}^{i,s}_0 = \frac{3\delta x_i}{2\Delta t} I + D_1(\mathcal{W}^s_i, \mathcal{W}^s_{i+1}) - D_2(\mathcal{W}^s_{i-1}, \mathcal{W}^s_i) \\ \mathcal{B}^{i,s}_1 = D_2(\mathcal{W}^s_i, \mathcal{W}^s_{i+1}) \\ \mathcal{S}^s_i = -\left(\frac{\delta x_i}{2\Delta t}(3\mathcal{W}^s_i - 4\mathbf{W}^n_i + \mathbf{W}^{n-1}_i) + \boldsymbol{\Phi}^s_{i+\frac{1}{2}} - \boldsymbol{\Phi}^s_{i-\frac{1}{2}}\right), \end{cases}$$

D_1 and D_2 being the generic matrices of the approximation (26). Only a few iterations are needed to obtain second-order accuracy in time (Bilanceri 2011; Martin and Guillard 1996).

3.2 Extension to 3D URANS Equations in a Rotating Frame

3.2.1 Convective Fluxes

The preconditioned LD-HLL scheme (24), previously defined for the 1D case, can be extended to 3D turbulent flows by defining $\mathbf{n}_{ij}$ as the integral over ∂C_{ij} of the outer unit normal to the cell boundary. Indeed, it is possible to approximate $\mathbf{\Phi}_{ij}$ by exploiting a 1D flux function between $\mathbf{W}_i$ and $\mathbf{W}_j$, along the direction $\mathbf{n}_{ij}$ and the fact that each additional variable can be considered as a passive scalar. In the turbulent flow case, the generalization of the scheme presented in Sect. 3.1 must be done by considering the sum of the laminar and turbulent convective fluxes $n_k\left(\mathbf{F}_k + \widetilde{\mathbf{F}}_k\right)$. As a result, it is possible to approximate $\mathbf{\Phi}_{ij}$ by the following preconditioned flux function (we refer to Bilanceri 2011 for details):

$$\mathbf{\Phi}_{ij} = \frac{n_{ij,k}\left(\mathbf{F}_k(\mathbf{W}_i) + \mathbf{F}_k(\mathbf{W}_j)\right)}{2} + \frac{n_{ij,k}\left(\widetilde{\mathbf{F}}_k(\mathbf{W}_i) + \widetilde{\mathbf{F}}_k(\mathbf{W}_j)\right)}{2} - \frac{1}{2} S_{ij}(\mathbf{W}_j - \mathbf{W}_i) \quad (38)$$

$$S_{ij} = \begin{pmatrix} \lambda_1^p & 0 & 0 & 0 & 0 & 0 \\ 0 & \left(\Delta^{32}\lambda^p\right) n_{ij,1}^2 + \lambda_3^p & \left(\Delta^{32}\lambda^p\right) n_{ij,1}n_{ij,2} & \left(\Delta^{32}\lambda^p\right) n_{ij,1}n_{ij,3} & 0 & 0 \\ 0 & \left(\Delta^{32}\lambda^p\right) n_{ij,2}n_{ij,1} & \left(\Delta^{32}\lambda^p\right) n_{ij,2}^2 + \lambda_3^p & \left(\Delta^{32}\lambda^p\right) n_{ij,2}n_{ij,3} & 0 & 0 \\ 0 & \left(\Delta^{32}\lambda^p\right) n_{ij,3}n_{ij,1} & \left(\Delta^{32}\lambda^p\right) n_{ij,3}n_{ij,2} & \left(\Delta^{32}\lambda^p\right) n_{ij,3}^2 + \lambda_3^p & 0 & 0 \\ 0 & 0 & 0 & 0 & \lambda_3^p & 0 \\ 0 & 0 & 0 & 0 & 0 & \lambda_3^p \end{pmatrix} \quad (39)$$

where $\Delta^{32}\lambda^p = \lambda_2^p - \lambda_3^p$, $\lambda_1^p = \theta^{-1}\lambda_{ij}$, $\lambda_2^p = \theta\lambda_{ij}$, $\lambda_3^p = \lambda_{ij}$; the parameters θ and λ_{ij} were defined in Sect. 3.1.

The second-order extension of the convective fluxes is carried on using a MUSCL reconstruction technique similar to the one described for the 1D case in Sect. 3.1. In particular the flux function $\mathbf{\Phi}_{ij}$ is not computed using the values $\mathbf{W}_i$ and $\mathbf{W}_j$ but considering two suitable states which, except in the presence of discontinuities, approximate with second-order accuracy the solution at the boundary between the two cells, that is,

$$\mathbf{\Phi}_{ij} = \mathbf{\Phi}_{ij}\left(\mathbf{W}_{ij}^-, \mathbf{W}_{ij}^+\right) \quad (40)$$

where

$$\mathbf{W}_{ij}^- = \mathbf{W}_i + \frac{1}{2}\nabla\mathbf{W}_i \cdot \mathbf{d}_{ij}, \quad \mathbf{W}_{ij}^+ = \mathbf{W}_j - \frac{1}{2}\nabla\mathbf{W}_j \cdot \mathbf{d}_{ij} \quad (41)$$

where $\mathbf{d}_{ij}$ is the vector joining the i-th node with the j-th one and $\nabla\mathbf{W}_i$ is an approximation of the gradient in the i-th cell, possibly taking into account slope limiters. The definition of $\nabla\mathbf{W}_j \cdot \mathbf{d}_{ij}$ in this work is based on the same approach considered in Camarri et al. (2004), Farhat et al. (1991) when limiters have to be included. First, a linear approximation $\nabla\mathbf{W}|_{T_h}$ of the gradient in each tetrahedron T_h is considered:

$$\nabla\mathbf{W}|_{T_h} = \sum_{k\in i(T_h)} \mathbf{W}_k \nabla\zeta_{T_h}^k, \quad (42)$$

where $i(T_h)$ is the set of nodes belonging to T_h, $\zeta_{T_h}^k$ are the barycentric coordinates associated with the kth-vertex of T_h, while $t(i)$ is the set of the tetrahedrons which share the ith vertex. Once $\nabla \mathbf{W}|_{T_h}$ is available, a centered approximation of the gradient in the i-th cell is derived as follows:

$$\nabla \mathbf{W}_i^c = \frac{\sum\limits_{h \in t(i)} \mathcal{V}_{T_h} \nabla \mathbf{W}|_{T_h}}{\sum\limits_{h \in t(i)} \mathcal{V}_{T_h}}, \tag{43}$$

$\mathcal{V}_{T_h}$ being the volume of the h-tetrahedron. Following Farhat et al. (1991), it is now possible to obtain a second-order MUSCL reconstruction operator with slope limiters by defining $\nabla \mathbf{W}_i \cdot \mathbf{d}_{ij}$ and $\nabla \mathbf{W}_j \cdot \mathbf{d}_{ij}$ by:

$$\begin{cases} \nabla \mathbf{W}_i \cdot \mathbf{d}_{ij} = \text{minmod}\{\mathbf{W}_j - \mathbf{W}_i, 2\nabla \mathbf{W}_i^c \cdot \mathbf{d}_{ij} - (\mathbf{W}_j - \mathbf{W}_i)\} \\ \nabla \mathbf{W}_j \cdot \mathbf{d}_{ij} = \text{minmod}\{\mathbf{W}_j - \mathbf{W}_i, 2\nabla \mathbf{W}_j^c \cdot \mathbf{d}_{ij} - (\mathbf{W}_j - \mathbf{W}_i)\}, \end{cases} \tag{44}$$

where minmod$\{a, b\}$ is defined as follows:

$$\text{minmod}\{a,b\} = \begin{cases} \min\{a,b\} & \text{if } a > 0, b > 0 \\ \max\{a,b\} & \text{if } a < 0, b < 0 \\ 0 & \text{otherwise.} \end{cases} \tag{45}$$

3.2.2 Viscous Fluxes

The discretization of the viscous fluxes is instead based on P1 finite elements, in which the test functions are linear on the tetrahedral element. The approximation of the viscous fluxes in each tetrahedron is the following (see Bilanceri 2011 for details on its derivation):

$$\mathbf{\Upsilon}_i = - \sum_{T_h \in t(i)} \mathcal{V}_{T_h} \, \mu_h^T \, (\mathbf{V}_j^{T_h} + \widetilde{\mathbf{V}}_j^{T_h}) \, \frac{\partial \phi^{(i,T_h)}}{\partial x_j} \bigg|_i, \tag{46}$$

where $\phi^{(i,T_h)}$ is the P1 finite-element test function associated to the vertex i restricted to the tetrahedron T_h and μ_h^T is given by

$$\mu^{T_h} = \sum_{L \in i(T_h)} \frac{\mu_L}{4}.$$

Finally, $\mathbf{V}_j^{T_h}$ and $\widetilde{\mathbf{V}}_j^{T_h}$ are P1 approximations of the laminar and turbulent viscous fluxes on the tetrahedron T_h. For instance, for laminar viscous fluxes they are given by

$$\begin{cases} \mathbf{V}_1^{T_h} = \left(\ 0, \sigma_{11}^{T_h}, \sigma_{12}^{T_h}, \sigma_{13}^{T_h} \ \right)^T \\ \mathbf{V}_2^{T_h} = \left(\ 0, \sigma_{21}^{T_h}, \sigma_{22}^{T_h}, \sigma_{23}^{T_h} \ \right)^T \\ \mathbf{V}_3^{T_h} = \left(\ 0, \sigma_{31}^{T_h}, \sigma_{32}^{T_h}, \sigma_{33}^{T_h} \ \right)^T, \end{cases} \tag{47}$$

where

$$\sigma_{ij}^{T_h} = \left(\sum_{L \in i(T_h)} -\frac{2}{3} \left(u_{L,k} \frac{\partial \phi^{(L,T_h)}}{\partial x_k} \right) \delta_{ij} \right. \\ \left. + \left(u_{L,i} \frac{\partial \phi^{(L,T_h)}}{\partial x_j} + u_{L,j} \frac{\partial \phi^{(L,T_h)}}{\partial x_i} \right) \right), \tag{48}$$

where $u_{L,i}$ is the i-th component of the velocity at node L.

3.2.3 Source Terms

The discretization of the turbulence source term appearing in (7), and, in particular, of the term $\mathcal{P}$, is carried out as previously described for the viscous terms.

The source term due to the frame rotation is discretized as follows:

$$\mathbf{S}_i := \begin{pmatrix} 0 \\ -2\,\boldsymbol{\omega} \wedge \rho_i \mathbf{u}_i + \rho_i\, \mathbf{r}_i \end{pmatrix} \quad \mathbf{r}_i := -\,\boldsymbol{\omega} \wedge \left(\boldsymbol{\omega} \wedge \mathbf{g}_i \right), \tag{49}$$

$\mathbf{g}_i$ being the position vector of the centroid of the ith cell.

3.2.4 Time Advancing

The time discretization for the 3D numerical method can be readily obtained from its 1D counterpart. Let us consider a second-order accurate in space and time approach (the modifications for the first-order case are straightforward). Then the implicit formulation of (14) is

$$\mathcal{V}_i \frac{3}{2\Delta^n t} \Delta^n \mathbf{W}_i + \sum_{j \in N(i)} \Delta^n \boldsymbol{\Phi}_{ij} + \Delta^n \boldsymbol{\Upsilon}_i - \Delta^n (\boldsymbol{\Omega}_i + \mathbf{S}_i) = \\ \mathcal{V}_i \frac{1}{2\Delta^n t} \left(-\mathbf{W}_i^{n-1} + \mathbf{W}_i^n \right) - \sum_{j \in N(i)} \boldsymbol{\Phi}_{ij}^n - \boldsymbol{\Upsilon}_i^n + \boldsymbol{\Omega}_i^n + \mathbf{S}_i^n. \tag{50}$$

It is now necessary to define a suitable approximation for $\Delta^n \boldsymbol{\Phi}_{ij}$, $\Delta^n \boldsymbol{\Upsilon}_i$, and $\Delta^n (\boldsymbol{\Omega}_i + \mathbf{S}_i)$.

Concerning the variation of the convective part, $\Delta^n \boldsymbol{\Phi}_{ij}$, the time linearizations proposed in Sect. 3.1 are applicable to the 3D case.

Concerning the term $\Delta^n \boldsymbol{\Upsilon}_i$, a simple linearization technique is sufficient since $\boldsymbol{\Upsilon}_i$ is linearly dependent on the velocity vector. In particular $\Delta^n \sigma_{ij}^{T_h}$ can be written as

$$\Delta^n \sigma_{ij}^{T_h} = \left(\sum_{L \in i(T_h)} -\frac{2}{3} \left(\Delta^n u_{L,k} \frac{\partial \phi^{(L,T_h)}}{\partial x_k} \right) \delta_{ij} + \left(\Delta^n u_{L,i} \frac{\partial \phi^{(L,T_h)}}{\partial x_j} + \Delta^n u_{L,j} \frac{\partial \phi^{(L,T_h)}}{\partial x_i} \right) \right). \tag{51}$$

By approximating $\Delta^n u_{L,k}$ as

$$\Delta^n u_{L,k} = -\frac{u_{L,k}^n}{\rho_L^n} \Delta^n \rho_L + \frac{1}{\rho_L^n} \Delta^n \rho u_{L,k} + O\left((\Delta^n t)^2 \right), \tag{52}$$

a second-order time linearization for the viscous fluxes is achieved. With a similar approach a second-order time linearization can be obtained also for the source terms. Once $\Delta^n \boldsymbol{\Phi}_{ij}$, $\Delta^n \boldsymbol{\Upsilon}_i$, and $\Delta^n(\boldsymbol{\Omega}_i + \mathbf{S}_i)$ are given, the second-order defect correction approach described in Sect. 3.1 can be applied also for the 3D case.

In order to reduce the memory and computational requirements, we adopted a weakly coupling between the mass and momentum balance equations and the equations for k and ε, i.e., in the implicit time advancing of the mass and momentum equations, we consider ρk and $\rho \varepsilon$ frozen at time step n, while when advancing the equations for k and ε, ρ, ρu, ρv, and ρw are frozen. This permits to successively solve two linear systems of smaller dimensions: one composed by 4×4 blocks, which updates the flow variables, ρ, ρu, ρv, ρw, and another one for the turbulent variables, ρk, $\rho \varepsilon$, composed by 2×2 blocks. Note that, even if this introduces an additional approximation to the discretization technique described in the previous sections, a linear system composed by modified blocks as previously described can be considered as a particular instance of the DeC approach (see also Bilanceri 2011). As a consequence, even if for this case we do not have theoretical results of a finite termination property, the second-order approximation should be recovered asymptotically when a full convergence of the DeC iteration is obtained.

4 Application to the Flow in a Turbopump Inducer

In this section the modeling and numerical tools previously described are applied to the simulations of the flow in a real three-blade axial inducer designed for liquid-propelled rockets (Torre et al. 2010). It is a three-blade inducer with a tip blade radius

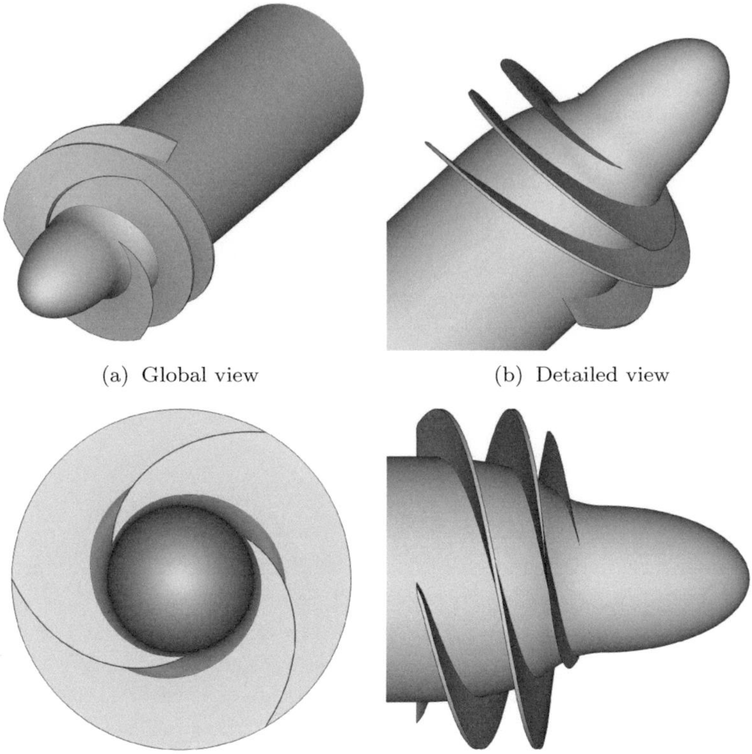

(a) Global view (b) Detailed view

(c) Front view (d) Lateral view

Fig. 1 Geometry of the inducer considered in this work

of 81-mm and 2-mm radial clearance between the blade tip and the external case (see Fig. 1).

Experimental data are available for all the numerical simulations described in the following. The pressure jump between two different stations has been measured for a wide range of working conditions: from small to large mass flow rates, non-cavitating and cavitating conditions, and different values of the rotational speed ω_z. The results are presented in terms of the mean adimensionalized pressure jump Ψ as a function of the adimensionalized discharge Φ:

$$\Psi = \frac{\Delta P}{\rho_L \omega_z^2 R_T^2} \qquad \Phi = \frac{Q}{\pi R_T^2 \omega_z R_T}, \tag{53}$$

where Q is the discharge, R_T is the radius of the tip of blade, ρ_L is the density of the liquid, and ω_z is the angular velocity. Note that the numerical pressure jump is averaged over one complete revolution of the inducer.

Table 1 Conditions of the numerical simulations and of the experiments

Benchmark	Ind1	Ind2	Ind3	Ind4	Ind5	Ind6
Φ	0.0584	0.0391	0.0185	0.0531	0.0531	0.0531
ω_z (rpm)	1500	1500	1500	3000	3000	3000
p_{out} (kPa)	125	125	125	60	85	82.5
T (°C)	25	25	25	16.8	16.8	16.8
σ	–	–	–	0.056	0.084	0.077

A cylindrical computational domain is used, whose external surface is coincident with the inducer case. The inlet is placed 249 mm ahead of the inducer nose and the outlet is placed 409 mm behind. A second computational domain, characterized by a larger streamwise length (the inlet 1120 mm ahead the inducer nose), has also been considered. Two different grids have been generated to discretize the shorter domain: the basic one G1 (1926773 cells) and G2 (3431721 cells) obtained from G1 by refining the region between the blade tip and the external case. The larger domain has been discretized by grid G1L (2093770 cells), which coincides with G1 in the original domain.

The working conditions considered in this work are shown in Table 1, where p_{out} is the outlet pressure of the flow and the cavitating number (only shown for cavitating simulations) is defined as follows:

$$\sigma = \frac{p - p_{Lsat}}{0.5\rho\omega_z^2 R_T^2}.$$

Note that, except when differently stated, the simulations do not include turbulence effects.

As shown in Table 1, all the simulations in non-cavitating conditions use the same rotational velocity of 1500 rpm. In the Φ-Ψ plane the experimental curves of the performances of the inducer are roughly independent from the rotational velocity ω_z (Torre et al. 2010). As a consequence, validating the numerical tool for a specific rotational velocity and different flow rates should validate the proposed numerical tool for a generic rotational velocity.

Table 2 shows the results for the non-cavitating simulations. It clearly appears that the lower is the discharge Φ, the worse are the results. Already with the coarsest grid G1, rather satisfactory results are obtained for intermediate and high discharge values, Ind2 and Ind1, respectively. The quantitative agreement is further improved considering the more refined grid G2 for the case Ind2. Conversely, for the low discharge case, Ind3, the simulations with the grid G1 and G2 greatly overestimate the pressure jump by, respectively, 41% and 30%. The magnitude of this error could be ascribed to the backward flow between the inducer blades and the external case. The correct resolution of this flow is of crucial importance for the determination of the performance of an inducer. Since the smaller is the mass flow rate the greater is the

Table 2 Pressure jump in non-cavitating conditions

	Experimental Ψ	Numerical Ψ	Error%
G1-Ind1	0.122	0.114	−6.6
G1-Ind2	0.186	0.204	+9.7
G2-Ind2	0.186	0.179	−3.8
G1-Ind3	0.214	0.302	+41
G2-Ind3	0.214	0.278	+30
G1L-Ind3	0.214	0.297	+39
G1L-Ind3-T	0.214	0.239	+12

Table 3 Numerical results for the cavitating simulations

	Experimental Ψ	Numerical Ψ	Error%
G1-Ind4	0.105	0.143	+36
G2-Ind5	0.143	0.130	−8.9
G2-Ind6	0.137	0.130	−5.0

backflow (see Fig. 2), we investigated two possible explanations of this behavior. The first one is the distance of the inlet from the inducer nose, which could be not large enough to avoid spurious effects on the solution; the second one is turbulence effect, which for this case could be important. The results of the first simulations for the longer computational domain, G1L-Ind3, show that even if there is a small effect, i.e., a decrease of the error from 41% to 39%, this is not the source of the error. Instead the results of the simulation G1L-Ind3-T, i.e., the one done considering the RANS model, show that in this case turbulence is a key issue. Indeed, in this case the error falls down to 12%, less than the error obtained with the refined grid G2 in laminar conditions. As expected the effects of turbulence are particularly important near the gap between the blades and the external case, as it is shown in Fig. 3 by considering the isocontours of k.

This strongly affects the backflow and, thus, the pressure jump. This also explains why for larger flow rates, for which the backflow is less important, the effects of turbulence are not so strong and a good agreement with experimental data can be obtained also in laminar simulations.

The mass flow rate for the cavitating cases is large enough to prevent the issues related to the backflow previously described, and thus only laminar simulations are considered. The results for the cavitating conditions, reported in Table 3, show that the first grid G1 is not enough refined to correctly describe cavitation for this case. The pressure jump is greatly overestimated: for these conditions the error is related to the underestimation of the cavitating region: the experimental data for $\sigma = 0.056$ show a large cavitating zone and consequently the performance of the inducer is significantly deteriorated. Instead, in the simulation with grid G1, the extension of the cavitating region is greatly underestimated and, as a consequence, the "numerical"

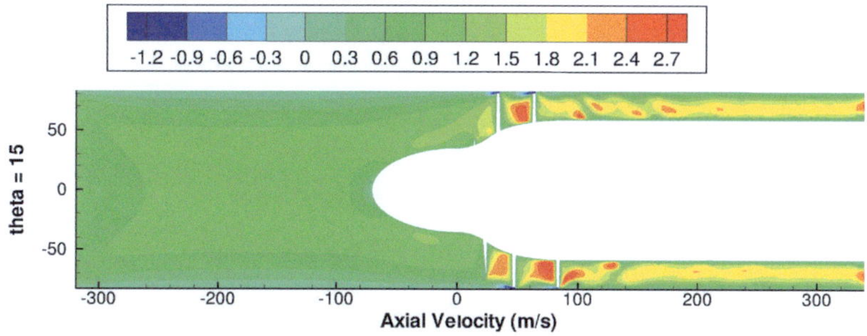

(a) Benchmark Ind1

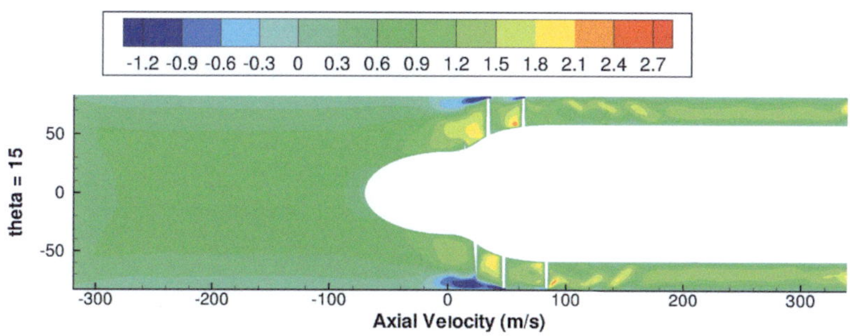

(b) Benchmark Ind2

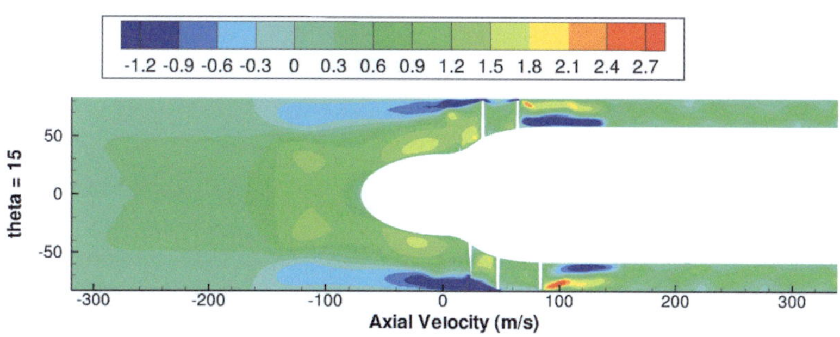

(c) Benchmark Ind3

Fig. 2 Comparison of the averaged axial velocity for the three non-cavitating benchmarks, grid G1

performance of the inducer is similar to the non-cavitating case. Grid refinement is particularly effective as shown by the results for the simulations, G2-Ind5 and G2-Ind6. The error in the prediction of the pressure jump is reduced and the extension of the cavitating region, even if it is still underestimated, is closer to the one found in experiments, as it is shown in Fig. 4 which plots the isocontours of the void fraction, corresponding to the cavitating region.

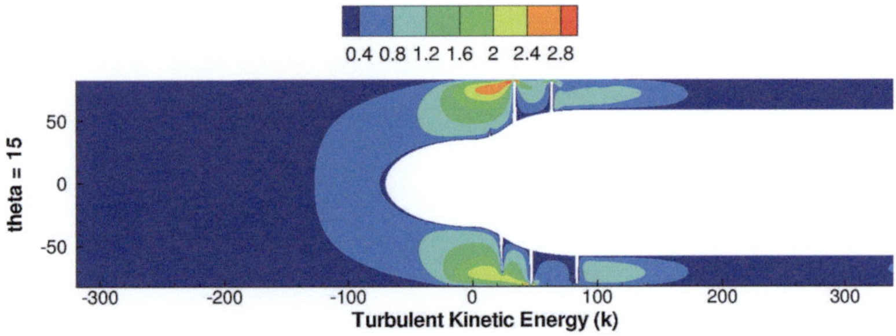

Fig. 3 Cross section of the averaged k field, simulation G1L-Ind3 (view of the shorter domain)

Fig. 4 Isocontours of the void fraction, α, for the simulation G2-Ind6

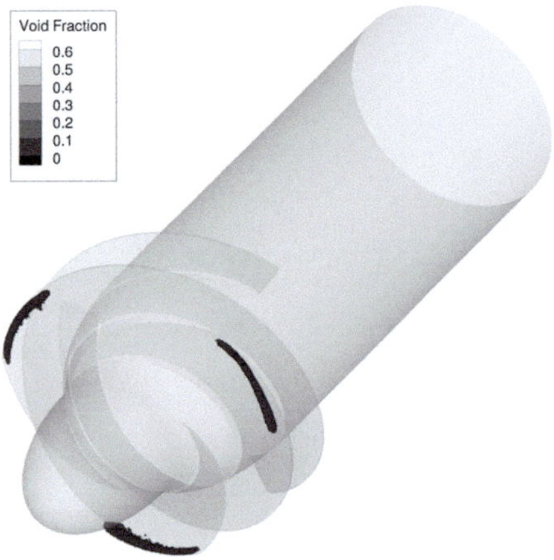

As a final remark, in this complex engineering application, the proposed numerical methodology proved to be robust and reliable in non-cavitating but also in cavitating conditions. The cavitating simulations have been carried out using CFL numbers up to 500 without encountering stability problems. Since in the definition herein adopted the time step is proportional to the mesh size and to the CFL and inversely proportional to the maximum local speed of sound, large values of CFL allow reasonable time steps to be used and thus simulations to be carried out at affordable computational costs to be carried out also for complex applications as the one herein considered.

5 Concluding Remarks

A numerical methodology has been presented for the simulation of cavitating flows in realistic engineering problems through a barotropic homogeneous-flow cavitation model. The key ingredients of this methodology are (i) the use of unstructured grids to deal with complex geometries and local refinement, (ii) a low-diffusive HLL scheme for convective fluxes, and (iii) implicit time advancing together with a linearization in time accounting for speed of sound variations and a defect correction approach to reduce the computational costs.

As an example of a real engineering problem, the application to the flow in a real turbopump inducer designed for liquid-propelled rockets in both cavitating and non-cavitating conditions. The obtained results showed a global good agreement with the experimental data and the proposed methodology appeared to be robust and efficient, leading to simulations requiring reasonable computational costs also in cavitating conditions.

Acknowledgements Marco Bilanceri and François Beux are gratefully acknowledged for their precious contribution to the work presented in this contribution.

References

Barberon, T., & Helluy, P. (2005). Finite volume simulation of cavitating flows. *Computers and Fluids*, *34*, 832858.

Bilanceri, M. (2011). Numerical simulations of barotropic flows in complex geometries. Ph.D. thesis, Aerospace Engineering, University of Pisa. https://etd.adm.unipi.it/theses/available/etd-03232011-094909/.

Bilanceri, M., Beux, F., & Salvetti, M. V. (2010). An implicit low-diffusive hll scheme with complete time linearization: Application to cavitating barotropic flows. *Computers and Fluids*, *39*, 19902006.

Camarri, S., Salvetti, M. V., Koobus, B., & Dervieux, A. (2004). A low diffusion MUSCL scheme for LES on unstructured grids. *Computer and Fluids*, *33*, 1101–1129.

Clerc, S. (2000). Numerical simulation of the homogeneous equilibrium model for two-phase flows. *Journal of Computational Physics*, *161*, 354–375.

Coutier-Delgosha, O., Fortes-Patella, R., & Delannoy, Y. (2003). Numerical simulation of the unsteady behavior of cavitating flow. *International Journal of Numerica Methids in Fluids*, *42*, 527–548.

Coutier-Delgosha, O., Fortes-Patella, R., Reboud, J. L., Hakimi, N., & Hirsch, C. (2005). Numerical simulation of cavitating flow in 2D and 3D inducer geometries. *International Journal for Numerical Methods in Fluids*, *48*, 135–167.

d'Agostino, L., Rapposelli, E., Pascarella, S., & Ciucci. A. (2011). A modified bubbly isenthalpic model for numerical simulation of cavitating flows. In *37th AIAA/ASME/SAE/ASEE Joint Propulsion Conference, Salt Lake City, UT, USA*.

Delannoy, Y., & Kueny, J. L. (1990). *Cavity flow predictions based on the Euler equations* (pp. 153–158). ASME Cavitation and Multiphase Flow Forum.

Delis, A. I., Skeels, C. P., & Ryrie, S. C. (2000). Implicit high-resolution methods for modelling one-dimensional open channel flow. *Journal of Hydraulic Research*, *38*(5), 369–382.

Farhat, S., Fezoui, L., & Lanteri, S. (1991). Computational fluid dynamics with irregular grids on the connection machine. INRIA Rapport de Recherche 1411, INRIA. http://hal.inria.fr/inria-00075149/PDF/RR-1411.pdf.

Goncalves, E., & Fortes, R. (2009). Numerical simulation of cavitating flows with homogeneous models. *Computers and Fluids*, *38*, 1682–1696.

Guillard, H., & Viozat, C. (1999). On the behaviour of upwind schemes in the low Mach number limit. *Computers and Fluids*, *28*, 63–86.

Harten, A., Lax, P. D., & van Leer, B. (1983). On upstreaming differencing and Godunov-type schemes for hyperbolic conservation laws. *SIAM Review*, *25*(1), 35–61.
Kunz, R. F., Boger, D. A., Stinebring, D. R., Chyczewski, T. S., Lindau, J. W., Gibeling, H. J., et al. (2000). A preconditioned Navier-Stokes method for two-phase flows application to cavitation prediction. *Computers and Fluids*, *29*, 849–875.
Launder, B., & Spalding, D. (1974). The numerical computation of turbulent flows. *Computer Methods in Applied Mechanics and Engineering*, *3*(2), 269–289.
Leveque, R. J. (1994). *Numerical methods for conservation laws*. Basel: Birkhäuser-Verlag.
Liu, T. G., Khoo, B. C., & Xie, W. F. (2004). Isentropic one-fluid modelling of unsteady cavitating flow. *Journal of Computational Physics*, *201*, 80–108.
Luo, H., Baum, J. D., & Löhner, R. (2001). An accurate, fast, matrix-free implicit method for computing unsteady flows on unstructured grids. *Computers and Fluids*, *30*, 137–159.
Martin, R., & Guillard, H. (1996). A second order defect correction scheme for unsteady problems. *Computers and Fluids*, *25*(1), 9–27.
Meng, H., & Yang, V. (2003). A unified treatment of general fluid thermodynamics and its application to a preconditioning scheme. *Journal of Computational Physics*, *189*, 277304.
Park, S. H., & Kwon, J. H. (2003). On the dissipation mechanism of Godunov-type schemes. *Journal of Computational Physics*, *188*(2), 524–542. doi:10.1016/S0021-9991(03)00191-8.
Qin, Q., Song, C. C. S., & Arndt, R. E. A. (2003). A virtual single-phase natural cavitation model and its application to CAV2003 hydrofoil. In *Proceedings of CAV2003—Fifth International Symposium on Cavitation, Osaka (Japan), November 2003*.
Roe, P. L. (1981). Approximate Riemann solvers, parameter vectors, and difference schemes. *Journal of Computational Physics*, *43*, 357–372. doi:10.1016/0021-9991(81)90128-5.
Rusanov, V. V. (1961). The calculation of the interaction of non-stationary shock waves with barriers. *Journal of Computational Mathematics and Physics USSR*, *1*, 267–279.
Saurel, R., Cocchi, J. P., & Butler, P. B. (1999). Numerical study of cavitation in the wake of a hypervelocity underwater projectile. *Journal of Propulsion and Power*, *15*(4), 513–522.
Senocak, I., & Shyy, W. (2002). A pressure-based method for turbulent cavitating flow computations. *Journal of Computational Physics*, *176*, 363–383.
Sinibaldi, E. (2006). Implicit preconditioned numerical schemes for the simulation of three-dimensional barotropic flows. Ph.D. thesis, Scuola Normale Superiore di Pisa.
Sinibaldi, E., Beux, F., & Salvetti, M. V. (2006). A numerical method for 3D barotropic flows in turbomachinery. *Flow, Turbulence and Combustion*, *76*(4), 371–381.
Srinivasan, V., Salazar, A. J., & Saito, K. (2009). Numerical simulation of cavitation dynamics using a cavitation-induced-momentum-defect (cimd) correction approach. *Applied Mathematical Modelling*, *33*, 15291559.
Toro, E. F. (1997). *Riemann solvers and numerical methods for fluid dynamics*. Springer.
Torre, L., Pace, G., Miloro, P., Pasini, A., Cervone, A., & d'Agostino, L. (2010). Flow instabilities on a three bladed axial inducer at variable tip clearance. In: *13th International Symposium on Transport Phenomena and Dynamics of Rotating Machinery, Honolulu, Hawaii, USA*.
van Leer, B. (1979). Towards the ultimate conservative difference scheme V: A second-order sequel to Godunov's method. *Journal of Computational Physics*, *32*(1), 101–136.
Ventikos, Y., & Tzabiras, G. (2000). A numerical method for the simulation of steady and unsteady cavitating flows. *Computers and Fluids*, *29*, 63–88.
Yee, H. C. (1987). Construction of explicit and implicit symmetric TVD schemes and their applications. *Journal of Computational Physics*, *68*(1), 151–179.

From Cavitating to Boiling Flows

Richard Saurel, Olivier Le Métayer and Pierre Boivin

Abstract A flow model is derived for the numerical simulation of interfacial flows with phase transition. The model arises from the classical multi-component Euler equations, but is associated to a non-classical thermodynamic closure: each phase is compressible and evolves in its own subvolume, with phases sharing common pressure, velocity and temperature, leading to non-trivial thermodynamic relations for the mixture. Phase transition is made possible through the introduction of Gibbs free energy relaxation terms in the equations. Capillary effects and heat conduction—essential in boiling flows—are introduced as well. The resulting multi-phase flow model is hyperbolic, valid for arbitrary density jumps at interfaces as well as arbitrary flow speeds. Its capabilities are illustrated successively through examples of nozzle induced cavitation and heated wall induced boiling.

1 Introduction

Cavitating, boiling and evaporating are three phenomena that involve phase transition in multi-phase flows. They appear in countless engineering applications: steam generators, marine propellers, liquid fuel combustion, etc. Yet, modelling these phenomena remains an unsettled problem.

R. Saurel (✉)
University Institute of France, Paris, France
e-mail: Richard.Saurel@univ-amu.fr

O. Le Métayer
Aix-Marseille University, CNRS IUSTI, Marseille, France
e-mail: Olivier.Lemetayer@univ-amu.fr

R. Saurel · P. Boivin
Aix-Marseille University, CNRS M2P2, Marseille, France
e-mail: Pierre.Boivin@univ-amu.fr

R. Saurel
RS2N, Marseille, France

L. d'Agostino and M.V. Salvetti (eds.), *Cavitation Instabilities and Rotordynamic Effects in Turbopumps and Hydroturbines*, CISM International Centre for Mechanical Sciences 575, DOI 10.1007/978-3-319-49719-8_10

Cavitation in a liquid is a phase change phenomenon created by a pressure drop driven by (fast) acoustic waves, generating—as it name suggests—cavities and bubbles. When pressure becomes lower than the saturation one at the local temperature, phase change appears as the liquid internal energy or temperature is greater than the saturated one: the liquid is overheated. Indeed, during a pressure drop, the liquid temperature varies weakly and at low pressure it becomes hot compared to the saturation temperature, this one being strongly dependant of pressure.

Boiling is yet another phase transition phenomenon created by heat deposition in a liquid, most times by heat conduction from a hot wall. The heating increases the liquid temperature and when it becomes greater than the saturation temperature at local pressure (most times uniform in the domain) phase change appears. Therefore, unlike cavitation, this process is governed by (slow) heat conduction.

In both instances, the liquid is overheated due to a departure from the saturation conditions, whether it comes from a pressure drop (cavitation) or from a temperature rise (boiling). Consequently, phase change occurs, provided that some impurities (nucleation sites) are present. In industrial and natural fluids impurities are always present in large enough concentrations (solute gases, trapped bubbles in wall roughness etc.). Cavitation is oftentimes assumed to be isothermal, with severe pressure gradients, whereas boiling and evaporation are roughly isobaric, with important temperature gradients (Sinibaldi et al. 2006; d'Agostino and Salvetti 2008; Goncalves and Patella 2009). Although they have different characteristic times, as pressure gradients are associated to acoustic waves (fast), and temperature gradients to heat conduction (slow), cavitation, boiling and evaporation are driven by phase transition and may therefore be modelled by the same approach.

However, in the literature, cavitation and boiling flows are considered through very different approaches. Most cavitation models consider liquid and two-phase mixture evolving at uniform temperature (Coutier-Delgosha et al. 2003; Barre et al. 2009). In these formulations the flow model is barotropic and the energy conservation principle as well as the second law of thermodynamics are omitted. In these models, the EOS is built to mimic some behaviour of two-phase mixtures, such as the mixture sound speed that evolves non-monotonically with respect to the volume fraction.

Boiling flows are considered by different approaches, most of them being based on Cahn and Hilliard (1958) approach of capillary fluids. Here the energy equation is considered as the importance of energetic effects are obvious. Contrarily to cavitation models, sound propagation is wrongly considered, as the square sound speed may become negative in the phase change domain (Menikoff and Plohr 1989). Indeed, the EOS is of cubic-type (van der Waals for example), with undefined sound speed in specific thermodynamic domain. Many other restrictions appear with the Cahn-Hilliard second gradient theory, such as for example the need to enlarge interfaces to make possible practical computations (Jamet et al. 2001). Fundamental issues also arise, such as shock wave existence in these media. The two most popular approaches for cavitating and boiling flows have consequently obvious limitations and restrictions.

In the present work a unified approach is provided and its ability to model and compute cavitating and boiling flows is shown with computational examples. Most of the scientific material results from former investigations by Saurel (2008), Le Martelot et al. (2013, 2014). The flow model is a hyperbolic system of partial differential equations with Gibbs free energy relaxation. The thermodynamic closure is built without ambiguity and results in a mixture EOS valid in pure liquid, pure vapour and two-phase mixture. The presence of non-condensable phase may be considered as well. The sound speed is defined in all space of variables. Nonlinear waves, such as shocks have a clear definition too. Last, the formulation is valid for any flow speed and any fluid density ratio at interfaces.

2 Basic Flow Model Specification

The flow model has to deal with:

(a) Liquid and gas compressibility: liquid compressibility consideration is mandatory in cavitating flows and gas compressibility must be addressed in most situations of flows with phase change.
(b) Pure liquid and pure gas dynamics as well as interfaces motions, these ones separating pure phases and those of two-phase mixtures.
(c) Mass exchange in mixtures and at interfaces, for both evaporation and condensation.
(d) Heat conduction, important in boiling flows.
(e) Capillary effects, important in boiling flows as well.

Obviously the flow model has to be in agreement with the basic principles of physics such as mass, momentum and energy conservation, second law of thermodynamics, frame invariance and thermodynamic consistency (convexity of the EOS and sound speed existence) resulting in hyperbolicity. In the basic flow model version we address items (a) and (b) that are the most challenging. The model we address considers mixtures and material interfaces in:

- velocity and pressure equilibrium (mechanical equilibrium),
- temperature equilibrium (thermal equilibrium).

This set of constraints is obviously valid in pure phases. In two-phase mixtures it assumes that velocity slip is absent and that the mixing is fine enough to reach temperature equilibrium. It means that the two-phase mixture is made of small drops, small bubbles or foams. Experimental observations of cavitating and flashing flows near macroscopic interfaces support these assumptions (Simoes-Moreira and Shepherd 1999). From these assumptions it is possible to reduce non-equilibrium two-phase flow models to mechanical and thermal equilibrium one. Such reduction method is addressed in Kapila et al. (2001), Murrone and Guillard (2005), Saurel et al. (2008), Lund (2012), Le Martelot et al. (2014) in various contexts. When both mechanical and thermal relaxation are assumed to be stiff, it results in the following system of partial differential equations:

$$\begin{aligned}
&\frac{\partial(\alpha\rho)_1}{\partial t} + div\big((\alpha\rho)_1\mathbf{u}\big) &&= 0, \quad \text{or alternatively} \quad && \frac{\partial\rho}{\partial t} + div(\rho\mathbf{u}) &&= 0\\
&\frac{\partial(\alpha\rho)_2}{\partial t} + div\big((\alpha\rho)_2\mathbf{u}\big) &&= 0 && \frac{\partial\rho Y_1}{\partial t} + div(\rho Y_1\mathbf{u}) &&= 0\\
&\frac{\partial\rho\mathbf{u}}{\partial t} + div\left(\rho\mathbf{u}\otimes\mathbf{u} + p\underline{\underline{I}}\right) &&= 0\\
&\frac{\partial\rho E}{\partial t} + div\left((\rho E + p)\mathbf{u}\right) &&= 0
\end{aligned} \tag{1}$$

where α_k, Y_k, ρ_k $(k = 1, 2)$ denote respectively the volume fraction, the mass fraction and the material density. ρ represents the mixture density ($\rho = \alpha_1\rho_1 + \alpha_2\rho_2$), u represents the centre of mass velocity, p denotes the pressure and E the mixture total energy ($E = e + u^2/2$). The mixture internal energy is defined as $e = Y_1e_1 + Y_2e_2$.

System (1) is clearly reminiscent of the reactive (or multi-component) Euler equations widely used in chemically reacting flows. However, the thermodynamic closure departs significantly of the one used in gas mixtures. Indeed, in the present context it results from the following algebraic system:

$$\begin{cases}
T_1 = T_2 = T,\\
e = Y_1e_1(T,p) + Y_2e_2(T,p),\\
p_1 = p_2 = p,\\
\alpha_1 + \alpha_2 = 1 \text{ or alternatively, } Y_1v_1(T,p) + Y_2v_2(T,p) = v
\end{cases} \tag{2}$$

In this algebraic system, the phases are in pressure equilibrium and each one occupies its own sub-volume or volume fraction α_k. System (2) corresponds to a nonlinear system of two equations with the two unknowns T and p. To determine its explicit solution, EOS for the phases have to be provided. This is addressed in the forthcoming section. This thermodynamic closure differs significantly from that of ideal gas mixtures. In ideal mixtures each fluid occupies the entire volume, this assumption replacing the last equality of System (2). Also, the pressure is defined by the Dalton's law $p = \sum_k p_k$ instead of the pressure equilibrium condition. In formulation (1) with thermodynamic closure (2) each phase occupies its own volume (and not the entire one) and evolves in temperature and pressure equilibrium with the other phase.

At interfaces separating pure liquid and pure gas the assumption of single temperature seems unrealistic as interface conditions, in the absence of mass transfer and heat diffusion, reduce to normal equal velocities and equal pressures, implying arbitrary temperature jumps. However, when heat diffusion is present an additional interface condition appears, corresponding to temperatures equality. Thus System (2) is valid for the computation of interfacial flows when:

- Heat diffusion is present and when it is possible to resolve the heat diffusion layer, as for example in the boiling flow configurations that will be considered latter. This is similar to laminar flames computations.
- A mushy zone is present at the interface. This is the case for example with flashing and super-cavitating flows where the interface is not a clear discontinuity but a sharp mixture layer in which thermodynamic relaxation and heat exchanges occur intensively. This is similar to turbulent flames computations, where it is not possible to resolve all heat diffusion and chemical relaxation layers at subscale, but for which 'turbulent heat conduction' imply global turbulent flame propagation.

System (1) with closure (2) is thus valid for the computation of:

- local interface dynamics when heat transfer is considered (this effect will be inserted later),
- global (or macro-scale) interface dynamics when a micro-scale or subscale structure such as a mushy zone is present. Indeed, at subscale, heat transfer is necessarily present, imposing temperature equality at the interface.

When phase transition is addressed Gibbs free energy relaxation terms have to be considered in one of the mass equations:

$$\frac{\partial \rho Y_1}{\partial t} + \frac{\partial \rho u Y_1}{\partial x} = \rho \nu (g_2 - g_1) \tag{3}$$

where $g_k = h_k - Ts_k$ denotes the phase k Gibbs free energy. h_k and s_k represent the specific enthalpy and specific entropy. ν represents a relaxation parameter that controls the rate at which thermodynamic equilibrium is reached. The way this relaxation parameter is estimated will be addressed later. The entropy equation associated to System (1)–(3) reads,

$$\frac{\partial \rho s}{\partial t} + \frac{\partial \rho s u}{\partial x} = \frac{\rho \nu (g_2 - g_1)^2}{T}, \tag{4}$$

with the following definition for the mixture entropy: $s = Y_1 s_1 + Y_2 s_2$.

The same type of remark as before with temperature equality at interfaces is needed to explain the validity of a single velocity model to compute interfacial flows with phase transition. When heat diffusion and Gibbs energy relaxation are addressed and solved at the interface, i.e., when the interface structure is solved, three velocities appear at the global scale even if a single local velocity is present in the flow model. The three velocities that appear at the global scale are the liquid one, the vapour one and the phase transition front one.

This is similar to flames computation in which the reactive Navier–Stokes equations are appropriate to compute the reacting and burnt gas dynamics as well as flame front dynamics.

The model has been shown to converge to exact sharp interface solutions in Le Martelot et al. (2014). When the interface has more complex structure with turbulent mixing at subscale, the same three velocities appear with different dynamics.

This is a consequence of turbulent heat diffusion and effective properties of fluid media. In the limit, when flash evaporation is considered, the front velocity doesn't exceed the acoustic wave speed of System (1) or its thermodynamic equilibrium analogue when the relaxation parameter ν in Eq. (3) tends to infinity. Indeed, metastable liquid is produced at a rate controlled by expansion waves, themselves propagating at the speed of acoustic waves (Saurel et al. 2008). In this limit, the deflagration speed of Chapman-Jouguet (Chaves 1984) is recovered as kinetic relation for the global phase change front velocity. To summarize the discussion on the validity of flow model (1), it is unable to compute accurately interfaces of simple mechanical contact but it becomes valid:

- when heat conduction is considered and the interface structure is resolved, as done for boiling flows (Le Martelot et al. 2014),
- when phase change occurs through interfaces with subscale structure, such as cavitating and flashing flows (Saurel et al. 2008; Le Martelot et al. 2013).

For practical use of System (1)–(2), EOSs have to be specified for the various phases. This is the aim of the next section.

2.1 *Equations of State*

Phase transition and equations of state is a long lasting challenge.

About the van der Waal EOS

As mentioned in the introduction, the Cahn-Hilliard (1958) approach is quite popular with boiling flows modelling, not with cavitating ones. It uses the van der Waals (VdW) equation of state (EOS) to compute the pressure. This EOS provides the pressure for the liquid phase, the gas phase and the two-phase mixture. It reads,

$$p = \frac{\rho RT}{1 - \rho b} - c\rho^2 \tag{5}$$

where R denote the specific gas constant ($R = \frac{r}{W}$) with r = 8.314 J/mol/K and W (kg/mol) the molar mass. b represents the specific covolume, i.e. the volume occupied by the molecules (m^3/kg) and c is a constant associated to attractive effects.

This EOS contains a fundamental drawback schematized in Fig. 1 where an isentrope for the VdW EOS is shown.

EOS (5) thus presents unphysical behaviour during the phase change process. The following interpretations follow:

- Equation (5) is aimed to close the balance equations of mass, momentum and energy of the mixture. Unlike System (1), the volume or mass concentrations are not considered in such formulations. A possible reason for choosing models with

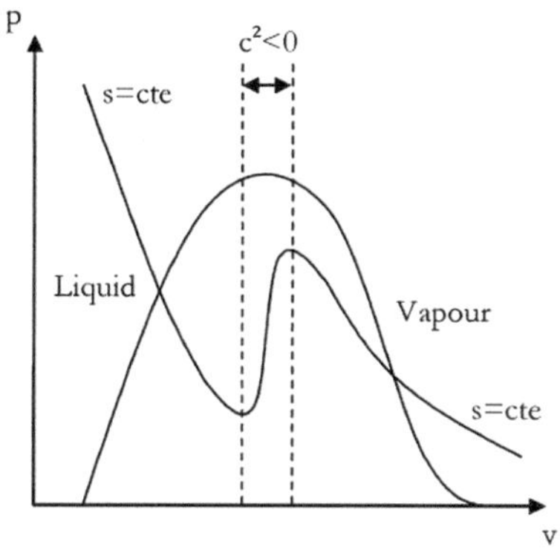

Fig. 1 Thermodynamic path along an isentrope showing an expansion process starting from a pure liquid to the pure vapour region. The square sound speed $c^2 = -v^2\left(\frac{\partial p}{\partial v}\right)_s$ is well defined in the pure liquid and pure gas but is undefined in the two-phase region, between the metastable liquid and the metastable gas

three equations only is that adding a concentration equation requires additional information, such as the thermo-chemical relaxation rate ν. These options, however, lead to the loss of hyperbolicity.

- In the VdW representation, phase transition is modelled as a thermodynamic path and not as a kinetic one. Consequently the time variable is obviously absent in the EOS. However, phase transition is a time delayed phenomenon and a thermodynamic path seems a crude representation.

These remarks naturally lead us to the following thermodynamic closure, based on three parameters $P(\rho, e, Y)$ instead of two $P(\rho, e)$ in the Van der Waals approach.

EOS for pure fluids

In the following we adopt an approach where each phase has its own EOS, each EOS being thermodynamically consistent (convex) with well defined sound speed. The connexion between the two phases is done through a kinetic path instead of a thermodynamic one. Determination of the kinetic relaxation rate will be addressed later. Schematic representation of the thermo-kinetic approach of phase transition is shown in Fig. 2. This idea was promoted by Saurel et al. (2008).

In this frame, liquid and gas require their own EOS, these ones being linked by the phase diagram. The building of such EOS has been addressed in Le Métayer et al. (2004) on the basis of the stiffened gas (SG) EOS. The main formulas for this EOS read for a given phase $k = 1, 2$:

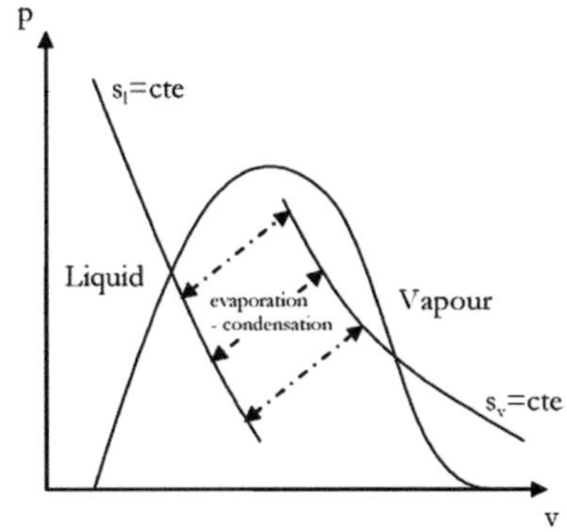

Fig. 2 Liquid and gas isentropes are connected through a kinetic path (Gibbs free energy relaxation process) during phase change

$$
\begin{aligned}
p_k(\rho_k, e_k) &= \rho_k(\gamma_k - 1)(e_k - q_k) - \gamma_k p_{\infty k},\\
T_k(p_k, \rho_k) &= \frac{p_k + p_{\infty k}}{\rho_k C_{vk}(\gamma_k - 1)},\\
g_k(p_k, T_k) &= (\gamma_k C_{vk} - q'_k)T_k - C_{vk} T_k \ln \frac{T_k^{\gamma_k}}{(p_k + p_{\infty k})^{\gamma_k - 1}} + q_k.
\end{aligned}
\tag{6}
$$

For a given phase, the following parameters are needed: γ_k, $p_{\infty k}$, C_{vk}, q_k, and q'_k. As shown by Le Métayer et al. (2004) there is no difficulty to determine these parameters once the saturation curves for the liquid and gas are known. For liquid water and steam, the fluid parameters, optimized in the [300–500 K] temperature range, are the following:

Liquid water:

$$\gamma_{liq} = 2.62 \quad P_{\infty,liq} = 9058.29 \times 10^5\,\text{Pa} \quad C_{v,liq} = 1606.97\,\text{J}\cdot\text{kg}^{-1}\cdot\text{K}^{-1}$$

$$q_{liq} = -1.150975\ 10^6\,\text{J}$$

Water vapour:

$$\gamma_{vap} = 1.38 \quad P_{\infty,vap} = 0\,\text{Pa} \quad C_{v,vap} = 1192.51\,\text{J}\cdot\text{kg}^{-1}\cdot\text{K}^{-1}$$

$$q_{vap} = 2.060759 \times 10^6\,\text{J}$$

These parameters are used in the computational examples of boiling and cavitating flows.

Expressing the thermal EOS of System (6) differently the following relation appears:

$$p_k(\rho_k, T_k) = \rho_k(\gamma_k - 1)C_{vk}T_k - p_{\infty k}. \tag{7}$$

The agitation part is $\rho_k(\gamma_k - 1)C_{vk}T_k$ and is similar to the ρRT part of (5). The attractive part $(-p_{\infty k})$ is here a constant while it is density varying with the VdW EOS. The repulsive part $(1 - \rho b)^{-1}$ is neglected in the SG formulation. Its consideration is addressed in Le Métayer and Saurel (2016).

The sound speed of a given phase reads,

$$c_k = \sqrt{\gamma_k \frac{p_k + p_{\infty k}}{\rho_k}}. \tag{8}$$

With this formulation the phases sound speed are always defined as $p_k > -p_{\infty k}$. Having in hands EOS (6) for each phase we now address building of the EOS for the mixture.

Mixture EOS

The mixture EOS is the one that closes System (1) with the help of mixture rules (2) on the basis of formulation (6) for each phase. The algebraic system to solve is thus,

$$\begin{aligned} e &= Y_1 e_1(T,p) + Y_2 e_2(T,p), \\ v &= Y_1 v_1(T,p) + Y_2 v_2(T,p), \end{aligned} \tag{9}$$

where e and v are obtained from the resolution of (1) and combination of the mass, momentum and energy equations. The unknowns in (9) are consequently the mixture temperature T and pressure P. Combining the caloric and thermal EOSs of (6) the energy for a given phase reads,

$$e_k(T_k, \rho_k) = C_{vk}T_k + \frac{p_{\infty k}}{\rho_k} + q_k.$$

Inserting this expression in the first equation of (6) the specific volume as a function of pressure and temperature is obtained:

$$v_k(T_k, p_k) = \frac{(\gamma_k - 1)C_{vk}T_k}{p_k + p_{\infty k}}. \tag{10}$$

Inserting this expression in one of the EOS (6) the internal energy as a function of pressure and temperature is obtained:

$$e_k(T_k, p_k) = \frac{(p_k + \gamma_k p_{\infty k})C_{vk}T_k}{p_k + p_{\infty k}} + q_k.$$

With these definitions, System (9) becomes,

$$T(e,p,Y_1) = (e-\bar{q})\left(\sum_i \frac{Y_i C_{vi}(p+\gamma_i p_{\infty i})}{p+p_{\infty i}}\right)^{-1}, \tag{11}$$

$$T(v,p,Y_1) = v\left(\sum_i \frac{(\gamma_i-1)Y_i C_{vi}}{p+p_{\infty i}}\right)^{-1}, \tag{12}$$

with $\bar{q} = Y_1 q_1 + Y_2 q_2$.

Eliminating the temperature from these two equations the EOS for the pressure is obtained as solution of the following quadratic equation:

$$a_2 p^2 + a_1 p + a_0 = 0$$

$$a_2 = Y_1 C_{v1} + Y_2 C_{v2}$$
$$a_1 = Y_1 C_{v1}(p_{\infty_2} + \gamma_1 p_{\infty 1} - (\gamma_1 - 1)Q) + Y_2 C_{v2}(p_{\infty_1} + \gamma_2 p_{\infty 2} - (\gamma_2 - 1)Q)$$
$$a_0 = -Q((\gamma_1 - 1)Y_1 C_{v1} p_{\infty 2} + (\gamma_2 - 1)Y_2 C_{v2} p_{\infty 1}) + p_{\infty 1} p_{\infty 2}(\gamma_1 Y_1 C_{v1} + \gamma_2 Y_2 C_{v2})$$

where

$$Q = \rho(e - \bar{q})$$

The pressure is given by the only positive root:

$$p = \frac{-a_1 + \sqrt{a_1^2 - 4a_0 a_2}}{2a_2} \tag{13}$$

Once the pressure is determined, the temperature is computed by either (11) or (12). The mixture sound speed has a non-monotonic behaviour versus the volume fraction as shown in Fig. 3.

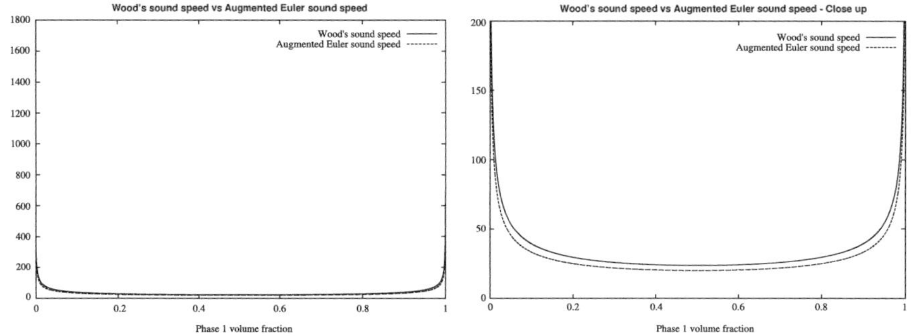

Fig. 3 Representation of the mechanical equilibrium mixture sound speed (in *lines*) and the pressure-temperature equilibrium mixture sound speed (in *dashed lines*). Both sound speeds present a non-monotonic behaviour versus volume fraction. The mechanical equilibrium sound speed is always slightly higher than the pressure-temperature equilibrium one

From the results shown in Fig. 3 it appears that there is no practical need to compute the mixture sound speed that is quite complex and computationally expensive (its explicit expression is given in Le Martelot et al. (2014)). As it is always lower than the mechanical equilibrium one, this one is preferred in the numerical computations, as better stability is guaranteed with this estimate. The mechanical equilibrium sound speed obeys the well known Wood (1930) formula:

$$\frac{1}{\rho c_w^2} = \frac{\alpha_1}{\rho_1 c_1^2} + \frac{\alpha_2}{\rho_2 c_2^2} \tag{14}$$

The volume fractions are determined from the resolution of System (1) and the specific volumes computed by (10) with the pressure given by (13) and temperature given by (12). The phase sound speeds are given by (8).

3 Kinetic Relaxation Rate—Thermo-Chemical Relaxation Solver

When dealing with phase transition the relaxation rate ν present in the concentration equation (3) has to be specified. Following Saurel et al. (2008), the guess $\nu \to \infty$ is appropriate in most situtations. In other words, pure liquid and pure gas are allowed to have any temperature during the hyperbolic step and instantly reach thermodynamic equilibrium at the saturation conditions where both liquid and gas are present (interface and mixture zones). This assertion is justified as follows:

- When dealing with the direct numerical simulation of boiling flows, the fluids become metastable as a consequence of heat diffusion. Heat conduction is slow and controls the global rate of phase change. Thus, one can assume that phase transition occurs at any rate, given it is greater than heat conduction. For the sake of simplicity, infinitely fast relaxation is adopted.
- When dealing with cavitating flows, it is assumed that the fluids contain enough impurities to have many nucleation sites. Around evaporating interfaces, phase transition happens and their collective effects result in macroscopic cavitation fronts surrounded by mushy zones. In these mushy zones the interfacial area is so large that heat and mass exchange are intense enough so that a flow model with a unique temperature and stiff Gibbs energy relaxation is appropriate.
- When none of the fluids is metastable (pure liquid state and pure gas state) no thermo-chemical relaxation occurs and the flow model reduces to single phase equations, with appropriate thermodynamics.

With this approach the thermodynamic path that the fluid follows during an isentropic expansion is shown in Fig. 4.

As the thermo-chemical solver is used as soon as metastable states appear, the effective thermodynamic path that the fluid follows is that of Fig. 4. Such a path could be reproduced by a reduced version of the flow model (1) with three equations only

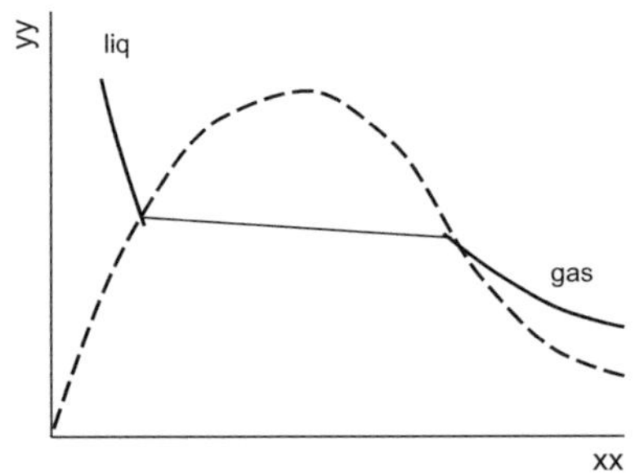

Fig. 4 As metastable states are transformed stiffly to equilibrium mixture states the effective thermodynamic path corresponds in the two-phase region to a line whose slope has been exaggerated in the figure. The slope is non zero but is weak, as sound propagates slowly in two-phase mixtures

(mixture mass, mixture momentum and mixture energy). However, the sound speed of such a model is not convenient, as it is costly to compute, and most importantly, discontinuous at the phase diagram boundaries.

Here, the flow model has a sound speed always defined and continuous. As mentioned earlier, Eq. (14) is a fair approximation. Also, extending System (1) to the presence of non-condensable gas is straightforward while it is non-trivial with a more reduced model.

The presence of a stiff relaxation term in (3) does not result in computational difficulties as integration of such term is never addressed: in the following, only the equilibrium state is required.

The relaxed solution corresponding to thermodynamic equilibrium is obtained considering the mixture mass and mixture energy definition:

$$v = \frac{1}{\rho} = Y_1 v_1 + Y_2 v_2 = cst = v_0 \tag{15}$$

$$e = Y_1 e_1 + Y_2 e_2 = cst = e_0 \tag{16}$$

where $Y_1 = \dfrac{\alpha_1 \rho_1}{\rho}$ and $Y_2 = \dfrac{\alpha_2 \rho_2}{\rho} = 1 - Y_1$ are the varying mass fractions of both phases. In the following the liquid and its vapour are denoted respectively by the subscripts '1' and '2'.

The specific volumes and internal energies are given by the SG EOS:

$$v_k = \frac{(\gamma_k - 1) C_{vk} T_k}{p_k + p_{\infty,k}}, \tag{17}$$

$$e_k = C_{vk} T_k \left(1 + \frac{(\gamma_k - 1) p_{\infty,k}}{p_k + p_{\infty,k}}\right) + q_k. \tag{18}$$

All parameters appearing in relations (17), (18) and (19) are computed in order to satisfy the experimental liquid/vapour saturation curves (Le Métayer et al. 2004).

The final relaxed state, denoted by the superscript '$*$' corresponds to the thermodynamic equilibrium state. The liquid and vapour phases have a common pressure, temperature and Gibbs free energy. Equality of Gibbs free energy of both phases,

$$g_k = h_k - T_k s_k = (\gamma_k C_{vk} - q_k')T_k - C_{vk} T_k ln \frac{T_k^{\gamma_k}}{(p + p_{\infty,k})^{\gamma_k - 1}} + b_k P_k + q_k, \tag{19}$$

provides a relation between the pressure and the temperature:

$$T^*(p^*) = T_{sat}(p^*) \tag{20}$$

This relation represents the evolution of the saturation temperature as a function of pressure. Thanks to (20), the relation (15) reads,

$$v_0 = Y_1^* v_1^*(p^*) + Y_2^* v_2^*(p^*) = Y_1^* v_1^*(p^*) + (1 - Y_1^*) v_2^*(p^*), \tag{21}$$

with,

$$v_k^*(p^*) = \frac{(\gamma_k - 1) C_{vk} T^*(p^*)}{p^* + p_{\infty,k}}. \tag{22}$$

Variables v_1^* and v_2^* thus correspond to the saturated specific volumes of the liquid and vapour phases respectively.

A first relation linking the liquid mass fraction and the pressure is obtained from (21):

$$Y_1^* = \frac{v_2^*(p^*) - v_0}{v_2^*(p^*) - v_1^*(p^*)}. \tag{23}$$

In relation (23) the existence of physical solution is fulfilled by the following condition:

$$0 < Y_1^* < 1 \Leftrightarrow v_1^*(p^*) < v_0 < v_2^*(p^*) \tag{24}$$

By using once more the saturation relation (20), the total energy equation (16) becomes,

$$e_0 = Y_1^* e_1^*(p^*) + Y_2^* e_2^*(p^*) = Y_1^* e_1^*(p^*) + (1 - Y_1^*) e_2^*(p^*), \tag{25}$$

where,

$$e_k^*(p^*) = C_{vk} T^*(p^*) \left(1 + \frac{(\gamma_k - 1) p_{\infty,k}}{p^* + p_{\infty,k}}\right) + q_k. \tag{26}$$

A second relation linking the liquid mass fraction and the final pressure is obtained,

$$Y_1^* = \frac{e_0 - e_2^*(p^*)}{e_1^*(p^*) - e_2^*(p^*)}. \tag{27}$$

It is more convenient to rewrite relation (27) in terms of specific enthalpies by combining (21) and (25),

$$e_0 + p^* v_0 = Y_1^* \left(e_1^*(p^*) + p^* v_1^*(p^*)\right) + Y_2^* \left(e_2^*(p^*) + p^* v_2^*(p^*)\right) = Y_1^* h_1^*(p^*) + Y_2^* h_2^*(p^*), \qquad (28)$$

which directly introduces the latent heat of vaporization $L_v(p^*)$,

$$L_v(p^*) = h_2^*(p^*) - h_1^*(p^*). \qquad (29)$$

Equation (27) then becomes:

$$Y_1^* = \frac{h_2^*(p^*) - (e_0 + p^* v_0)}{h_2^*(p^*) - h_1^*(p^*).} \qquad (30)$$

In relation (30) a second existence condition appears,

$$0 < Y_1^* < 1 \Leftrightarrow h_1^*(p^*) < e_0 + p^* v_0 < h_2^*(p^*). \qquad (31)$$

Equating relations (23) and (30) leads to an equation where the final pressure p^* is the only unknown,

$$\frac{h_2^*(p^*) - (e_0 + p^* v_0)}{h_2^*(p^*) - h_1^*(p^*)} - \frac{v_2^*(p^*) - v_0}{v_2^*(p^*) - v_1^*(p^*)} = 0. \qquad (32)$$

Once the solution of (32) is obtained—through Newton's method for instance—the other thermodynamic variables are easily obtained by the preceding relations presented above.

However Eq. (32) may not provide a physical solution depending on the initial energy e_0 and specific volume v_0. This is the case when conditions (24) and (31) are not fulfilled. In such instances the liquid/vapour system tends towards a final state where a single phase is present. Total evaporation or condensation thus occurs during the relaxation process, and the corresponding thermodynamic state is computed with Eq. (13) where the mass fraction of one of the phases has been set to 1.

4 Hyperbolic Solver

For the sake of simplicity, the presentation of the solver is one-dimensional, and limited to first order. Details about higher order extensions may be found in Toro (2009).

System (1) is written in compact form as,

$$\frac{\partial U}{\partial t} + \frac{\partial F(U)}{\partial x} = 0, \tag{33}$$

where $U = \left[\rho \ \rho u \ \rho E \ \rho Y\right]^t$ is the conservative variables vector, and $F = \left[\rho u \ \rho u^2 + p \ (\rho E + p)u \ \rho u Y\right]^t$ represents the associated flux vector.

Godunov scheme

System (33) is a hyperbolic system of conservation laws, with wave speeds u, $u - c_{eq}$ and $u + c_{eq}$.

The first-order Godunov method reads,

$$U_i^{n+1} = U_i^n - \frac{\Delta t}{\Delta x}(F^*_{i+1/2} - F^*_{i-1/2}), \tag{34}$$

under the CFL stability condition,

$$\Delta t \leq \frac{\Delta x}{\max(u \pm c_{eq})}. \tag{35}$$

The cell boundary fluxes $F_{i\pm1/2}$ are computed with the following approximate solver.

HLLC solver

The HLLC solver (Toro et al. 1994) is an approximate Riemann solver for the Euler equations that is easy to adapt to the present "real gas" context. In the HLLC context, each wave is considered as a discontinuity:

- a contact discontinuity wave, with characteristic speed S_M, through which pressure and velocity is constant
- two discontinuities, with characteristic speed S_L and S_R (left facing and right facing waves), through which mass fractions Y_i are constant.

As shown in Fig. 5, each of these waves separate the states: $(\bar{U}_L, \bar{U}_L^*, \bar{U}_R^*, \bar{U}_R)$, from which it is possible to compute the appropriate flux between the two cells during the time step.

In the example depicted in Fig. 5, U_L^* is the appropriate state to compute the inter-cell flux F, since it is the state which persists at the physical position of the cell discontinuity during the time step. In a more general context, there are four possible cases depending on the signs of the three wave velocities:

case a $S_L > 0 \rightarrow F = F(U_L)$
case b $S_R < 0 \rightarrow F = F(F_R)$
case c $S_M > 0, S_L < 0 \rightarrow F = F(U_L^*)$
case d $S_M < 0, S_R > 0 \rightarrow F = F(U_R^*)$

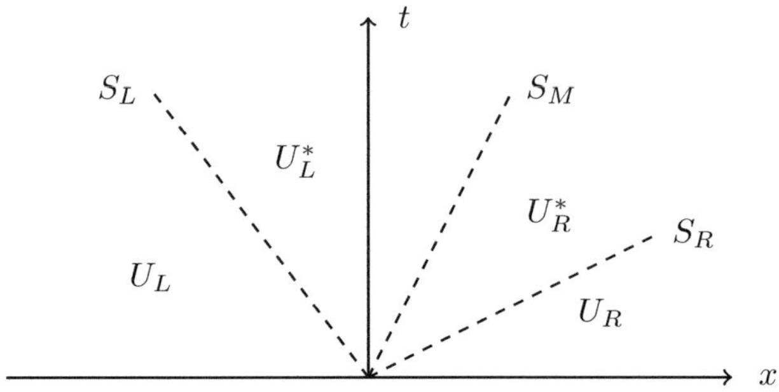

Fig. 5 Schematic representation of the Riemann problem emerging at each cell boundary. Three nonlinear waves are emitted, each one being considered as a discontinuity

Wave speeds estimates and fluxes computation

Following Davis (1988), the right and left velocities are approximated as,

$$S_R = \max((u + c_w)_R, (u + c_w)_L), \tag{36}$$

$$S_L = \min((u - c_w)_R, (u - c_w)_L). \tag{37}$$

Knowledge of c_{eq} is not required as it is replaced everywhere by its approximation c_w, given by (14), in the solver that follows.

Each discontinuity obeys the Rankine-Hugoniot conditions given by:

$$[F] - S_k[U] = 0 \tag{38}$$

where S_k denotes the speed of the kth discontinuity and [...] denotes the jump of a given variable across the discontinuity.

Under HLL approximation, the intermediate wave speed is given by

$$S_M = u^* = \frac{p_R - p_L + \rho_R u_R(u_R - S_R) - \rho_L u_L(u_L - S_L)}{\rho_R(u_R - S_R) - \rho_L(u_L - S_L)} \tag{39}$$

Using again the Rankine–Hugoniot jump conditions, one can derive the following relations for the intermediate state:

$$\begin{cases} \rho_R^* = \rho_R \frac{u_R - S_R}{S_M - S_R} \\ u_R^* = u^* = S_M \\ p_R^* = p_R + \rho_R(u_R - S_R)(u_R - S_M) \\ E_R^* = E_R + \frac{p_R(u_R - S_M)}{\rho_R(u_R - S_R)} - S_M(u_R - S_M) \\ Y_{R,i}^* = Y_{R,i} \end{cases} \tag{40}$$

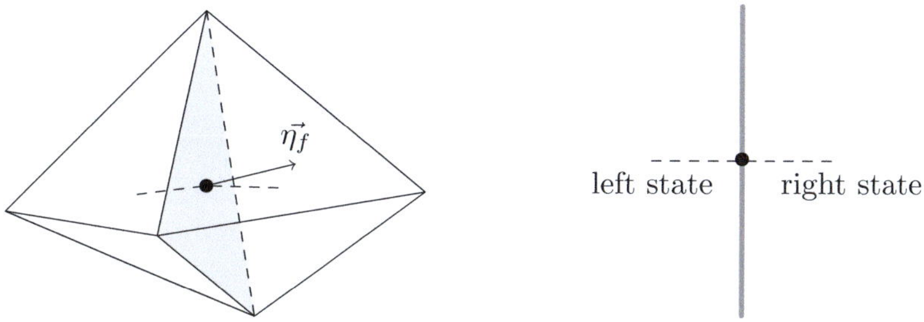

Fig. 6 3D Godunov method applied to computational control volumes made of tetrahedrons. The code is cell-centered, and the Riemann problem is solved at each face. ▲ tetrahedron center. • face center

These relations fully define the intercell set of conservative variables U_R^*, and consequently the flux $F(U_R^*)$ given by,

$$F_R^* = F_R - S_R(U_R - U_R^*). \tag{41}$$

Substituting the L index to the R index in the above formula leads to a similar expression for the remaining flux $F(U_L^*)$.

The computational examples that follow were achieved with the DALPHADT code based on tetrahedron meshes. In this frame, the Godunov method reads,

$$U_{k,i}^{n+1} = U_{k,i}^{n+1} - \frac{\Delta t}{V_i} \sum_{f=1}^{4} (F_k^* \cdot \vec{\eta})_f S_f, \tag{42}$$

where S_f is the face area, and $(F_k^* . \vec{\eta})_f$ represents the flux solution of the Riemann problem solved along the face normal vector $\vec{\eta}_f$ of a given face of the tetrahedron, as illustrated in Fig. 6.

We now address computational examples.

5 Cavitating Flows

In this section, 2D two-phase flow computations in Venturi nozzle are addressed. The configuration studied corresponds to the experimental facility built at LEGI Laboratory in Grenoble, France by the group leaded by S. Barre. The test section corresponds to a Venturi channel with a nozzle divergent inclined at an angle of 8°. The geometry is shown in Fig. 7.

The fluids considered correspond to liquid and vapour water, with the equation of state parameters given in Sect. 2.1. The boundary conditions correspond to imposed mass inflow with both imposed stagnation enthalpy and entropy at the inlet and

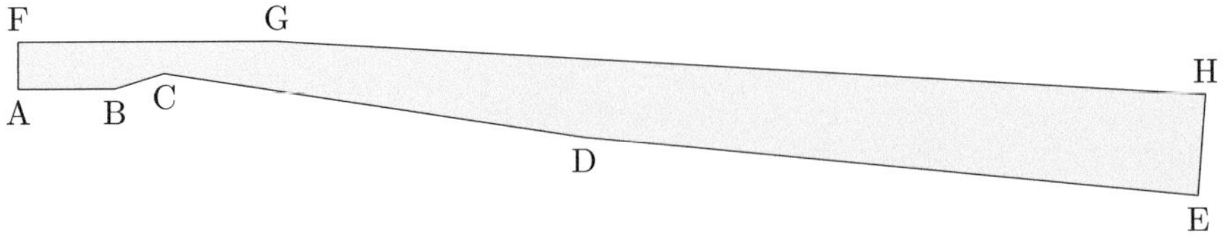

	X (abscissa) (m)	Y (m)		X(abscissa) (m)	Y (m)
A	0	0	E	1.225	-0.114
B	0.1	0	F	0	0.0488
C	0.153	0.0157	G	0.271	0.0488
D	0.588	-0.0517	H	1.233	-0.00845

Fig. 7 Geometrical data of the Venturi 8° nozzle of LEGI, France

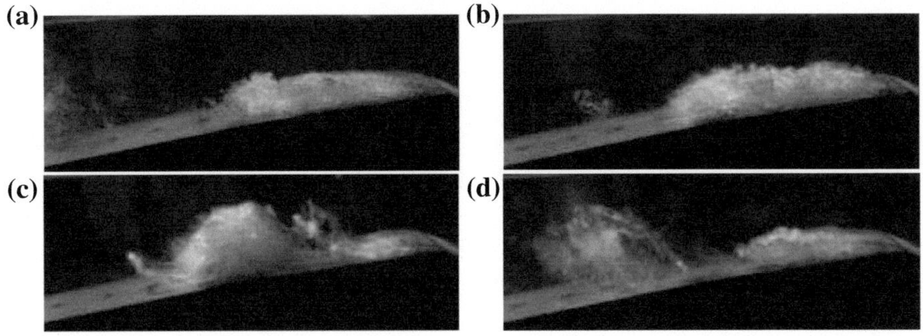

Fig. 8 Experimental photographs of the break off cycle observed in the 8° Venturi nozzle with the boundary conditions aforementioned. Flow direction is from right to left. A cavitation pocket appears, extends and separates in two sub-pockets, one transported with the mean flow and another one collapsing close to the nozzle. Courtesy of S. Barre (LEGI)

prescribed pressure at the outlet. The imposed conditions at the inlet are the following,

$$\begin{cases} m = 7514.917\,\text{kg.m}^{-2}.\text{s}^{-1} \\ \rho_{liq} = 1067.56\,\text{kg.m}^{-3} \\ \rho_{vap} = 0.387\,\text{kg.m}^{-3} \\ \alpha_{liq} = 0.999 \\ P = 51825\,\text{Pa} \end{cases}$$

while, at the right outlet, the prescribed pressure is $P = 72025$ Pa.

With these boundary conditions a periodic flow has been observed experimentally as shown in the Fig. 8.

In the first stage of the cycle (a), a cavitation sheet is attached to the throat and grows. In a second stage, the sheet reaches its maximum length (b) and breaks into two main parts (c). At the end, the downstream part is swept along within the

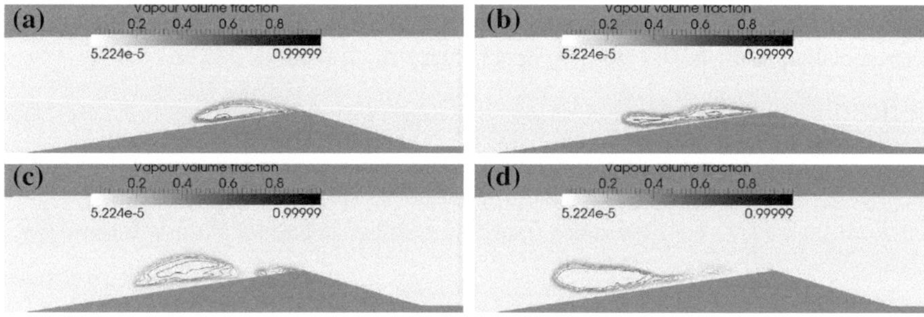

Fig. 9 Computed volume fraction of water vapour. This example shows the same four different parts as those observed during the experimental studies and shown in Fig. 8. The mean attached cavity length is about 45 mm, in perfect agreement with the experiments

stream and starts to collapse while the attached part starts another cycle (d). The mean attached cavity length value is 45 ± 5 mm while the quasi-periodic vapour cloud shedding frequency is about 45 Hz.

To compute this unstable flow a 2D unstructured mesh containing 52450 cells is used. The grid is refined at the throat in order to capture the cavitation pocket. The average cell size is 0.13 mm at throat and 0.8 mm elsewhere.

The explicit scheme summarized in Sect. 4 is extended to time implicit integration and the Riemann problem is preconditioned with the method detailed in Le Martelot et al. (2013). The flow is computed during 1.8 s of physical time, this time being long enough to obtain a quasi-stationary flow with quasi-periodic vapour clouds shedding. An example of the obtained cloud shedding is shown in the volume fraction contours of Fig. 9. Examining the water vapour volume fraction contours oscillations, we were able to determine a vapour pocket shedding frequency of about 43 Hz, in excellent agreement with the measured frequency for the pressure fluctuations (45 Hz.).

By performing measurements during every cycle, an average attached cavity length of about 45 mm has been measured from the computations. These results show very good agreement with the experiments. Indeed, experimental measurements gave a mean attached cavity length equals to 45 ± 5 mm.

These results show that it is possible to reproduce the large structures of such cavitating flows with a flow model free of parameters. In particular, no turbulence model is used.

6 Boiling Flows

To deal with DNS-type of boiling flows, System (1) has to be enhanced by introducing additional physical effects:

- buoyancy,
- surface tension,
- heat conduction.

Surface tension effects are considered through the Continuum Surface Force (CSF) method of Brackbill et al. (1992). The capillary force is modelled as,

$$\mathbf{F}_\sigma = \sigma\kappa\overrightarrow{\nabla Y_1} \tag{43}$$

where σ represents the surface tension coefficient (N.m^{-1}), κ represents the local curvature (m^{-1}) and Y_1 is the mass fraction of phase 1. The local curvature reads,

$$\kappa = -div\left(\frac{\overrightarrow{\nabla Y}}{|\overrightarrow{\nabla Y}|}\right). \tag{44}$$

The gravity force is modelled as,

$$\mathbf{F}_g = \rho\mathbf{g},$$

where $\mathbf{g}$ represents the gravity constant.

Heat conduction is inserted in the total energy equation of the model through the Fourier law $\mathbf{q} = -\lambda_c\overrightarrow{\nabla T}$ where the "mixture" thermal conductivity is given by $\lambda_c = \alpha_1\lambda_1 + \alpha_2\lambda_2$ and λ_k represents the thermal conductivity of phase k. Details are given in Le Martelot et al. (2014).
Inserting these extra effects, the flow model now becomes:

$$\begin{aligned}
&\frac{\partial \rho Y_1}{\partial t} + div\left(\rho Y_1\mathbf{u}\right) = \rho\nu(g_2 - g_1)\\
&\frac{\partial \rho}{\partial t} + div\left(\rho\mathbf{u}\right) = 0\\
&\frac{\partial \rho\mathbf{u}}{\partial t} + div\left(\rho\mathbf{u}\otimes\mathbf{u} + P\underline{\underline{I}}\right) = \sigma\kappa\overrightarrow{\nabla Y_1} + \rho\mathbf{g}\\
&\frac{\partial \rho E}{\partial t} + div\left((\rho E + P)\mathbf{u}\right) = div\left(\lambda_c\overrightarrow{\nabla T}\right) + \sigma\kappa\overrightarrow{\nabla Y_1}\bullet\mathbf{u} + \rho\mathbf{g}\bullet\mathbf{u}
\end{aligned} \tag{45}$$

A conservative form is available as well,

$$\begin{aligned}
&\frac{\partial \rho Y_1}{\partial t} + div\left(\rho Y_1\mathbf{u}\right) = \rho\nu(g_2 - g_1)\\
&\frac{\partial \rho}{\partial t} + div\left(\rho\mathbf{u}\right) = 0\\
&\frac{\partial \rho\mathbf{u}}{\partial t} + div\left(\rho\mathbf{u}\otimes\mathbf{u} + P\underline{\underline{I}} - \sigma\underline{\underline{\mathbf{m}}}\right) = \rho\mathbf{g}\\
&\frac{\partial \rho E + \sigma|\overrightarrow{\nabla Y_1}|}{\partial t} + div\left((\rho E + P + \sigma|\overrightarrow{\nabla Y_1}|)\mathbf{u} - \sigma\underline{\underline{\mathbf{m}}}\bullet\mathbf{u} - \lambda_c\overrightarrow{\nabla T}\right) = \rho\mathbf{g}\bullet\mathbf{u}
\end{aligned} \tag{46}$$

where $\underline{\underline{\mathbf{m}}} = \left(|\overrightarrow{\nabla Y_1}| \underline{\underline{I}} - \frac{\overrightarrow{\nabla Y_1} \otimes \overrightarrow{\nabla Y_1}}{|\overrightarrow{\nabla Y_1}|} \right)$.

This system is closed by the EOS (13). The entropy equation associated to System (45) reads,

$$\frac{\partial \rho S}{\partial t} + div\left(\rho S\mathbf{u} - \frac{\lambda_c \overrightarrow{\nabla T}}{T} \right) = \frac{\rho\nu(g_2 - g_1)^2}{T} + \lambda_c \frac{(\nabla T)^2}{T^2} \tag{47}$$

with $S = Y_1 s_1 + Y_2 s_2$, and shows agreement with the second law of thermodynamics, as the right hand side of (47) is positive.
System (46) is considered hereafter to compute boiling flow examples.

A closed and adiabatic rectangular domain (12 cm × 7 cm) in which the lower half is filled with saturated liquid water and the upper half is filled with saturated vapour is considered.

The surface tension coefficient is set to $\sigma = 73.10^{-3}$ N.m^{-1}, the contact angle is taken constant and equal to $\theta = 45°$ and the gravity acceleration is set to $|\mathbf{g}| = 9.81$ m.s^{-2}.

At start, the initial volume fraction of vapour is $\alpha_{vap} = 0.0001$ in the lower half domain and $\alpha_{vap} = 0.9999$ in the upper part. Moreover, the initial pressure and temperature are initialised with the hydrostatic gravity profile with the constraint $T = T_{SAT}(P)$ in each cell. The bottom wall temperature is set constant at $T_{SAT}(P_{atm}) + 15K$.

The numerical scheme presented in the former section is rendered time implicit and the Riemann solver is low Mach pre-conditioned (Le Martelot et al. 2014). A mesh made of 960 × 560 cells is used in the computations that follow.

Three vapour bubbles (radius = 3 mm) are set initially at the bottom wall as shown in the first image of Fig. 10. The volume fraction of vapour inside these bubbles is $\alpha_{vap} = 0.9999$. The computed vapour mass function is shown in the same figure.

The first instants show the three first bubbles moving toward the surface due to buoyancy effects while, as the bottom wall of the box is heated, liquid water begins to boil, creating a vapour film. The boiling phenomenon appears as consequence of wall heating effect that renders the liquid locally slightly overheated. Indeed, the assumption made is that the liquid contains enough impurities to not accept overheating, as pure liquids are able to become metastable while real liquids are not. Therefore, using the phase transition solver of Sect. 4 after checking that the liquid is not in stable state, the equilibrium state is computed on the basis of Eq. (32) and a mixed cell appears. From this "nucleation cell site" and merging effects due to surface tension, convection and inertia, new bubbles appear. Once created, theses new bubbles begin to rise and, as there is now again liquid in contact with the bottom wall, new bubbles appear behind them and begin to grow.

It is worth to mention that the bottom wall of the box is a perfect surface, where uniform temperature and constant contact angle are set. According to the velocity

Fig. 10 Vapour mass fraction contours at times $t = 0\,\mathrm{s}$, $t = 50\,\mathrm{ms}$, $t = 100\,\mathrm{ms}$, $t = 200\,\mathrm{ms}$, $t = 300\,\mathrm{ms}$ and $t = 400\,\mathrm{ms}$. A vapour film appears at the first instants. As a result of wall effects combined with capillary ones a big elongated bubble is created at the centre and breaks up in several bubbles. New bubbles appear at the bottom wall as a consequence of wall heating

profiles, the first bubbles seem to be created between the convective rolls, where the velocity is the lowest, as shown in Fig. 10.

7 Conclusions

Boiling, evaporation and cavitation are essentially the same phenomenon, driven by phase transition, but in different pressure and temperature conditions. Consequently, they ought to be described by the same model, and deriving such a model is the object of the present work. The flow model is essentially a hyperbolic system with relaxation terms. Coping with boiling, evaporation and cavitation, it is—in the authors knowledge—the first approach showing such a wide range of applicability. It is also a very robust computational approach of phase transition compared to existing alternatives. Although not shown here, our preliminary investigations show that the flow model can be extended to cope not only with a liquid–vapour couple, but also with multi-component gas mixtures.

Acknowledgements Part of this work has been carried out in the framework of the Labex MEC (ANR-10-LABX-0092) and of the A*MIDEX project (ANR-11-IDEX-0001-02), funded by the Investissements d'Avenir French Government program managed by the French National Research Agency (ANR). We also acknowledge funding from ANR through project ANR-14-CE22-0014. Pr. Stéphane Barre (LEGI) is also gratefully acknowledged for providing the photographs of Fig. 8.

References

Barre, S., Rolland, J., Boitel, G., Goncalves, E., & Fortes Patella, R. (2009). Experiments and modeling of cavitating flows in Venturi: Attached sheet cavitation. *European Journal of Mechanics-B/Fluids, 280*(3), 444–464.

Brackbill, J. U., Kothe, D. B., & Zemach, C. I. (1992). A continuum method for modeling surface tension. *Journal of Computational Physics, 1000*(2), 335–354.

Cahn, J. W., & Hilliard, J. E. (1958). Free energy of a nonuniform system. I. Interfacial free energy. *The Journal of Chemical Physics, 280*(2), 258–267.

Chaves, H. (1984). Changes of phase and waves on depressurization of liquids with high specific heat. *NASA STI/Recon Technical Report N, 84*, 25003.

Coutier-Delgosha, O., Fortes-Patella, R., & Reboud., J.-C. (2003). Evaluation of the turbulence model influence on the numerical simulations of unsteady cavitation. *Journal of Fluids Engineering, 1250*(1), 38–45.

d'Agostino, L., & Salvetti, M. V. (2008). *Fluid dynamics of cavitation and cavitating turbopumps* (Vol. 496). Springer Science & Business Media.

Davis, S. F. (1988) Simplified second-order Godunov-type methods. *SIAM Journal on Scientific and Statistical Computing, 90*(3), 445–473.

Goncalves, E., & Patella, R. F. (2009). Numerical simulation of cavitating flows with homogeneous models. *Computers & Fluids, 380*(9), 1682–1696.

Jamet, D., Lebaigue, O., Coutris, N., & Delhaye, J. M. (2001). The second gradient method for the direct numerical simulation of liquid–vapor flows with phase change. *Journal of Computational Physics, 1690*(2), 624–651.

Kapila, A. K., Menikoff, R., Bdzil, J. B., Son, S. F., & Stewart, D. S. (2001). Two-phase modeling of deflagration-to-detonation transition in granular materials: Reduced equations. *Physics of Fluids (1994–present), 130*(10), 3002–3024.

Le Martelot, S., Nkonga, B., & Saurel, R. (2013). Liquid and liquid-gas flows at all speeds. *Journal of Computational Physics*, *255*, 53–82.

Le Martelot, S., Saurel, R., & Nkonga, B. (2014). Towards the direct numerical simulation of nucleate boiling flows. *International Journal of Multiphase Flow*, *66*, 62–78.

Le Métayer, O., & Saurel, R. (2016). The Noble-Abel—Stiffened-Gas equation of state. *Physics of Fluids 28*(4), 046102.

Le Métayer, O., Massoni, J., & Saurel, R. (2004). Elaboration des lois d'état d'un liquide et de sa vapeur pour les modèles d'écoulements diphasiques. *International Journal of Thermal Sciences, 430*(3), 265–276.

Lund, H. (2012). A hierarchy of relaxation models for two-phase flow. *SIAM Journal on Applied Mathematics, 720*(6), 1713–1741.

Menikoff, R., & Plohr, B.J. (1989). The Riemann problem for fluid flow of real materials. *Reviews of Modern Physics, 610*(1), 75.

Murrone, A., & Guillard, H. (2005). A five equation reduced model for compressible two phase flow problems. *Journal of Computational Physics, 2020*(2), 664–698.

Saurel, R., Petitpas, F., Abgrall, R., et al. (2008). Modelling phase transition in metastable liquids: Application to cavitating and flashing flows. *Journal of Fluid Mechanics, 6070*(1), 313–350.

Simoes-Moreira, J. R., & Shepherd, J. E. (1999). Evaporation waves in superheated dodecane. *Journal of Fluid Mechanics, 382*, 63–86.

Sinibaldi, E., Beux, F., & Salvetti, M. V. (2006). A numerical method for 3D barotropic flows in turbomachinery. *Flow, Turbulence and Combustion, 760*(4), 371–381.

Toro, E. F. (2009). *Riemann solvers and numerical methods for fluid dynamics: A practical introduction*. Springer Science & Business Media.

Toro, E. F., Spruce, M., & Speares, W. (1994). Restoration of the contact surface in the HLL-Riemann solver. *Shock Waves, 40*(1), 25–34.

Wood, A. B. (1930). *A textbook of sound*. London: G. Bell and Sons Ltd.

MIX
Papier aus verantwortungsvollen Quellen
Paper from responsible sources
FSC® C105338
FSC
www.fsc.org

If you have any concerns about our products,
you can contact us on
ProductSafety@springernature.com

In case Publisher is established outside the EU,
the EU authorized representative is:
Springer Nature Customer Service Center GmbH
Europaplatz 3, 69115 Heidelberg, Germany

Printed by Libri Plureos GmbH
in Hamburg, Germany